Mobile WiMAX

Mobile WiMAX

Edited by

Kwang-Cheng Chen
National Taiwan University, Taiwan

J. Roberto B. de Marca
Pontifical Catholic University, Brazil

IEEE PRESS
IEEE Communications Society, Sponsor

WILEY
John Wiley & Sons, Ltd.

Other Wiley Editorial Offices

John Wiley & Sons Inc., 111 River Street, Hoboken, NJ 07030, USA

Jossey-Bass, 989 Market Street, San Francisco, CA 94103-1741, USA

Wiley-VCH Verlag GmbH, Boschstr. 12, D-69469 Weinheim, Germany

John Wiley & Sons Australia Ltd, 42 McDougall Street, Milton, Queensland 4064, Australia

John Wiley & Sons (Asia) Pte Ltd, 2 Clementi Loop #02-01, Jin Xing Distripark, Singapore 129809

John Wiley & Sons Canada Ltd, 6045 Freemont Blvd, Mississauga, ONT, L5R 4J3, Canada

Wiley also publishes its books in a variety of electronic formats. Some content that appears in print may not be
available in electronic books.

IEEE Communications Society, Sponsor. COMMS-S Liaison to IEEE Press, Mostafa Hashem Sherif

Library of Congress Cataloging-in-Publication Data

Mobile WiMAX / Edited by Kwang-Cheng Chen, J Roberto B. de Marca.
    p. cm.
  Includes index.
  ISBN 978-0-470-51941-7 (cloth)
  1. Wireless metropolitan area networks.   I. Chen, Kwang-Cheng.   II. Marca, J. Roberto B. de.
TK5105.85.M63 2008
621.384–dc22

                           2007039298

British Library Cataloguing in Publication Data

A catalogue record for this book is available from the British Library

ISBN 978-0-470-51941-7 (HB)

Typeset in 10/12pt Times by Aptara Inc., New Delhi, India
Printed and bound in Great Britain by Antony Rowe Ltd, Chippenham, England.
This book is printed on acid-free paper responsibly manufactured from sustainable forestry
in which at least two trees are planted for each one used for paper production.

Contents

18   Business Model for a Mobile WiMAX Deployment in Belgium   353

*Bart Lannoo, Sofie Verbrugge, Jan Van Ooteghem, Bruno Quinart,
Marc Casteleyn, Didier Colle, Mario Pickavet, and Piet Demeester*

Contributors

Fausto Andreotti, Italtel

Enrico Angori, Elsag-Datamat

Giuseppe Baruffa, University of Perugia

Anand Bedekar, Motorola Inc.

Yan Q. Bian, University of Bristol

Ezio Biglieri, Universitat Pompeu Fabra

Eugen Borcoci, University Politehnica of Bucharest

Marc Casteleyn, Strategy and Business Development, Belgacom

Tsz Ho Chan, The Hong Kong University of Science and Technology

Chih-Wei Chang, SoC Tech. Center, Indus. Tech. Research Inst., Taiwan

Kwang-Cheng Chen, National Taiwan University

Chui Ying Cheung, The University of Washington, Seattle

Didier Colle, Ghent University – IBBT

Marilia Curado, University of Coimbra

Piet Demeester, Ghent University – IBBT

Fabrizio Frescura, University of Perugia

I-Kang Fu, National Chiao Tung University

Stephan Göbbels, RWTH Aachen University

Andrea Goldsmith, Stanford University

Eren Gonen, Motorola Inc.

Emiliano Guainella, University of Rome 'La Sapienza'

Mounir Hamdi, The Hong Kong University of Science and Technology

Christian Hoymann, RWTH Aachen University

Chia-Chi Huang, National Chiao Tung University

Chih-Wei Huang, University of Washington

Jenq-Neng Hwang, University of Washington

Prakash Iyer, Intel Corp.

Marcos Katz, Technical Research Centre of Finland

Hyunpyo Kim, KT Corp.

Bart Lannoo, Ghent University – IBBT

Byeong Gi Lee, Seoul National University

Jaekon Lee, Samsung Electronics

Yong-Hwan Lee, Seoul National University

Yi-Ching Liao, MediaTek Inc.

Longsong Lin, INTEL Corp.

Shiang-Jiun Lin, National Chiao Tung University

Yi-Hsueh Lin, RealTek Inc.

Maode Ma, Nanyang Technological University, Singapore

Joseph P. McGeehan, Toshiba Research Europe Limited

Luís Geraldo Pedroso Meloni, State University of Campinas – Unicamp

Paolo Micanti, University of Perugia

Bertrand Muquet, SEQUANS Communications

Nat Natarajan, Motorola Inc

Pedro Neves, Portugal Telecom Inovaçao

Andrew R. Nix, University of Bristol

Mario Pickavet, Ghent University – IBBT

Bruno Quinart, Ghent University – IBBT

Hikmet Sari, SUPELEC and SEQUANS Communications

Mohanty Shantidev, Intel Corp.

Wern-Ho Sheen, National Chiao Tung University

Yong Sun, Toshiba Research Europe Limited

Zhifeng Tao, Mitsubishi Electric Research Laboratories

Koon Hoo Teo, Mitsubishi Electric Research Laboratories

Jan Van Ooteghem, Ghent University – IBBT

Muthaiah Venkatachalam, Intel Corp.

Sofie Verbrugge, Ghent University – IBBT

Xiangying Yang, Intel Corp.

Jae-Heung Yeom, Seoul National University

Chung-Kei Yu, National Taiwan University

Jinyun Zhang, Mitsubishi Electric Research Laboratories

Preface

The Worldwide Interoperability for Microwave Access technology, under its trade name of WiMAX, has been the talk of the world in the wireless communications industry for the past five years. It is a technology that aims to provide wireless long-distance broadband access for a variety of applications. It all started in 1999 when the IEEE Standards Association authorized the start of the working group known as 802.16, also referred to as the Wireless MAN (Metropolitan Area Network) working group. Although some results were produced by this group in 2002 with the initial standards for line-of-sight operation in frequencies in the range of 11–66 GHz, the first comprehensive standard that encompasses also non-line-of-sight operation was released at the end of 2004. The IEEE 802.16-2004 standard (developed by group 802.16d), was developed for point-to-point and point-to-multi-point operations and includes profiles for operations in the 2–11 GHz spectrum. The other important development that took place in this period was the creation in 2001 of the industry partnership called the WiMAX Forum. The WiMAX Forum defines itself as an industry-led non-profit organization comprising more than 470 companies including 141 operators (as of October 2007) committed to promoting and certifying interoperable WiMAX products. Their web-site also states that 'WiMAX products are designed to deliver wireless broadband services to both residential customers and businesses by creating economies of scale made possible by standards-based, interoperable products and services'. There is no question that this Forum is playing and will continue to play an important role if WiMAX technology is to become an operational success. On the other hand, they are also responsible for the great news hype surrounding this technology. However, media hype and aggressive marketing campaigns with possibly over-optimistic claims have been a constant feature whenever a new communication technology has been developed in the past 20 years.

Meanwhile the efforts within IEEE 802.16 continued, aiming at a new version of the technology that was suitable to provide services to mobile terminals. The corresponding standard was approved at the end of 2005 and is known as IEEE802.16e-2005 leading to what is often called Mobile WiMAX or m-WiMAX. The excitement about Mobile WiMAX is not only hype but is also due to the great flexibilities that this technology offers. It also results from the fact that it is an open standard family of solutions that has the potential to compete with 3G technologies (and their evolutions). This excitement is also due to the roster of novel and efficient techniques included in its specification. These novelties include a scalable OFDMA access mode which is very well suited to operate with MIMO (multiple input, multiple output) schemes, the possible use of low-density parity error correcting codes and an all-IP structure. When all this wealth of knowledge is put together, there is the justified expectation that the resulting performance in terms of spectral efficiency and achieved throughput will surpass the existing options.

There is a myriad of applications envisioned for both 802.16-2004 and Mobile WiMAX. One of the immediate applications is point-to-point communications backhaul usage. A simple point-to-multipoint application is the interconnection of wireless LANs access points. Once mobility is added, the spectrum of applications significantly enlarges to include, as described in Chapter 16, telemedicine and accident prevention services, internet access to the general population in developing countries as well as a full-blown public cell service that can be offered also by non-incumbent operators.

Of course, there are still significant challenges ahead before the great promise and hype can become a real market success, a technology that has been truly adopted by society. As we all know, it is not unknown for a great technology on paper never to really enjoy widespread adoption. Some of the challenges are business-related, others are technical in nature. One of the business challenges is the competition from the evolution of 3G systems. It seems clear that the incumbent operators will want to continue offering their services, in the existing spectrum through upgrades of their current infrastructure. In the recent past the deployment of open standard technology (WiFi) has offered an option for data transmission that has slowed the deployment and acceptance of 3G technologies, since users could satisfy some of their needs through this fixed alternative. Of course, WiFi had a great advantage due to the low cost of access points and the large number of WiFi-enabled laptops. It is not clear yet whether the cost of the Mobile WiMAX infrastructure will offer a significant (or any at all) advantage over options arising from the traditional royalty-based cellular industry. Another related challenge is that the size of the terminal market will guarantee that sufficient low-cost advanced terminals will be available as an attractive option to swing customers to Mobile WiMAX. Of course, the decision by some countries, such as South Korea with its WiBro project (see Chapter 14), to provide full support to its vendor and operational industries to allow them to adopt and develop the new technology will play an important role in helping disseminate its use and encourage the lowering of the terminal cost.

There are also technical challenges. Some of these challenges will be discussed in detail in this book and they arise from the flexibilities afforded by the standard approved and that has also contributed greatly to the excitement about this technology in the engineering and scientific communities. As an example, the scheduling algorithm at the MAC layer that is critical in offering differentiated quality of services is not defined in the IEEE standard. Similarly, the standard, as typical in the 802 series, does not address network topology and protocols. This is being done now under the sponsorship of the WiMAX Forum. Mobility management requires a definition of handoff algorithms that should work not only between WiMAX base stations but also across technologies. The Mobile WiMAX specification allows ample opportunities to optimize performance through radio resource management techniques. There is also a lot to be learned in terms of frequency planning of WiMAX systems. Finally, the efficiency and performance of the Mobile WiMAX technology in rendering the envisioned services still are not fully understood and we hope this book will contribute to answering these questions.

Structure of the Book

This book is organized into four parts, attempting to cover the broad scope of issues essential to the success of Mobile WiMAX, ranging from physical layer developments to existing field trials and business model discussions.

The book kicks off with a tutorial by Roger Marks (the IEEE 802.16 Chair), L. Lin and K.C. Chen that describes the main features of the 802.15 family of standards. In particular, they focus on the MAC layer characteristics, the Mobile WiMAX physical layer (PHY) properties and the current state of the development of the network structure.

In Part One, Physical Layer Transmission, the following four chapters deal with performance, optimization and improvement opportunities of the Mobile WiMAX physical layer. As already mentioned, the Mobile WiMAX PHY standard includes many different features and options to make the best use of the wireless channel characteristics. One of these features is the use of multiple input, multiple output (MIMO) techniques such as transmit/receive diversity and spatial multiplexing. It is well known that multiple antenna schemes can be used to improve the performance of wireless systems by increasing the transmitted data rate through spatial multiplexing, and/or reducing interference from other users. This is the topic of Chapter 2, written by the top-notch team of Muquet, Bigileri, Goldsmith and Sari, where they initially present a general description of MIMO systems. Next, the authors review the multi-antenna profiles adopted for WiMAX systems, discuss their relative merits, and address implementation issues.

Chapter 3 by Yeom and Lee discusses the use of interference cancellation techniques to improve quality of service at the edge of a cell. This is a problem that also affects most 3G systems which can result in severe unfairness if the user is stationary. There is a desire to operate Mobile WiMAX without frequency reuse and therefore this problem becomes critical. On the other hand, improving service to terminals at the cell border will greatly reduce the overall throughput. In this chapter Yeom and Lee first describe conventional inter-cell interference (ICI) mitigation techniques for OFDMA systems and briefly describe how such mitigation techniques can be applied to the Mobile WiMAX system. The authors also offer a new strategy to resolve the problem.

Another feature of the Mobile WiMAX physical layer is the use of adaptive modulation and coding (AMC) to better match instantaneous channel and interference conditions. However, policies on how to select the most appropriate modulation and coding scheme that should be used under various link conditions are not specified in the IEEE standard. In Chapter 4, Chan, Cheung, Ma and Hamdi, in addition to offering a comprehensive overview of the IEEE 802.16e MAC layer, investigate rate adaptation algorithms suitable for use in conjunction with MIMO techniques. The authors also propose a framework in which the PHY layer metrics can be passed into the MAC layer in a practical simulation environment that is required to evaluate the performance of rate adaptation procedures.

Mobile WiMAX uses an Orthogonal Frequency Division Multiple Access (OFDMA) scheme which has several advantages in dealing with multipath fading and in providing high spectral efficiency. However, poor phase noise spectrum can be very detrimental to the overall uplink performance if not properly compensated for. Yu, Liao, Lin and Chen, in Chapter 5, describe several models of phase noise sources and their effect in OFDM and OFDMA systems. They also show how to mitigate multiple phase noise in OFDMA uplink for two different sub-carrier assignment schemes.

Part Two, Medium Access Control and Network Architecture, comprising Chapters 6–8, is devoted to issues related to layer 3 and above.

TCP-based applications such as web browsing, email, and FTP are among the most popular internet applications and should be supported by Mobile WiMAX with good performance. The main focus of Chapter 6, by Yang, Venkatachalam and Yang, is to show that the flexible MAC framework of WiMAX is the key to optimizing system-level application performance.

The group of authors formed by Yang, Venkatachalam and Shantidev show in Chapter 7 that different schedulers have particular impacts on TCP performance in terms of throughput and fairness. It is observed that MAC layer enhancement alone is not sufficient to improve the application of end-to-end performance, particularly in Mobile WiMAX networks. Joint optimization of physical layer parameters and MAC layer algorithms can significantly improve overall throughput without compromising the performance of individual flows and fairness among users. Optimized hard handover as well as related sleep/idle mode operations should be carefully studied to guarantee a seamless mobile computing experience. A related topic also associated to mobility is how to locate a mobile station when there is a need to establish a connection to that station. Furthermore, the paging procedure adopted must be energy-efficient to exchange battery charge life. Another requirement critical to most applications is an upper bound on the paging latency. Chapter 7 considers the trade-off between paging latency and signaling message overhead. The same authors of the previous chapter initially offer an overview of idle mode and paging operation in Mobile WiMAX networks and then proceed to describe a novel algorithm that strikes a good balance between signaling load and paging latency.

The specifications contained in the IEEE 802.16e-2005 standard, as well as the IEEE 802.16-2004, are limited to physical layer and the medium access control (MAC) sub-layer. A Convergence Sub-layer (CS) was added to the standards, allowing multiplexing of various types of network traffic into the MAC layer. In January 2005, the WiMAX Forum constituted a working group to specify the complementary end-to-end interoperable network architecture. This network specification targets an end-to-end all-IP architecture optimized for a broad range of IP services. Chapter 8 is devoted to a brief description of the main concepts and functions of the network architecture that is currently being developed within the WiMAX Forum. Natarajan, Iyer, Venkatachalam, Bedekar and Gonen examine the network design principles underlying the architecture and introduce the network reference model (NRM), which identifies key functional entities and reference points over which a network interoperability framework is defined. The chapter also addresses messaging and procedures that are being developed to provide network support of mobility.

Due to significant loss of signal strength along the propagation path and the transmit power constraint of IEEE 802.16/16e mobile stations, the sustainable coverage area for a specific high data rate is often of limited geographical size. This observation is also valid regarding 3G cellular technologies. The performance can certainly be improved by deploying additional base stations. The drawbacks are increased infrastructure and maintenance costs and a more difficult interference management scenario. An alternative approach is to use low cost relay stations, introduced into the network to help extend the range, improve quality of service (QoS), boost network capacity, and eliminate dead spots, all in a cost-effective fashion. In March 2006, a new task group, IEEE 802.16j, was officially established, which attempts to improve the current IEEE 802.16e-2005 standard defining a minimal set of functional enhancements to support mobile multi-hop relay (MMR) operation. Recently a baseline document was issued to this effect. Part Three, Multi-hop Relay Networks, comprising the next four chapters in this book (Chapters 9–12) are devoted to this new exciting development in the area of Mobile WiMAX. The first chapter in this part is authored by Tao, Teo and Zhang, and they start by explaining the current view of the IEEE 802.16j MMR network and the challenges faced in advancing this new technology. They follow by introducing a new scheme called *tunneling*, which is designed specifically to leverage the inherent notion of 'aggregation' in relay links.

These authors argue that the tunneling mechanism can significantly simplify the routing, QoS management and relay station (RS) handover at the intermediate RSs along the relay path, while still maintaining backward compatibility. Chapter 10 is authored by Lin, Sheen, Fu, Huang, and focuses on new resource scheduling methods when directional antennas equip both the base station and the relay stations in a Manhattan-like environment. Results show that the overall system throughput can be dramatically increased by the new methods, as compared to the system with omni-directional antennas. Chapter 11 pursues a similar line proposing another approach to increase the efficiency of the relays in a Mobile WiMAX environment. Sun, Bian, Nix and McGeehan provide a thorough analysis of relay efficiency in the context of Mobile WiMAX. A directional distributed relaying architecture is introduced for highly efficient radio resource sharing. This architecture is based on both interference cancellation and interference avoidance. The results presented demonstrate that resource sharing has the potential to double the system efficiency compared to relay systems without resource sharing. Furthermore, it is noted that relay deployment extends the applicability of adaptive antenna systems to control and mitigate interference. Chapter 12 by Hoymann and Göbbels offers an extremely interesting and comprehensive exercise on dimensioning a cellular multi-hop WiMAX network. It takes into account the effects of sectorization and clustering and discusses in detail time and space division multiplexing of relay sub-cells. In the end, they compute the capacity of an IEEE 802.16e-2005 both for single hop as well as multi-hop configurations. As a result, the authors draw very enlightening conclusions regarding the advantages and suitability of each solution.

Part Four, Multimedia Applications, Services, and Deployment, comprising Chapters 13–18, deals with applications, the actual commercial deployment of Mobile WiMAX and with business aspects.

A special feature of this book is Chapter 14 by H. Kim, J. Lee and B.G. Lee that describes in great detail the WiBro (*Wireless Broadband*) that since early 2007 has been in commercial operation in the Seoul area. WiBro has been fully harmonized with the IEEE 802.16e-2005. The chapter provides a wealth of information about the Korean system, including network architecture, planning aspects, terminal characteristics and service options.

Chapters 13 and 16 focus on the potential of video applications to be offered using WiMAX. Video streaming over Mobile WiMAX is the subject of Chapter 13, authored by Hwang, Huang and Chang. The authors show that the advanced QOS features in WiMAX can afford very reliable wireless transmission. They contend also that the use of a cross-layer design that considers both the WiMAX MAC functionality as well as an end-to-end mechanism can greatly contribute to the observed benefits. The focus of Chapter 16 is a very interesting application for both 802.16-2004 and 802.16e-2005. Micanti, Baraffa and Frescura consider the distribution of digital cinema from studios to one or more regional theaters and also to end users with access to broadband infrastructure. The choice of Mobile WiMAX as one of the distribution technologies allows the destination of the video content to be an audience located, for example, on a bus or in a high speed train. Micanti, Baruffa and Frescura then present a technique for encapsulating Digital Cinema JPEG 2000 compressed sequences into a reliable multicasting protocol, for the purpose of distribution among a main production site and the projection theaters or end users.

The potential social benefits of a versatile and efficient technology such as Mobile WiMAX are described in Chapters 15 and 17. L.G.P. Meloni in Chapter 15 considers the use of Mobile WiMAX as the return channel technology of a digital TV system in a developing nation. In this scenario, the return channel could be the best way to provide access to modern information services to an underprivileged segment of the population. There are specific requirements for this application including a very high number of users in dense urban areas, fairness considerations when users are located at cell edges and high volume of simultaneous access in some peak short periods caused by live audience programs. The author, through simulation experiments, evaluates the sector capacity as well as delay numbers for different traffic combinations and propagation scenarios. Chapter 17, by Guainella, Borcoci, Katz, Mendes, Curado, Andreotti and Angori, illustrates the adoption of WiMAX technology in support of environmental monitoring, accident prevention and telemedicine in rural areas. The work was performed within the scope of the WEIRD project funded by the European Commission. The authors describe the key technologies adopted by the project and the open system architecture specified, fulfilling the requirements of mobility and Quality of Service. They describe also how the results will be validated with the use of four testbeds.

Business models and rollout scenarios is the topic of our final chapter. A team of Belgian authors (Lanoo, Verbrugge, van Ooteghen, Quinart, Casteleyn, Colle, Pickavert and Demeester) developed a detailed business model to investigate the potential model of Mobile WiMAX to offer broadband services in their country. The model includes different business and rollout cases and relies on a planning tool developed by the authors using several technical features of Mobile WiMAX.

We hope you will enjoy the breadth of coverage as well as the quality of the contributions that were compiled for this book. We hope that by reading it you will get a better understanding of the potential of this new technology and the issues that are keeping engineers and scientists busy trying to make it a market success and to further improve its performance.

Acknowledgments

Last but not the least, we appreciate the IEEE ComSoc staff support organization of the 2007 Mobile WiMAX Symposium, and Ms G.L. Pai at the National Taiwan University who helped tremendously in preparing the manuscripts for this book. Kwang-Cheng (K.-C.) Chen would like to especially thank the Dr Irving T. Ho Foundation and the National Taiwan University for their supporting appointment as *Irving T. Ho Chair Professor* from January 2007, to allow him to focus more on developing new communication and networking technology and serve the research community related to mobile WiMAX and its future evolution.

1

Introduction to Mobile WiMAX

Longsong Lin, and Kwang-Cheng Chen

1.1 IEEE 802.16

In order to introduce WiMAX, we must start from the IEEE 802.16. IEEE 802 defines international standards (more precisely, to be recognized by the ISO later) for local area networks (LAN) and metropolitan area networks (MAN), such as IEEE 802.3 well known as Ethernet. IEEE 802 projects generally deal with the physical layer transmission (PHY) and medium access control (MAC), and leave the network layer and above to other international standards such ISO. Since 1990, there have been a few wireless standards in IEEE Project 802:

- IEEE 802.11 wireless LANs (WLAN);
- IEEE 802.15 wireless personal area networks (WPAN);
- IEEE 802.16 wireless metropolitan area networks (WMAN);
- IEEE 802.20 and several others.

With popular WiFi applications (i.e. wireless LANs) especially after hot-spot deployment, more reliable wireless broadband technology for Internet access attracts great interest. The concept for wireless metropolitan area networks (WMAN) has therefore been introduced in recent years. Of the many efforts, the IEEE 802.16 standard originally defining fixed broadband wireless (FBW) is widely considered a new generation technology to replace the past wireless local loop (WLL) in telecommunications, and to deliver performance comparable to traditional cable, T1, xDSL, etc. The advantages of IEEE 802.16 include:

- quick deployment, even in those areas where it is difficult for wired infrastructure to reach;
- the ability to overcome physical limitation of traditional wired infrastructure;
- reasonable installation costs to support high rate access.

Mobile WiMAX   Edited by Kwang-Cheng Chen and J. Roberto B. de Marca.
© 2008 John Wiley & Sons, Ltd

In other words, standardized FBW can support flexible, cost-effective, broadband access services in a wide range of devices. WiMAX (Worldwide Interoperability for Microwave Access) Forum is a non-profit corporation formed by equipment and component suppliers to promote the adoption of IEEE 802.16-compliant equipment by operators of broadband wireless access systems, which is comparable to the WiFi Alliance in promoting IEEE 802.11 wireless LANs. WiMAX is establishing 'System Profiles' for all compliant equipment, which can also address regulatory spectrum constraints faced by operators in different geographical regions. The WiMAX forum is also developing higher-layer specifications to match IEEE 802.16. In the meantime, WiMAX-defining conformance tests in conjunction with interoperability enable service providers to choose multiple vendors. WiMAX is working with the ETSI (European Telecommunications Standards Institute) to develop the HIPERMAN standard.

In April 2002, IEEE 802.16 was published for 10–66G Hz operations, while line-of-sight transmission is considered a primary application. To promote immediate wider applications, IEEE 802.16a was published in January 2003, which aims at 2–11G Hz operations for non-line-of-sight performance.

Fixed broadband wireless (FBW) access applications based on point-to-multipoint network topology primarily include:

- cellular (or Fixed-Network) backhaul;
- broadband on demand;
- residential broadband;
- underserved areas services;
- nomadic wireless services.

As a consequence, FBW (later refined as Broadband Wireless Access, BWA, for the IEEE 802.16) systems and networks supports:

- high throughput;
- high degree of scalability;
- quality-of-service (QoS) capability;
- high degree of security;
- excellent radio coverage.

IEEE 802.16 Wireless MAN has a connection-oriented MAC and PHY is based on non-line-of-sight radio operation in 2-11 GHz. For licensed bands, channel bandwidth will be limited to the regulatory provisioned bandwidth divided by any power of 2, no less than 1.25M Hz. Three technologies have been defined:

- single carrier (SC);
- orthogonal frequency division multiplexing (OFDM);
- orthogonal frequency division multiple access (OFDMA).

The communication of frame-based IEEE 802.16 is based on the fundamental concept by defining burst profiles in each BS-SS communication link. To better reflect the new application scenarios, IEEE 802.16 is now known as Wireless Broadband Access.

IEEE 802.16 had a revision published in October 2004, which is known as IEEE 802.16-2004. The mobile version of IEEE 802.16 has been developed in the IEEE 802.16e (official name, 'Physical and Medium Access Control Layers for Combined Fixed and Mobile Operation in Licensed Bands'), which is commonly known as *Mobile WiMAX*, especially considering its OFDMA (orthogonal frequency division multiple access) PHY. Such a mobile enhancement of IEEE 802.16e is primarily specified for licensed bands and Korean *WiBro* provides mobile services based on IEEE 802.16-2004 and IEEE 802.16e. Chapter 14 introduces WiBRO systems and applications. At the ITU-R May 2007 meeting in Japan, Mobile WiMAX was recommended as OFDMA TDD WMAN (though still subject to further formal approval), thus leaving 50M Hz bandwidth internationally available at 2.57–2.62 GHz from 3G TDD spectrum, on a per nation basis.

Since December 2006, IEEE 802.16m has started as a new amendment project to study the IEEE 802.16 WirelessMAN-OFDMA specification to provide an advanced air interface for operation in licensed bands, and to meet the cellular layer requirements for IMT-Advanced for the next generation of mobile networks, of course, with continuing support for legacy WirelessMAN-OFDMA equipment and devices. The target speed for IEEE 802.16m is 100M bps, with supporting high mobility, so that it can serve as a candidate for IMT-Advanced. Consequently, 3G LTE (long-term evolution) from 3GPP, UMB (ultra-mobile broadband) from 3GPP2, and IEEE 802.16e and 802.16m, are all adopting OFDMA-based technology.

1.2 IEEE 802.16 MAC

IEEE 802.16 Medium Access Control (MAC), which IEEE 802.16e MAC generally follows, has a network topology of point to multi-point (PMP), with support for mesh network topology. Its backhaul can be either ATM (asynchronous transfer mode) or packet-based (such as IP networks). From the reference model as illustrated in Figure 1.1, there are three sub-layers in the MAC:

- *Service Specific Convergence Sub-layer (CS)*: providing any transformation or mapping of external network data through CS SAP (CS service access point).
- *MAC Common Part Sub-layer (MAC CPS)*: classifying external network service data units (SDUs) and associating these SDUs to proper MAC service flow and Connection Identifier (CID). Multiple CS specifications are provided to interface with various protocols.
- *Privacy (or Security) Sub-layer*: supporting authentication, secure key exchange, and encryption.

Different from typical MACs using random access techniques in the IEEE 802, IEEE 802.16 MAC is connection oriented, and similar to time division multiple access (TDMA). Once a subscriber station (SS) enters the network, it creates one or more connections to communicate with the base station (BS). It also performs link adaptation and automatic repeat request (ARQ) functions to maintain the target bit error rate. To further support multimedia traffic, IEEE 802.16 MAC may have to use radio resources, and provide quality-of-service (QoS) differentiation in services, which are not considered typical MAC functions. To support OFDMA PHY, the MAC layer is responsible for assigning frames to the proper zones and exchanging this structure

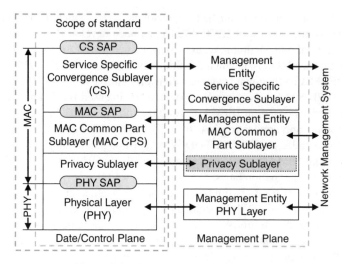

Figure 1.1 Reference model of IEEE 802.16

information with the SSs in the DL and UL maps. Transmit diversity and adaptive antenna system (AAS), as well as MIMO zone, are included.

IEEE 802.16 MAC is connection-oriented. As BS controls the access to the medium, band-width is granted to SSs on demand. At the beginning of each frame, the BS schedules the uplink and downlink grants to meet the negotiated QoS requirements. Each SS learns the boundaries of its allocation under current uplink sub-frame via the UL-MAP message. The DL-MAP delivers the timetable of downlink grants in the downlink sub-frame.

The IEEE 802.16e MAC provides QoS differentiation for different types of applications, and defines four types of services:

- *Unsolicited Grant Services (UGS)*: UGS is designated for constant bit rate (CBR) services, such as T1/E1 emulation and VoIP without silence suppression.
- *Real-Time Polling Services (rtPS)*: rtPS is designated for real-time services that generate variable size of data packets on a periodic basis, such as MPEG video and VoIP with silence suppression.
- *Non-Real-Time Polling Services (nrtPS)*: nrtPS is designated for non-real-time services that require variable size data grant burst types on a regular basis.
- *Best Effort Services (BE)*: It counts typical data traffic such as Internet web browsing and FTP file transfer.

There are different bandwidth-request mechanisms in WiMAX. For unsolicited granting, a fixed amount of bandwidth on the periodic basis is requested at the set-up phase of uplink. Then, bandwidth is never explicitly requested. The unicast poll allocates necessary bandwidth for a polled uplink connection. The broadcast polls are issued by the BS to all uplink connections, while a truncated exponential back-off algorithm is employed to resolve possible collisions in polling. Based on the bandwidth requested and granted, the BS uplink scheduler estimates the residual backlog at each uplink connection, and allocates future grants. An SS scheduler must be implemented with each SS MAC, in order to re-distribute the granted capacity to all its connections. However, note that IEEE 802.16 does not specify scheduling algorithms that are left to manufacturers.

Similar to the concept of cellular layer-2/3, IEEE 802.16 MAC has the radio link control (RLC) to control PHY transition from one burst profile to another, in addition to traditional power control and ranging.

Another important sub-layer in the IEEE 802.16 MAC is the security sub-layer, and an improved version has been developed for the IEEE 802.16e. The Privacy and Key Management Protocol version 2 (PMKv2) is the basis of Mobile WiMAX security. Device and user authentication adopts IETF EAP protocol. The traffic encryption follows the IEEE 802.11i using AES-CCM to protect traffic data. The keys used to derive the ciphertext are generated from the EAP authentication. To avoid further attacks and hostile analysis, a periodic key (TEK) refreshing mechanism enables improved protection. A three-way handshake scheme in Mobile WiMAX optimizes the re-authentication mechanism for fast handover by preventing man-in-the-middle-attacks.

To deal with mobility in Mobile WiMAX, IEEE 802.16e the MAC specifies MAC layer handover procedure, while the exact handover decision algorithm is not specifically defined. Handover happens in two possible situations:

- when the mobile station (MS) moves and needs (due to signal fading, interference level, etc.) to change the BS that is currently connected, in order to provide a better signal quality;
- when MS can be served with higher QoS at another BS.

Prior to handover, network topology acquisition must be achieved in three steps:

1. *Network topology advertisement*: A BS broadcasts information regarding the network topology (typically using MOB_NBR-ADV), which might be obtained from the backbone.
2. *MS scanning the neighboring BSs*: For the purpose of MS seeking and monitoring suitability of neighboring BSs as handover candidates, BS can allocate time interval(s) to MS and such a scanning duration is known as a scanning interval. Once a BS is identified through scanning, MS may attempt to synchronize with its downlink transmission and estimate the quality of physical channel. The serving BS may buffer the incoming data during the scanning interval until the exit of the scanning mode.
3. *Association*: Association is an optional initial ranging procedure during the scanning interval with respect to one of the neighboring BSs. The function of association is to enable the MS to acquire and to record ranging parameters and service availability information for the purpose of proper selection of the handover target.

After network topology acquisition, the handover process proceeds for a MS migrating from the air-interface (or radio resource) provided by one BS to that provided by another BS, as the following stages:

- *Cell re-selection*: MS may use neighboring BS information, or may request to schedule scanning intervals to scan/range, in order to evaluate MS interests in handover to neighboring BS.

- *Handover decision and initiation*: A handover begins with a decision for an MS to switch from a serving BS to a target BS. Such a decision can be originated by either MS or serving BS.
- *Synchronization to target BS downlink*: MS in handover process first synchronizes to downlink transmissions of target BS to obtain DL and UL transmission parameters (such as MAP). If the target BS has previously received handover notification from serving BS through backbone, the target BS may allocate a non-contention-based initial ranging opportunity.
- *Ranging*: The target BS may get information from the serving BS through the backbone network. The MS and target BS will conduct either initial ranging or handover ranging to set up the correct communication parameters.
- *Termination of MS context*: This is the final step in handover, to terminate service at the serving BS. An MS may terminate handover at any time prior to termination.

1.3 IEEE 802.16e Mobile WiMAX

Mobile WiMAX is generally considered to be the IEEE 802.16e-2005 adopting OFDMA PHY. In this book, we shall describe recent advances in mobile WiMAX from technology to services and applications. First, we shall briefly introduce Mobile WiMAX in this section.

The IEEE 802.16e-2005 supports both time division duplexing (TDD) and frequency division duplexing (FDD) modes. However, the initial release of Mobile WiMAX profiles only considers the TDD mode of operation for the following reasons:

- It enables dynamic allocation of downlink (DL) and uplink (UL) radio resources to effectively support asymmetric DL/UL traffic that is common in Internet applications. The allocation of radio resources in DL and UL is determined by the DL/UL switching point(s).
- Both DL and UL are in the same frequency channel to yield better channel reciprocity and to better support link adaptation, multi-input-multi-output (MIMO) techniques, and closed-loop advanced antenna technique such as beam-forming.
- A single frequency channel in DL and UL can provide more flexibility for spectrum allocation.

To further alleviate spectrum allocation efforts, mobile WiMAX adopts simple frequency reuse schemes by reusing 1 and 3 with PUSC (Figure 1.2).

As pointed out in [3], time division DL/UL in multiple-cell wireless networks can create up-down collisions (or interference) resulting in performance loss. Fractional frequency reuse (FFR) as shown in Figure 1.3 can be applied by utilizing frequency reuse $1 \times 3 \times 1$ near the center and frequency reuse $1 \times 3 \times 3$ near the cell edges. There is no need for frequency planning and it is very flexible to configure the networks. Frequency reuse factor 1 at the center of the cell maximizes the network spectral efficiency, while higher reuse factor at cell edges alleviates (co-channel) interference.

When we design a mobile WiMAX system, we normally use the wide-sense stationary uncorrelated scattering (WSSUS) to stochastically model the time-varying fading wireless channels in time and frequency domains. Two main factors from this model are used in developing the system parameters: Doppler spread and thus coherence time of the channel, and multipath

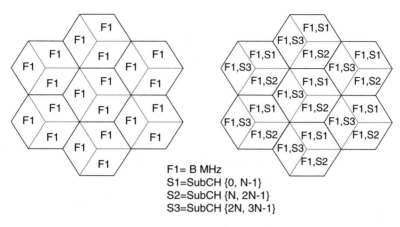

Reuse 1x3x1   Reuse 1x3x3

Figure 1.2 Frequency reuse schemes (a) $1 \times 3 \times 1$ (b) $1 \times 3 \times 3$

delay spread and thus coherence bandwidth. Stanford University Interim (SUI) channel models are widely accepted in the study of WiMAX systems.

One very special feature in (mobile) WiMAX is ranging while SS at initial entry and also periodically in normal operation, in which the mobile subscriber station (MS) acquires frequency, time, and power adjustments, so that all MS transmissions can align with the UL sub-frame received by the base station (BS). The ranging process proceeds by MS transmitting a signal and BS responding with required adjustments, which is a closed loop control process critical to OFDMA communications in (mobile) WiMAX. Ranging happens in three ways: initial/handoff ranging, periodic ranging, and BW request ranging.

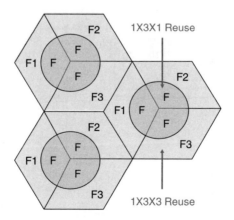

F=F1+F2+F3   F1: F,S1 F2: F,S2 F3: F,S3

Figure 1.3 Fractional frequency reuse

Mobile WiMAX OFDMA PHY adopts scalable OFDMA with 1.25×2^n MHz bandwidth, $n = 0, 1, 2, 3, 4$, at fixed sub-carrier spacing. There are three types of OFDMA sub-carriers:

- data sub-carriers for data transmission;
- pilot sub-carriers for estimation and synchronization purposes;
- null sub-carriers for guard band and DC carriers, without transmission at all.

The pilot sub-carrier allocation can be performed in different modes. For DL Fully Used Subchannelization (FUSC), the pilot tones (or sub-carriers) are allocated first and then the remaining sub-carriers are arranged for data sub-channels. For DL Partially Used Subchannelization (PUSC) and all UL modes, the set of all used sub-carriers (pilot and data) is partitioned into sub-channels, and then pilot sub-carrier(s) are allocated within each sub-channel.

Adaptive modulation and coding (AMC) is adopted by using QPSK, 16QAM, 64QAM (optional in UL) as modulation, and convolutional codes (mandatory), turbo codes, low-density parity check codes for forward error correcting codes (FEC). Of most interest, space–time codes (STC) and spatial multiplexing (SM) are used to enhance PHY transmission speed and reception quality of signals from mobile stations. Along with adaptive antenna systems (AAS) using beamforming technique, STC and SM form the foundation of multi-input multi-output (MIMO) processing for the mobile WiMAX and Chapter 2 of this book has an excellent introduction to this.

STC originated in pioneer work of transmit diversity coding by S. Alamouti [9], and Figure 1.4 depicts the realization of closed-loop STC for mobile WiMAX, and Alamouti code as shown. There are three kinds of STC in the IEEE 802.16e OFDMA by using two or three antennas, while Alamouti code is one of them.

In addition to STC, spatial multiplexing can be further used, which is depicted in Figure 1.5. Spatial multiplexing transmits data streams via different spatial domains (typically multiple antennas). STC and spatial multiplexing can form the foundation of IEEE 802.16e MIMO

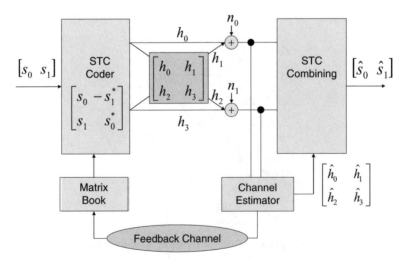

Figure 1.4   Closed-loop space time coding (Alamouti codes as an example)

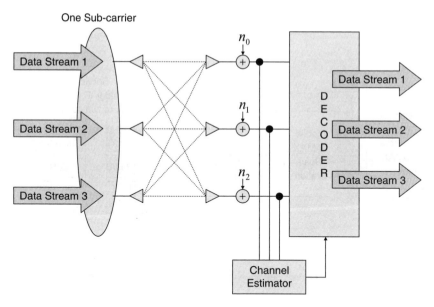

Figure 1.5 Spatial multiplexing

processing. The BS can send control messages to indicate if the subsequent allocation should use a certain permutation with a specific transmit diversity mode, and to describe DL allocations assigned to MIMO-enabled SSs by defining one of the three STC matrices.

To support STC and spatial multiplexing, pilots for multiple transmission antennas should be separate to avoid inter-stream interference. It is worth noting that OFDM particularly fits MIMO and adaptive antenna techniques, compared with CDMA and single carrier transmission.

The final issue in physical layer transmission is the radio resource allocation of OFDMA associated with appropriate pilot sub-carrier allocation as another important dimension of OFDMA communication. Pilot sub-carrier allocation can be found in [10], however, data sub-carrier, bits, and power allocation can be found in the literature without a detailed description as in [10].

1.4 Mobile WiMAX End-to-End Network Architecture

IEEE 802.16e only defines PHY and MAC. However, in light of the needs of interfaces at higher layers to allow multi-vendor supply as typical wireless communication standards, WiMAX Forum has working groups beyond the IEEE 802.16. The mobile WiMAX End-to-End Network Architecture is developed on an all-IP platform with all packet technology and without any legacy circuit telephony.

Figure 1.6 depicts an IP-based WiMAX network architecture, which consists of three major parts: (1) user terminals (i.e. subscriber/mobile stations); (2) an access service network (ASN); and (3) a connectivity service network (CSN). ASN defines a logical boundary to describe the aggregation of functional entities and corresponding message flows associated with the access services. The connectivity service network (CSN) represents a set of network functions

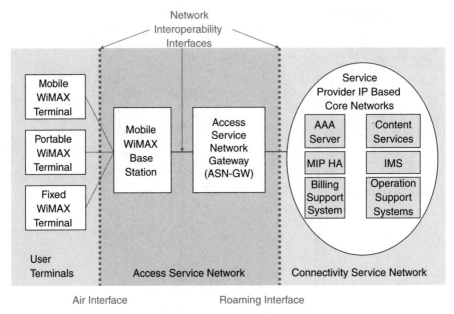

Figure 1.6 WiMAX network IP-based architecture

providing IP connectivity services to WiMAX subscribers. A CSN may compromise network elements such as AAA proxy and servers, routers, user database, and internetworking gateway.

The end-to-end WiMAX network architecture extensively supports mobility and handover, which includes

- vertical or inter-technology handovers under multi-mode operation;
- IPv4- and IPv6-based mobility management;
- roaming between network service providers (NSPs);
- seamless handover up to vehicular speed satisfying bounds of service disruptions.

WiMAX network architecture certainly has provisions to support QoS via differentiated levels of QoS, admission control, bandwidth management, and other appropriate policies.

For more details of mobile WiMAX, please consult [10] and the other useful references listed.

References

[1] C. Eklund, R.B. Marks, K.L. Stanwood, and S. Wang, 'IEEE Standard 802.16: A Technical Overview of the Wireless MAN Air Interface for Broadband Wireless Access', *IEEE Communications Magazine*, June 2002, 50–63.

[2] *INTEL Technology Journal*, special issue on WiMAX, **8**(3), 2004.

[3] K.C. Chen, 'Medium Access Control of Wireless Local Area Networks for Mobile Computing', *IEEE Networks*, 1994, 50–63.

[4] X. Fu, Y. Li, and H. Minn, 'A New Ranging Method for OFDMA Systems', *IEEE Trans. on Wireless Communications*, **6**(2), Feb. 2007, 659–669.

[5] C. Cicconetti, et al., 'Quality of Service Support in IEEE 802.16 Networks', *IEEE Network*, March/April 2006, 50–55.

[6] J. Wang, M. Venkatachalam, and Y. Fang, 'System Architecture and Cross-Layer Optimization of Video Broadcast over WiMAX', *IEEE Journal on Selected Areas in Communications*, **25**(4), May 2007, 712–721.

[7] Q. Ni, et al., 'Investigation of Bandwidth Request Mechanisms under Point-to-Multipoint Mode of WiMAX Networks', *IEEE Communications Magazine*, May 2007, 132–138.

[8] K. Lu, Y. Qian, and H-H. Chen, 'A Secure and Service-Oriented Network Control Framework for WiMAX Networks', *IEEE Communications Magazine*, May 2007, 124–130.

[9] S. Alamouti, 'Simple Transmit Diversity Technique for Wireless Communications', *IEEE Journal on Selected Areas in Communications*, **16**(8), October 1998, 1451–1458.

[10] 'Air Interface for Fixed and Mobile Broadband Wireless Access Systems, Amendment 2: Physical and Medium Access Control Layers for Combined Fixed and Mobile Operation in Licensed Bands', IEEE Std 802.16e-2005, February 2006.

Part One

Physical Layer Transmission

2

An Analysis of MIMO Techniques for Mobile WiMAX Systems

Bertrand Muquet, Ezio Biglieri, Andrea Goldsmith, and Hikmet Sari

2.1 Introduction

Mobile WiMAX systems are based on the IEEE 802.16e-2005 specifications [1] which define a physical (PHY) layer and a medium access control (MAC) layer for mobile and porequation broadband wireless access systems operating at microwave frequencies below 6 GHz. The IEEE 802.16e-2005 specifications actually define three different PHY layers: single-carrier transmission, orthogonal frequency-division multiplexing (OFDM), and orthogonal frequency-division multiple access (OFDMA). The multiple access technique used in the first two of these PHY specifications is pure TDMA, but the third mode uses both the time and frequency dimensions for resource allocation. From these three PHY technologies, OFDMA [2] has been selected by the WiMAX Forum as the basic technology for porequation and mobile services. Compared to TDMA-based systems, it is known that OFDMA leads to a significant cell range extension on the uplink (from mobile stations to base station). This is due to the fact that the transmit power of the mobile station is concentrated in a small portion of the channel bandwidth and the signal-to-noise ratio (SNR) at the receiver input is increased. Cell range extension is also achievable on the downlink (from base station to mobile stations) by allocating more power to carrier groups assigned to distant users. Another interesting feature of OFDMA is that it eases the deployment of networks with a frequency reuse factor of 1, thus eliminating the need for frequency planning.

Since radio resources are scarce and data rate requirements keep increasing, spectral efficiency is a stringent requirement in present and future wireless communications systems. On the other hand, random fluctuations in the wireless channel preclude the continuous use of highly bandwidth-efficient modulation, and therefore adaptive modulation and coding (AMC) has become a standard approach in recently developed wireless standards, including WiMAX. The idea behind AMC is to dynamically adapt the modulation and

Mobile WiMAX   Edited by Kwang-Cheng Chen and J. Roberto B. de Marca.
© 2008 John Wiley & Sons, Ltd

coding scheme to the channel conditions to achieve the highest spectral efficiency at all times [3, Chapter 9].

An additional dimension to modulation and coding aimed at increasing spectral efficiency (data rate normalized by the channel bandwidth) is the space dimension, i.e., the use of multiple antennas at the transmitter and receiver. More generally, multiple-antenna techniques can be used to increase diversity and improve the bit error rate (BER) performance of wireless systems, increase the cell range, increase the transmitted data rate through spatial multiplexing, and/or reduce interference from other users. The WiMAX Forum has selected two different multiple antenna profiles for use on the downlink. One of them is based on the space–time code (STC) proposed by Alamouti for transmit diversity [4], and the other is a simple 2x2 spatial multiplexing scheme. These profiles can also be used on the uplink, but their implementation is only optional.

This chapter discusses the use of multiple-antenna techniques in mobile WiMAX systems. We first present antenna array techniques, which primarily reduce interference and enhance the useful signal power. Next, we give a general description of multi-input multi-output (MIMO) systems, which can be used for different purposes including diversity, spatial multiplexing and interference reduction. Then, we focus on the multi-antenna profiles adopted for WiMAX systems, discuss their relative merits, and address the implementation issues.

2.2 Multiple Antenna Systems

The performance improvement that results from the use of diversity in wireless communications is well known and often exploited. On channels affected by Rayleigh fading, the BER is known to decrease proportionally to SNR^{-d}, where SNR designates the signal-to-noise ratio and d designates the system diversity obtained by transmitting the same symbol through d independently faded channels. Diversity is traditionally achieved by repeating the transmitted symbols in time, in frequency or using multiple antennas at the receiver. In the latter case, the diversity gain is compounded to the array gain, consisting of an increase in average receive SNR due to the coherent combination of received signals, which results in a reduction of the average noise power even in the absence of fading.

If, in addition to multiple receive antennas, one includes multiple transmit antennas, a MIMO system is obtained (see Figure 2.1, for a general block diagram).

Here, the situation is more complex, with a greater deal of flexibility in the design and potential advantages at the cost of a larger system complexity. In fact, in addition to array gain and diversity gain, one can achieve spatial multiplexing gain, realized by transmitting independent

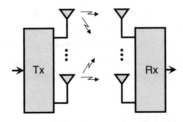

Figure 2.1   General block diagram of MIMO systems

information from the individual antennas, and interference reduction. The enormous values of the spatial multiplexing gain potentially achieved by MIMO techniques have had a major impact on the introduction of MIMO technology in wireless systems.

2.2.1 Antenna Array Techniques

Multiple antennas at the transmitter and the receiver can provide diversity gain as well as increased data rates through space-time signal processing. Alternatively, sectorization or smart (adaptive) antenna array techniques can be used to provide directional antenna gain at the transmitter or at the receiver. This directionality can increase the cell range, reduce channel delay spread and flat-fading, and suppress interference between users. Indeed, interference typically arrives at the receiver from different directions, and directional antennas can exploit these differences to nullify or attenuate interference arriving from given directions, thereby increasing system capacity. Exploiting the reflected multipath components of the signal arriving at the receiver requires an analysis of multiplexing/diversity/ directionality tradeoff. Whether it is best to use the multiple antennas to increase data rates through multiplexing, increase robustness to fading through diversity, or reduce channel delay spread and interference through directionality is a complex tradeoff decision that depends on the overall system design as well as on the environment (urban, semi-urban, rural).

The most common directive antennas are switched-beam or phased (directional) antenna arrays, as shown in Figure 2.2. In these systems, there are multiple fixed antenna beams formed by the array, and the system switches between these different beams to obtain the best performance, i.e., the strongest signal-to-interference-plus-noise-ratio (SINR) of the desired signal. Switched-beam antenna arrays are designed to provide high gain across a range of signal arrival angles, and can also be used to sectorize the directions that signals arrive from. In particular, sectorization is commonly used at base stations to cut down on interference: If different sectors are assigned different frequencies or time slots, then only those users within the same sector interfere with each other, thereby reducing the average interference by a factor

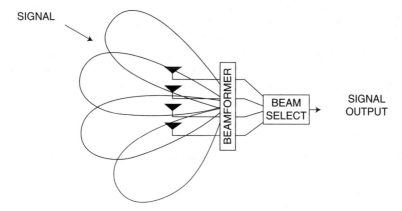

Figure 2.2   Switched-beam (sectorized) array

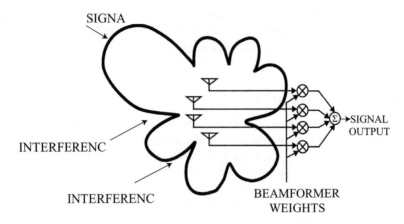

SIGNA

INTERFERENC

INTERFERENC

BEAMFORMER
WEIGHTS

SIGNAL
OUTPUT

Figure 2.3 Smart antenna (phased array)

equal to the number of sectors. For example, if a 360° angular range is divided into three sectors to be covered by three 120° sectorized antennas, then the interference in each sector is reduced by a factor of 3 relative to an omni-directional base station antenna. The price paid for this reduced interference is the increased complexity of sectorized antennas, including the need to switch a user's beam as it moves between sectors. The benefits of directionality that can be obtained with multiple antennas must be weighed against the potential diversity or multiplexing benefits of the antennas.

Adaptive (smart) antenna arrays typically use phased-array techniques to provide directional gain, which can be tightly controlled with a sufficient number of antenna elements. Phased-array techniques work by adapting the phase of each antenna element in the array, which changes the angular locations of the antenna beams (angles with large gain) and nulls (angles with small gain), as shown in Figure 2.3. For an antenna array with N antennas, N nulls can be formed to significantly reduce the received power of N separate interferers. If there are $N_I < N$ interferers, then the N_I interferers can be cancelled out using N_I antennas in a phased array, and the remaining $N - N_I$ antennas can be used for diversity or multiplexing gain. Note that directional antennas must know the angular location of the desired and interfering signals to provide high or low gains in the appropriate directions, and tracking of user locations can be a significant impediment in highly mobile systems.

The complexity of antenna array processing, along with the size of a large antenna array, makes the use of smart antennas in small, lightweight, low-power handheld devices challenging. However, base stations and access points already use antenna arrays in many cases.

2.2.2 *Performance Tradeoffs*

An adaptive array with N antennas can provide the following performance benefits:

1. a higher antenna gain for extended battery life, extended range, and higher throughput;
2. multipath diversity gain for improved reliability, including more robust operation of services;
3. interference suppression;

4. reduced interference into other systems on transmission;
5. higher link capacity through the use of MIMO with spatial multiplexing.

More specifically, an antenna array with N_t transmit antennas and N_r receiver antennas provides an array gain (average SNR increase) of $N_t + N_r$ and a diversity gain (BER slope reduction) of $N_t N_r$. Alternatively, in rich scattering it provides a min(N_t, N_r) multiplexing gain (data rate increase) or it can nullify N_r interferers on the receive end. For example, a 4-element antenna array can provide up to a 13 dB SNR gain (7 dB array gain plus a 6 dB diversity gain), or a four-fold increase in data rate assuming four antennas at both the transmitter and receiver, or a cancellation of up to three interfering signals. However, these improvements cannot all be obtained simultaneously (e.g., suppression of N_r–1 interferers and a diversity gain of N_r are mutually exclusive) yet, each adaptive array in a system can optimize its performance in different combinations of (1)–(5) depending on its situation.

The performance tradeoffs between diversity and multiplexing for antenna arrays are well known [3, 5], and recent developments in space-time codes achieve the fundamental tradeoff performance bounds. However, the tradeoff between interference cancellation (IC) and diversity gain is not well understood. Recent work [6] has explored this tradeoff to obtain the best use of multiple receive antennas in fading channels with interference. This work obtains closed-form expressions for the performance analysis of different antenna array processing schemes based on the outage probability under maximal ratio combining (MRC), optimum combining (OC) [7], and interference cancellation through beam steering. Though OC is known to be the optimum technique in the presence of interference, providing diversity and interference cancellation simultaneously, its implementation complexity is high. Therefore, it may be best to use combined MRC (to provide diversity) and IC (to suppress the strongest interferers). The results in [6] show that IC yields significantly better performance than MRC if the system is interference limited and the number of dominant interferers is lower than the number of receive antennas. When these conditions are not fulfilled, IC is better than MRC if the output SINR is low; and MRC yields better performance otherwise. In fact, at the extreme, optimal combining reduces to either MRC or IC: When interference dominates SINR degradation, OC reduces to IC, and when fading dominates the SINR, OC reduces to MRC to optimally mitigate fading.

A complete performance analysis of MRC and OC in MIMO systems with fading and interference, assuming multiple receive antennas and a single transmit antenna was undertaken in [8]. While the same techniques can be used to analyze performance under multiple transmit antennas, the mathematics become more involved. The main idea behind the analysis is to investigate the optimal weights for the received signal at all antennas to maximize SNR or SINR. The received signal vector across all antennas after weighting is given by

$$r = H_D w_t b_s + \sum_{i=1}^{L} \sqrt{\Omega_i} h_i b_i + n \tag{2.1}$$

where H_D is the vector of receive antenna channel gains for the desired signal, w_t is the vector of weights at the transmitter, b_s is the transmitted symbol of interest, b_i is the symbol of the ith interfering signal, h_i is the gain of the ith interfering signal, and Ω_i is the power of the ith interference signal relative to the desired signal. The combiner output is then

$$y = w_r^H r \tag{2.2}$$

where w_r are the antenna weights at the transmitter. In MRC, the weights w_r yield the maximum SNR of y, and in OC the weights maximize the SINR of y. For MRC the weights are well known to be $w_t = \sqrt{\Omega_D} u$ and $w_r = H_D u$.

It can be shown [9] that the SINR of y assuming weights associated with MRC is given by

$$\gamma = \frac{\Omega_D \lambda}{\sum_{i=1}^{L} \Omega_i \chi_i + \sigma^2} \tag{2.3}$$

where λ is the maximum eigenvalue of the matrix $H_D^H H_D$ and the χ_i are exponential random variables with unit mean. The SINR distribution thus depends on the distribution of λ and the power of the interferers.

In [8], a closed-form expression for the outage probability of γ is obtained based on the moment-generating function (MGF) of the sum of the interferers $\chi = \sum_i \Omega_i \chi_i$. Differentiating this outage probability yields the distribution of γ. This distribution is then used to obtain the probability of bit error via an MGF analysis assuming any fading distribution on both the desired signal and the interferers. For OC the received signal is given by

$$y_r = w^H c_s b_s + \sqrt{P_I} \sum_{i=1}^{L} w^H c_i b_i \tag{2.4}$$

where c_s is the fading on the symbol b_s of interest, c_i is the fading on the symbol b_i of the ith interferer, and P_I is the weighted power of the interferers. From [8] the optimal weights for OC are given by the vector

$$w = g R^{-1} c_s \tag{2.5}$$

where g is an arbitrary constant and $R = \sum_i c_i c_i^H$ is a Wishart distributed matrix, resulting in SINR $\gamma = P_i^{-1} c_s^H R^{-1} c_s$. The distribution of outage probability associated with this SINR, conditioned on the fading values for the desired and interfering signals, is shown in [9] to be gamma-distributed. The unconditional distribution is obtained in [7] via a MGF analysis, similar to the case of MRC.

The third technique is interference cancellation through beam steering, where array processing under N antennas can ideally nullify $N - 1$ interferers. If we assume perfect cancellation of the strongest $N - 1$ interferers, then performance analysis reduces to finding the outage and bit error probabilities for the residual $L - N - 1$ interferers that remain after cancellation. These distributions first require the order statistics for the strongest interferers, which are obtained in [10]. The MGF for the received signal and its corresponding pdf is then obtained in closed form, from which outage probability can be obtained. More details can be found in [7].

A performance comparison between OC, MRC, and IC is shown in Figure 2.4. These numerical results are based on an interference-dominated environment where noise is negligible, and equal-power Rayleigh-fading interferers. Figure 2.4 shows the outage probability as a function of SIR at each antenna for 2, 3, and 4 receive antennas. Note that as expected, OC has the best performance, since it generalizes both MRC and IC. We also see that IC does worse than MRC except at low SIR, where interference dominates performance degradation and hence canceling interference is the correct strategy. At high SIR values, performance degradation

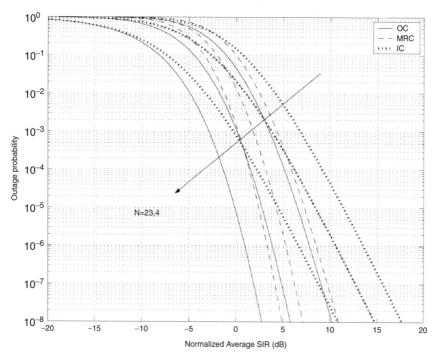

Figure 2.4   Performance comparison of optimum combining, maximum-ratio combining and interference cancellation [17]

due to multipath fading causes more degradation than interference and hence MRC leads to better performance than IC.

2.2.3 MIMO Systems

In this section, we discuss in more detail two fundamental tradeoffs mentioned in the previous section: The first one is between the diversity gain and the multiplexing gain [11]–[12], and the second one between performance and complexity. Focusing for simplicity on 2x2 MIMO systems, two limiting transmission schemes are as follows. One could transmit the same symbol, say s, from the two transmit antennas. In this case, the signal traverses four propagation paths, and, if these are affected by independent fading, the diversity achieved is 4. On the other hand, since only one signal is transmitted per channel use, one has *no multiplexing gain* with respect to single-antenna transmission. If two independent signals are transmitted simultaneously, then each one of them traverses two independent paths, thus achieving diversity 2, but every channel use transmits *two* signals, thus achieving a twofold multiplexing gain. One may also look for an intermediate situation, where multiplexing gain and diversity gain are traded off: A conceptually simple way of achieving this consists of introducing a certain amount of correlation between the symbols transmitted over the MIMO channel, which is achieved by coding across space and time (space–time codes). These codes can be generated by suitably combining good codes designed for single-antenna schemes (e.g., turbo or LDPC codes), or by using ad hoc designs (e.g., the Golden Code [13]).

The second tradeoff – that between performance and complexity – is crucial for the receiver design. As optimum receivers are in general very complex to implement, there is a considerable amount of research activity devoted to the design of suboptimum receivers. To motivate this point, consider a MIMO system with an equal number N of receive and transmit antennas, where we denote by $s_1, \ldots, s_N$ the transmitted symbols, and by h_{ij} the fading gain along the propagation path joining transmit antenna j to receive antenna i. These fading gains are organized in a square matrix $\mathbf{H}$, and the transmitted symbols in a vector $\mathbf{s}$. The received vector $\mathbf{r}$ can be expressed as

$$\mathbf{r} = \mathbf{H}\mathbf{s} + noise \tag{2.6}$$

and the receiver's goal consists of detecting the N transmitted signals. The simple device of solving the above system of equations, whereby $\mathbf{s}$ is the unknown vector, albeit simple, may not be (and in general is not) the best solution, as the presence of noise degrades performance whenever $\mathbf{H}$ is an ill-conditioned matrix, i.e., a matrix whose largest to the smallest eigenvalue ratio is large. Optimum (maximum-likelihood) detection of the transmitted signals should operate by minimizing, with respect to $s_1, \ldots, s_N$, the metric

$$\sum_{i=1}^{N} |r_i - \sum_{j=1}^{N} h_{ij}s_j|^2 \tag{2.7}$$

However, brute-force minimization of the above requires an exhaustive search among the M^N possible transmitted signal vectors, where M is the signal constellation size, i.e., the number of values taken on by each component of vector $\mathbf{s}$. For a 64QAM constellation and $N = 2$, the number of signal pairs to be enumerated amounts to $64^2 = 4096$, which may easily exceed the processing capability of the receiver. Of the possible ways out of this impasse, sphere detection plays a central role: This consists of enumerating only a subset of possible signal pairs, after making sure that the optimum pair is not excluded from consideration [11], [12].

A further cause of complexity in MIMO receivers comes from the observation that minimizing the above metric involves the knowledge of the N^2 fading gains (the elements of $\mathbf{H}$) appearing in it. This knowledge requires operations of channel estimation.

The WiMAX standard includes some profiles in order to exploit the benefits of MIMO in broadband wireless access systems. These profiles and the main challenges related to their implementation are described in the next section.

2.3 M Multiple Antennas in WiMAX Systems

2.3.1 Transmit Diversity

One of the WiMAX system profiles is the simple STC scheme proposed by Alamouti [4] for transmit diversity on the downlink. In the IEEE 802.16e-2005 specifications, this scheme is referred to as Matrix A. Originally, Alamouti's transmit diversity was proposed to avoid the use of receive diversity and keep the subscriber stations simple. This technique is applied subcarrier by subcarrier and can be described as shown in Figure 2.5.

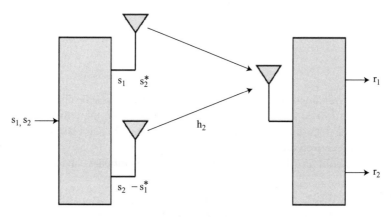

Figure 2.5 Schematic block diagram of Alamouti's transmit diversity

Suppose that (s_1, s_2) represent a group of two consecutive symbols in the input data stream to be transmitted. During a first symbol period t_1, transmit (Tx) antenna 1 transmits symbol s_1 and Tx antenna 2 transmits symbol s_2. Next, during the second symbol period t_2, Tx antenna 1 transmits symbol s_2^* and Tx antenna 2 transmits symbol $-s_1^*$. Denoting the channel response (at the subcarrier frequency at hand) from Tx1 to the receiver (Rx) by h_1 and the channel response from Tx2 to the receiver by h_2, the received signal samples corresponding to the symbol periods t_1 and t_2 can be written as:

$$r_1 = h_1 s_1 + h_2 s_2 + n_1 \tag{2.8}$$

$$r_2 = h_1 s_2^* - h_2 s_1^* + n_2 \tag{2.9}$$

where n_1 and n_2 are additive noise terms.

The receiver computes the following signals to estimate the symbols s_1 and s_2:

$$x_1 = h_1^* r_1 - h_2 r_2^* = \left(|h_1|^2 + |h_2|^2 \right) s_1 + h_1^* n_1 - h_2 n_2^* \tag{2.10}$$

$$x_2 = h_2^* r_1 + h_1 r_2^* = \left(|h_1|^2 + |h_2|^2 \right) s_2 + h_2^* n_1 + h_1 n_2^* \tag{2.11}$$

These expressions clearly show that x_1 (resp. x_2) can be sent to a threshold detector to estimate symbol s_1 (resp. symbol s_2) without interference from the other symbol. Moreover, since the useful signal coefficient is the sum of the squared moduli of two independent fading channels, these estimations benefit from perfect second-order diversity, equivalent to that of Rx diversity under maximum-ratio combining (MRC).

Alamouti's transmit diversity can also be combined with MRC when two antennas are used at the receiver. In this scheme, the received signal samples corresponding to the symbol periods t_1 and t_2 can be written as:

$$r_{11} = h_{11} s_1 + h_{12} s_2 + n_{11} \tag{2.12}$$

$$r_{12} = h_{11} s_2^* - h_{12} s_1^* + n_{12} \tag{2.13}$$

for the first receive antenna, and

$$r_{21} = h_{21}s_1 + h_{22}s_2 + n_{21} \tag{2.14}$$

$$r_{22} = h_{21}s_2^* - h_{22}s_1^* + n_{22} \tag{2.15}$$

for the second receiver antenna. In these expressions, h_{ji} designates the channel response from Tx i to Rx j, with $i, j = 1, 2$, and n_{ji} designates the noise on the corresponding channel. This MIMO scheme does not give any spatial multiplexing gain, but it has 4th-order diversity, which can be fully recovered by a simple receiver.

Indeed, the optimum receiver estimates the transmitted symbols s_1 and s_2 using:

$$x_1 = h_{11}^* r_{11} - h_{12} r_{12}^* + h_{21}^* r_{21} - h_{22} r_{22}^*$$

$$= \left(|h_{11}|^2 + |h_{12}|^2 + |h_{21}|^2 + |h_{22}|^2 \right) s_1 + h_{11}^* n_{11} - h_{12} n_{12}^* + h_{21}^* n_{21} - h_{22} n_{22}^*$$

$$x_2 = h_{12}^* r_{11} + h_{11} r_{12}^* + h_{22}^* r_{21} + h_{21} r_{22}^*$$

$$= \left(|h_{11}|^2 + |h_{12}|^2 + |h_{21}|^2 + |h_{22}|^2 \right) s_1 + h_{12}^* n_{11} + h_{11} n_{12}^* + h_{22}^* n_{21} + h_{21} n_{22}^*$$

and these equations clearly show that the receiver fully recovers the fourth-order diversity of the 2x2 system. It is worth noting that the MRC in this scheme can be modified to take into account the presence of some interferers and thus trade off diversity for interference cancellation.

2.3.2 Spatial Multiplexing

The second multiple antenna profile included in WiMAX systems is the 2x2 MIMO technique based on the so-called matrix $\mathbf{B} = (s_1, s_2)^T$. This system performs spatial multiplexing and does not offer any diversity gain from the Tx side. But it does offer a diversity gain of 2 on the receiver side when detected using maximum-likelihood (ML) detection.

To describe the 2x2 spatial multiplexing, we omit the time and frequency dimensions, leaving only the space dimension. The symbols transmitted by Tx1 and Tx2 in parallel are denoted as s_1 and s_2, respectively. Denoting by h_{ji} the channel response from Tx i to Rx j ($i, j = 1, 2$), the signals received by the two Rx antennas are given by

$$r_1 = h_{11}s_1 + h_{12}s_2 + n_1 \tag{2.18}$$

$$r_2 = h_{21}s_1 + h_{22}s_2 + n_2 \tag{2.19}$$

which can be written in a matrix form as

$$\begin{pmatrix} r_1 \\ r_2 \end{pmatrix} = \begin{pmatrix} h_{11} & h_{12} \\ h_{21} & h_{22} \end{pmatrix} \begin{pmatrix} s_1 \\ s_2 \end{pmatrix} + \begin{pmatrix} n_1 \\ n_2 \end{pmatrix} \tag{2.20}$$

The ML detector makes an exhaustive search of all possible values of the transmitted symbols and decides in favor of (s_1, s_2) which minimizes the Euclidean distance:

$$D(s_1, s_2) = \left\{ |r_1 - h_{11}s_1 - h_{12}s_2|^2 + |r_2 - h_{21}s_1 - h_{22}s_2|^2 \right\}$$

The complexity of the ML detector grows exponentially with the size of the signal constellation, and this motivates the use of simpler suboptimum detectors in practical applications. Among those are [5], [14], [15]:

1. Zero-forcing (ZF) detectors, which invert the channel matrix. The ZF receiver has a very small complexity that does not depend on the modulation. However, it does not completely exploit the system diversity and suffers from bad performance at low SNR.
2. Minimum mean-square error (MMSE) detectors, which reduce the combined effect of interference between the two parallel channels and additive noise. The MMSE receiver slightly improves the performance of the ZF receiver, but it requires knowledge of the SNR, which can be impractical. Besides, it does not completely exploit the channel diversity either.
3. Decision-feedback receivers, which make a decision on one of the symbols and subtract its interference of the other symbol based on that decision. These receivers offer improved performance when compared to ZF and MMSE receivers, but they are prone to error propagation and still lack optimality, which may lead to large performance losses.
4. Sphere detectors, which reduce the number of symbol values used in the ML detector. Note that this type of detectors may preserve optimality while reducing implementation complexity.

2.3.3 Comparison of MIMO Options

Since the Alamouti/MRC scheme and the 2x2 spatial multiplexing scheme have a diversity order of 4 and 2, respectively, the former obviously has better BER performance when the same modulation and coding schemes are used in both systems. Consequently, the Alamouti/MRC scheme can use a higher-level modulation if the two schemes are required to give the same BER performance. Of utmost interest is a performance comparison between the two MIMO schemes when they are used at the same spectral efficiency. (Note that the Alamouti/MRC technique with a modulation scheme transmitting $2m$ bits per symbol has the same spectral efficiency as the MIMO spatial multiplexing scheme with a modulation transmitting m bits per symbol.)

We undertook such a performance comparison using both uncoded and coded systems and different types of channels. Figure 2.6 shows the results on an uncorrelated Rayleigh fading channel when the Alamouti/MRC scheme uses 16-QAM and the spatial multiplexing scheme uses QPSK (4 bits per symbol period in both cases). It can be observed that the ZF receiver does not exploit the diversity of the spatial multiplexing scheme and that the slope of its BER curve is only half that of the ML receiver. The other major observation is that the slope of the Alamouti/MRC scheme is twice as large as that of the spatial multiplexing ML receiver, which is due to the diversity factor of 4 for the former and of 2 for the latter. These results, originally reported in [16] and [17] are in agreement with those reported in [18].

As predicted by the respective diversity gains of the two schemes, the results displayed in Figure 2.6 confirm that at high SNR values, the simple Alamouti/MRC scheme with 16-QAM achieves better performance than the 2x2 spatial multiplexing MIMO system with ML detection. This suggests that the best MIMO scheme to use in practice depends on the channel

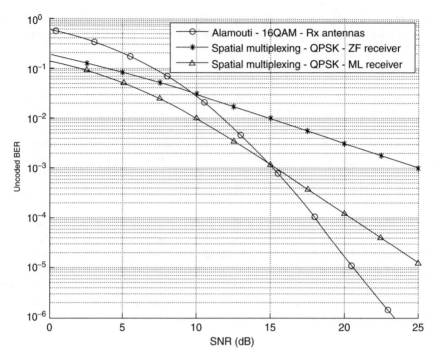

Figure 2.6   Comparison of Alamouti/MRC with 2×2 spatial multiplexing

SNR and the required throughput as well as on other considerations such as the interference
cancellation capability.

To be more specific on the choice between the two MIMO profiles, we summarize in
Table 2.1 the modulation and coding schemes available in WiMAX systems. (Note that Table
2.1 is restricted to the convolutional coding schemes included in the standard, and optional
interleaving and other coding schemes such as convolutional turbo codes are not considered.)
The spectral efficiency which appears in this equation is for single-antenna systems, and it is
of course doubled when spatial multiplexing is used.

In single-antenna systems, the throughput is optimized through link adaptation, which selects a constellation and a code rate as a function of the channel. This concept is called
adaptive modulation and coding (AMC). The basic idea is to measure the channel quality

Table 2.1   Constellations and convolutional coding schemes in WiMAX systems

Constellation	QPSK	QPSK	16QAM	16QAM	64QAM	64QAM
Code rate	$^1/_2$	$^3/_4$	$^1/_2$	$^3/_4$	$^1/_2$	$^3/_4$
Spectral efficiency (bits/symbol)	1	1.5	2	3	3	4.5

(for instance, by estimating the received power or the received SNR) at the mobile station. If the channel variations are sufficiently slow so that they are essentially constant, the channel quality measurement can be fed back to the base station with estimation error and delay that do not significantly degrade performance. The BS can then adapt the modulation and coding schemes to the channel and optimize the overall spectral efficiency subject to some performance criterion (for instance, the outage probability for a given packet error rate will be smaller than a predetermined value). Note that dedicated mechanisms such as the Fast Feedback Channel have been incorporated specifically in the standard for the purpose of link adaptation.

Figure 2.7 illustrates the AMC concept when the performance criterion is that the forward error correction (FEC) block error rate (FBER) must be smaller than 10^{-3}. For different combinations of the modulation and coding options of Table 2.1, Figure 2.7 shows the SNR thresholds above which the performance criterion is met. (The SNR thresholds are computed for a system using MIMO matrix A at the transmitter, two antennas with MRC at the receiver, and the ITU Pedestrian Channel A corresponding to a speed of 3 km/hour.) For instance, 16QAM with code rate 1/2 cannot be used for SNR values below 7 dB, because it yields an FEC block error rate greater than 10^{-3}. Above this threshold, the modulation meets the performance criterion and leads to a spectral efficiency of 2 bits per symbol. Further, Figure 2.7 shows that for SNR values exceeding 11dB, 16QAM can also be used with code

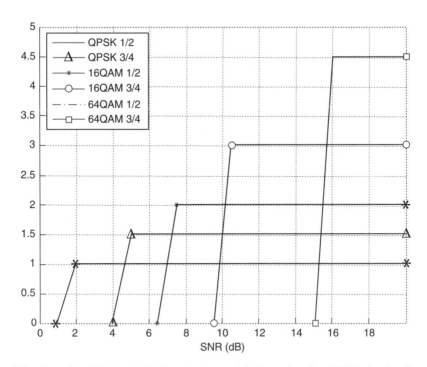

Figure 2.7 Operating SNR thresholds for adaptive modulation and coding (ITU Pedestrian Channel A, speed = 3 km/h, FBER = 10^{-3})

rate 3/4 and this increases the spectral efficiency from 2 to 3 bits per symbol. Based on the SNR thresholds shown, AMC consists of using the modulation/coding combination that leads to the highest spectral efficiency. Figure 2.7 shows that some combinations of modulation and coding schemes are not useful on the considered channel for the performance criterion used. For instance, it is meaningless to use 64QAM with code rate 1/2, because 16QAM with code rate 3/4 gives the same spectral efficiency and has a lower SNR threshold.

Returning now to the MIMO schemes in WiMAX systems, the best way to handle them is to add the MIMO dimension to modulation and coding, and select the best MIMO/Modulation/Coding combination through link adaptation. Figure 2.8 depicts the seven useful combinations for link adaptation over a pedestrian channel. Based on the results of this figure, MIMO matrix B (spatial multiplexing) will be usable with 16QAM and code rate 3/4 at SNR values higher than 22 dB yielding a spectral efficiency of 6 bits per symbol. Furthermore, at SNR values higher than 30 dB, this system can use 64QAM and code rate 3/4 leading to a spectral efficiency of 9 bits per symbol. This represents a significant increase of throughput compared to a MIMO matrix A system whose spectral efficiency is limited to 4.5 bits per symbol. It should be pointed out, however, that, in practice, the channel correlation due to the small distance between the receive antennas on the mobile station may seriously affect these results, and more particularly the Matrix B performance. Interference can also significantly impact the performance tradeoffs between Matrices A and B.

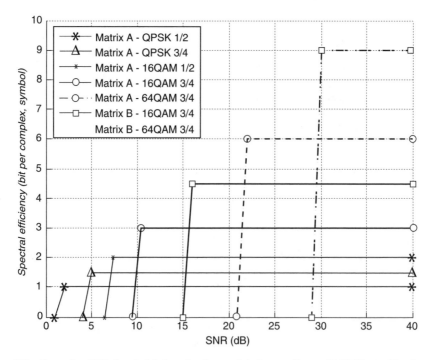

Figure 2.8 Operating SNR thresholds for adaptive modulation, coding and MIMO combinations (ITU Pedestrian Channel A, speed = 3 km/h, FBER = 10^{-3})

2.4 Conclusion

The Mobile WiMAX standard includes many different features and options to make the best use of the wireless channel characteristics. These include adaptive modulation and coding, and multiple antenna (MIMO) techniques such as transmit/receive diversity and spatial multiplexing. In this chapter, we first discussed the use of multiple antenna techniques in a general context and the tradeoffs between diversity, multiplexing gain and interference cancellation. Next, we described the two MIMO schemes included in the mobile WiMAX system specifications and analyzed their performance using the ITU pedestrian B channel model with a pedestrian speed of 3 km/h and assuming perfect channel state information and uncorrelated channels. It was first observed that at high SNR values, Alamouti's STC with MRC at the receiver significantly outperforms Spatial Multiplexing when the two systems employ modulation schemes leading to the same spectral efficiency. Next, for different modulation, coding and MIMO schemes, the SNR values leading to a BER of 10^{-3} were computed and the achievable spectral efficiency vs. SNR was plotted indicating which scheme can be used in which SNR region. The results indicated that MIMO Matrix A must be used except at very high SNR values, where Matrix B can lead to an increased spectral efficiency.

References

[1] IEEE 802.16-2005: IEEE Standard for Local and Metropolitan Area Networks – Part 16: Air Interface for Fixed and Mobile Broadband Wireless Access Systems – Amendment 2: Physical Layer and Medium Access Control Layers for Combined Fixed and Mobile Operation in Licensed Bands, February 2006.

[2] H. Sari and G. Karam, 'Orthogonal Frequency-Division Multiple Access and its Application to CATV Networks', *European Transactions on Telecommunications and Related Technologies* (*ETT*), **9**(6), November–December 1998, 507–516.

[3] A. Goldsmith, *Wireless Communications,* Cambridge: Cambridge University Press, 2005.

[4] S.M. Alamouti, 'A Simple Transmit Diversity Technique for Wireless Communications', *IEEE Journal on Selected Areas in Communications*, **16**(8), October 1998, 1451–1458.

[5] D. Tse and P. Viswanath, *Fundamentals of Wireless Communications*, Cambridge: Cambridge University Press, 2005.

[6] J. Romero and A. Goldsmith, 'Optimizing Antenna Array Processing in CCI Channels: Is It Better to Cancel or Combine?', in *Proc. 2006 IEEE Intl. Conf. Commun.*, June 2006.

[7] J.H. Winters, 'Optimum Combining in Digital Mobile Radio with Cochannel Interference', *IEEE Journal of Select. Areas Comm*, **2**(4), July 1984, 28–539.

[8] J. Perez and A.J. Goldsmith, 'Optimizing Antenna Array Processing in CCI Channels: Is it Better to Cancel or Combine?', submitted to *IEEE Trans. Wireless Commun.*, Aug. 2006.

[9] M.K Simon and M.-S. Alouini, *Digital Communications over Fading Channels,* Hoboken, NJ: John Wiley & Sons, Ltd, 2005.

[10] G. Sarfraz and A. Annamalai, 'Performance Evaluation of Cellular Mobile Radio Systems with Successive Cancellation of Nonidentically Distributed Co-channel Interferers in a Rayleigh Fading Environment', in *Proc. IEEE Wireless Commun. Net. Conf,* March 2003, 579–584.

[11] A. Paulraj, R. Nabar, and D. Gore, *Introduction to Space–Time Wireless Communications,* University Press, 2006.

[12] E. Biglieri, R. Calderbank, T. Constantinides, A. Goldsmith, A. Paulraj, and H.V. Poor, *MIMO Wireless Communications*, Cambridge: Cambridge University Press, 2006.

[13] J.-C. Belfiore, G. Rekaya, and E. Viterbo, 'The Golden Code: A 2x2 Full-Rate Space–Time Code with Nonvanishing Determinants', *IEEE Transactions on Information Theory*, **51**(4), April 2005, 1432–1436.

[14] E. Viterbo and E. Biglieri, 'A Universal Lattice Decoder', in *Proc. 14th GRETSI Symposium*, September 1993, Juan-les-Pins, France.

[15] M.O. Damen, A. Chkeif, and J.-C. Belfiore, 'Lattice Codes Decoder for Space-Time Codes', *Electronics Letters*, 4, May 2000, 161–163.

[16] B. Muquet, E. Biglieri and H. Sari, 'MIMO Link Adaptation in Mobile WiMAX Systems', in *Proc. WCNC 2007*, March 2007, Hong Kong.

[17] E. Biglieri, A. Goldsmith, B. Muquet and H. Sari, 'Diversity, Interference Cancellation and Spatial Multiplexing in MIMO Mobile WiMAX Systems', in *Proc. IEEE Mobile WiMAX Symposium 2007*, March 2007, Orlando, FL.

[18] R.W. Heath, Jr. and A.J. Paulraj, 'Switching Between Diversity and Multiplexing in MIMO Systems', *IEEE Transactions on Communication*, **53**(6), June 2005, 962–968.

3

Mitigation of Inter-Cell Interference in Mobile WiMAX

Jae-Heung Yeom and Yong-Hwan Lee

3.1 Introduction

Mobile WiMAX (m-WiMAX) has been proposed to support wireless data services at high rates comparable to wire-line schemes such as the digital subscriber line (DSL) and it is being considered as a migration path toward next generation wireless systems [1]. Recently, a version of Mobile WiMAX, called WiBro, has been used in Korea.

Since m-WiMAX considers the use of universal frequency reuse, it may suffer from inter-cell interference (ICI) near the cell boundary. In fact, the spectral efficiency in the worst channel condition can be reduced up to one-120th of that in the best channel condition (i.e., QPSK-1/12 with 3-time re-transmissions versus 64QAM-5/6 with a single transmission) [2]. As a consequence, when m-WiMAX provides real-time traffic services at a rate of 512 kbps near the cell boundary, it may need to allocate the whole downlink resources to a single user, which is practically unacceptable to service providers. Unless m-WiMAX can significantly improve the spectral efficiency particularly near the cell boundary, it may not be distinguishable from incumbent 3G systems. Thus, it is very important to solve the ICI problem in the m-WiMAX system.

First, we consider cell planning to mitigate ICI. The cellular structure was originally motivated from the signal propagation characteristics associated with the distance between the transmitter and receiver. The frequency can be reused in geographically separated areas where the signal is sufficiently attenuated due to the distance. However, the target signal cannot be perfectly isolated from ICI. There have been many attempts to alleviate the effect of ICI [3].

A simple approach is to sufficiently separate the cells using the same frequency band so as to maintain a minimum signal-to-interference power ratio (SIR) required in the whole coverage area. This can be done by dividing the whole system bandwidth into a large number of frequency sub-bands and allocating each frequency sub-band so that the co-channel cells are isolated as much as possible. This process is often called frequency reuse or cell planning. In the cellular

Mobile WiMAX   Edited by Kwang-Cheng Chen and J. Roberto B. de Marca.
© 2008 John Wiley & Sons, Ltd

system, the frequency reuse factor is one of the most important factors in the improvement of system capacity. The ICI can be reduced with the use of a low frequency reuse factor, improving the link performance. However, the overall spectral efficiency will be decreased due to partial use of the available band. When a high frequency reuse factor is used as in Figure 3.1(b), the overall spectral efficiency can be increased at the cost of degraded link performance.

ICI mitigation is closely related to multiple access schemes. Circuit-switched code division multiple access (CDMA) systems accurately control the transmit power to give more power to users near the cell boundary and use spread pseudo-noise codes to whiten the ICI. They can also obtain an ICI averaging effect since they support users on a single carrier. Moreover, they can achieve an inter-cell diversity gain through the use of soft handover. Thus, CDMA systems can achieve universal frequency reuse [4]. On the other hand, packet-based orthogonal frequency division multiple access (OFDMA) systems consider the use of fast adaptive modulation and coding (AMC) while providing all users with the same transmit power, which may cause users

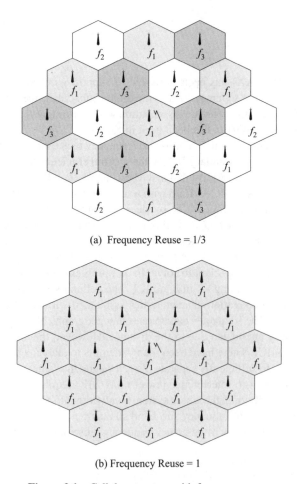

(a) Frequency Reuse = 1/3

(b) Frequency Reuse = 1

Figure 3.1   Cellular structure with frequency reuse

near the cell boundary to experience poor received signal strength. Moreover, users near the cell boundary may experience a large variation in ICI.

The use of hard handover can cause a severe ICI problem. As a consequence, the use of universal frequency reuse causes severe ICI problems to users near the cell boundary.

In this chapter, we first describe conventional ICI mitigation techniques for OFDMA systems and briefly describe how such ICI mitigation techniques can be applied to the m-WiMAX system. Then, we describe a new strategy to mitigate the ICI in m-WiMAX and verify its performance by computer simulation. Finally, we offer a conclusion.

3.2 ICI Mitigation Techniques for OFDMA Systems

OFDMA systems support intra-cell orthogonality and thus they may suffer from ICI as the main interference source. The effect of ICI is particularly detrimental to users near the cell boundary. To provide the desired service quality independent of user location, it is important to mitigate the ICI near the cell boundary. To this end, four prominent techniques have been proposed, called interference avoidance (IA), interference randomization, interference cancellation, and inter-sector cooperation [5].

3.2.1 ICI Avoidance

OFDMA systems can mitigate ICI by avoiding having resources overlapping with neighboring cells/sectors. That is, IA schemes dynamically allocate the channel resource to avoid the ICI by exchanging the information among adjacent cells [6]. Reuse partitioning techniques determine the reuse factor according to the interference condition for IA in OFDMA systems [6–8]. Fractional frequency reuse (FFR) techniques can be applied to avoid overlapping use of frequency resource by adjacent cells in response to ICI condition. Figure 3.2 shows that users

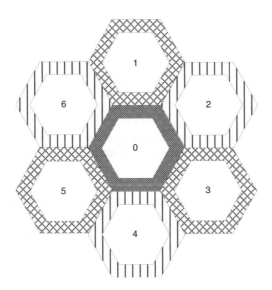

Figure 3.2 Frequency resource allocation based on FFR

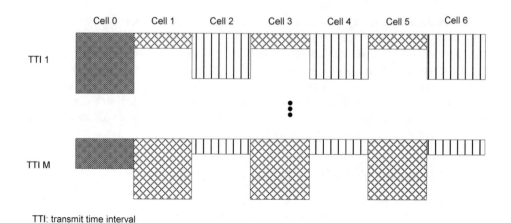

TTI: transmit time interval

Figure 3.3   Semi-static coordination of reserved frequency resource among cells

in the cell center are assigned to all subcarriers, and users near the cell boundary are assigned to one-third of all subcarriers. It may not be optimum to fix the pattern of partial frequency use. To improve the efficiency of frequency use, the size of reserved frequency can adaptively be determined according to the traffic load near the cell boundary [9]. Figure 3.3 illustrates the principle of semi-static coordination of reserved frequency sub-band, where adjacent cells are allocated to different reserved frequency resources. However, ICI avoidance may limit the peak transmission rate due to reduced reuse factors. Besides it has not been considered with combined use of other ICI mitigation techniques such as frequency hopping (FH) and ICI cancellation in the receiver.

3.2.2 ICI Randomization

ICI randomization techniques consider the use of cell-specific scrambling and FH techniques to whiten the ICI. Without the use of a cell-specific scrambling code, the mobile station (MS) decoder is matched not only to the target signal, but also to user signals in other cells. Thus, the use of cell-specific scrambling codes is useful to whiten the ICI seen by the receiver [10]. The ICI suppression capability is proportional to the processing gain of the turbo encoder in Figure 3.4. Note that this cell-specific scrambling does not affect the bandwidth, i.e., it is not a conventional spreading operation. As an alternative, the use of cell-specific interleaving, also known as IDMA, has been proposed to whiten the inter-cell interference and to improve the performance of advanced ICI cancellation [11]. FH provides a frequency diversity gain in addition to the ICI averaging gain. It can be employed without channel state information (CSI). It can use a hopping pattern changed once or more times during a data symbol time (i.e., fast FH) or once during several data symbols time (i.e., slow FH). The FH pattern utilizes the whole system bandwidth by changing the subcarrier in a random manner, providing an interference averaging effect. It is desirable to employ a strong channel code to sufficiently obtain the interference averaging effect because the information codewords are spread in the whole bandwidth. The m-WiMAX supports FH in two modes; partial usage of subchannels

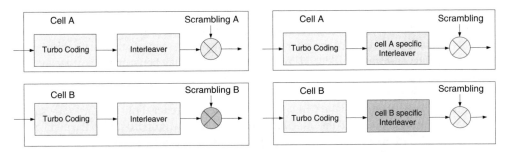

Figure 3.4 Cell-specific scrambling and cell-specific interleaving

(PUSC) and full usage of subchannels (FUSC). However, the use of FH may not be effective in full loading environments since ICI still remains without suppression.

3.2.3 ICI Cancellation

ICI cancellation techniques can increase the SIR even in full loading environments by canceling out the ICI based on inter-cell CSI [12]. The received signal **y** through the antenna array can be represented as [13]

$$\mathbf{y} = \mathbf{h}x + \mathbf{z} \tag{3.1}$$

where **h** is the received spatial signature of the user of interest, x is the target user signal and **z** is interference plus noise vector with covariance matrix $\mathbf{K}_z$. Assuming that the ICI is only from a single adjacent cell and the base station (BS) has an estimate of the received spatial signature **g** of the interference, the covariance matrix can be represented as

$$\mathbf{K}_z = \mathbf{g}\mathbf{g}^* + N_0\mathbf{I} \tag{3.2}$$

where the superscript * denotes transpose conjugate and N_0 denotes the spectral density of additive white Gaussian noise. Then, the target signal can be estimated as [13]

$$\hat{x} = \mathbf{v}_{mmse}^*\mathbf{y} = \mathbf{h}^*\mathbf{K}_z^{-1}\mathbf{y} \tag{3.3}$$

where $\mathbf{v}_{mmse}$ denotes the coefficient of a linear minimum mean squared error (MMSE) filter. However, this interference canceller may not work properly when the receiver does not experience dominant interferers and/or does not get their CSI. Also, the number of dominant interfering signals from other cells should be less than the number of receiver antennas. The performance of interference cancellation based on a joint detection is also significantly degraded when the carrier-to-interference-and-noise power ratio (CINR) is low [14].

3.2.4 Inter-Sector Cooperation

Inter-sector cooperation techniques originated in softer handoff in the CDMA, providing a diversity gain [15]. They are applicable to real-time traffic services because the information between the sectors can be exchanged in real time. By making two sectors transmit the same signal, the MS can obtain a delay diversity gain using an inter-sector diversity (i.e., macro diversity) technique [16]. Since the inter-sector diversity is the same as the cyclic shift transmit diversity, the performance depends on the propagation delay between the sectors and the channel coder rate [17]. Fully joint encoding schemes (e.g., dirty paper coding) can achieve the capacity without reducing the reuse factor even in multi-cellular systems. Since the participating BSs should share both the CSI and the information streams [18, 19], they may suffer from communication overhead that may increase exponentially proportional to the number of cooperative BSs. Moreover, the 3GPP long-term evolution sets a constraint on the maximum cooperative interval of an order of ten seconds to minutes [5]. Note that the IEEE 802.16e-2005 does not support joint encoding scheme [2]. Although transmit null beamforming can reduce inter-cell communication overhead [20], it still requires the CSI of interfering users, requiring inter-cell cooperation. It can work properly when the number of transmit antennas is larger than the number of interfering users.

3.3 Combined Use of ICI Mitigations in Mobile WiMAX

Previous attempts to mitigate ICI in the m-WiMAX system did not consider the combined use of these techniques [21]. The lack of conventional ICI mitigation techniques can be alleviated by employing them in a collaborative manner [22]. We consider the application of such an ICI mitigation strategy to the downlink of the m-WiMAX system in response to the change in the ICI condition.

3.3.1 Combined Use of IA and FH

This section presents combined use of IA and FH techniques for better mitigation of ICI in m-WiMAX. The m-WiMAX can support multiple zones, each of which utilizes subcarrier permutation and multi-antenna techniques according to the operation environments [2]. Subcarrier permutation is deeply associated with ICI mitigation techniques. In the PUSC mode, subcarriers are ranged in a major group that comprises several clusters and then ranged in the clusters [2]. The FH pattern of neighboring cells can be controlled using parameter *DL_PermBase* as

$$Renumbering\ Sequence\ (PhysicalCluster + 13*DL\_PermBase)\quad mod\ 60).\qquad(3.4)$$

Although parameter *Use_All_SC_indicator* is set to 1, IA techniques can be employed when the neighboring cells use the same *DL_PermBase* and the resource is allocated in what follows.

Consider the resource allocation for IA in the PUSC mode as illustrated in Figure 3.5. The whole frequency resource is divided into three bands; F_0, F_1 and F_2. The frequency reuse set φ_u is defined by

$$\varphi_0 = \{m_0,\ m_1,\ m_2\},\quad \varphi_1 = \{m_0,\ m_1\},\quad \varphi_2 = \{m_0,\ m_2\},\quad \varphi_3 = \{m_0\}\qquad(3.5)$$

where m_0, m_1 and m_2 denote the band index corresponding to m, $(m+1)\%3$ and $(m+2)\%3$, respectively. Here, $a\%b$ denotes a modulo b (e.g., when $m = 2$, $\varphi_0 = \{2, 0, 1\}$, $\varphi_1 = \{2, 0\}$, etc.). The reuse set φ_0 indicates the use of band F_{m_0}, F_{m_1} and F_{m_2} (i.e., the frequency reuse factor of 1), and the reuse set φ_1 indicates the use of band F_{m_0} and F_{m_1} (i.e., the frequency

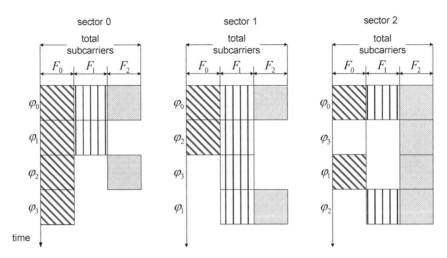

Figure 3.5 Resource allocation according to the frequency reuse factor in the PUSC mode

reuse factor of 2/3). The divided frequency resources are allocated to users according to the ICI condition. For example, when the MS in sector 0 receives strong interference from sector 2, the frequency resource in reuse set φ_1 is allocated to users in sector 0 and 1, not to users in sector 2. If the MS in sector 0 receives strong interference from sector 1 and 2, the frequency resource in reuse set φ_3 (i.e., the frequency reuse factor of 1/3) is allocated to users in sector 0, not to users in sector 1 and 2.

The IEEE802.16e-2005 specification recommends the use of two frequency reuse factors (i.e., 1 and 1/3). It is possible to use other reuse factors by making the neighboring cells use the same *DL_PermBase* as illustrated in Figure 3.5. IA and FH schemes can simultaneously be applied to the PUSC mode, enabling both ICI avoidance gain and frequency diversity gain.

3.3.2 Combined Use of ICI Cancellation and IA

The receiver can cancel out the ICI in the spatial domain, using multiple receive antennas when the signal is transmitted using less spatial dimension than the receiver [13]. The use of transmit beamforming can compensate for the signal level reduction due to the interference cancellation in the receiver and also increase the received level of the dedicated pilot signal, enhancing the channel estimation performance. The receiver may cancel out the ICI at each frame by decorrelating the desired signal with the ICI when the full CSI is available. The transmitter needs to use inter-cell scrambling patterns nearly orthogonal to each other by properly managing parameter *PRBS_ID* [2].

As illustrated in Figure 3.6, major interference is generated from two adjacent cells. When the number of receiver antennas of the user is equal to the number of ICI sources, the receiver does not have a large enough spatial degree of freedom to cancel out these two interferers. Moreover, it is hard for the MS to estimate the CSI of three cells since the inter-cell scrambling patterns can be orthogonal only between two cells. Therefore, the use of ICI cancellation may not provide desired performance. However, combined use of IA with a frequency reuse factor

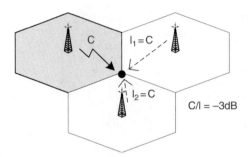

Figure 3.6   Cell edge environment

of 2/3 can successfully cancel out the interference: one interference can be suppressed by an
ICI cancellation technique and the other by an IA technique.

3.3.3 Inter-Sector Cooperation Using TDD Reciprocity

The use of inter-sector beamforming can alleviate the shortcomings of the inter-sector diversity
by exploiting the reciprocity of time division duplex (TDD) systems.

When the uplink sounding signal is transmitted near the sector boundary, it can be received
at least by two sectors, enabling the use of inter-sector beamforming. The inter-sector beamforming can yield an array gain at least twice that of the single-sector beamforming. The MS
can also avoid dominant inter-sector interference since it receives the same signals from two
sectors.

Consider the use of a (2×2) MIMO in the downlink and a (1×2) SIMO in the uplink
as suggested in the m-WiMAX operation profile [23]. Figure 3.7 illustrates an inter-sector
beamforming scheme, where sector 0 and 1 can generate a beamforming weight based on the
received uplink sounding signal from antenna 0 of the MS. The beamforming weight of sector

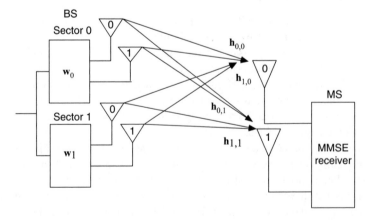

Figure 3.7   Inter-sector beamforming with two sector antennas

i can be represented as

$$\mathbf{w}_i \approx \frac{\mathbf{h}_{i,0}^*}{\|\mathbf{h}_{i,0}\|} \tag{3.6}$$

where $\mathbf{h}_{i,j}$ denotes the (1×2) channel vector from sector i to receive antenna j of the MS. Define $\mathbf{H}_i$ as the (2×2) channel matrix from sector i to the MS, given as,

$$\mathbf{H}_i = \begin{bmatrix} \mathbf{h}_{i,0} \\ \mathbf{h}_{i,1} \end{bmatrix} \tag{3.7}$$

The ICI represented by $\mathbf{H}_k x_k$ can be suppressed in the downlink by using a linear MMSE filter with coefficient vector [14]

$$\mathbf{v}_{\text{opt}} = \mathbf{K}_z^{-1}\mathbf{g} \tag{3.8}$$

where

$$\mathbf{g} = \mathbf{H}_0\mathbf{w}_0 + \mathbf{H}_1\mathbf{w}_1 \approx \begin{bmatrix} \|\mathbf{h}_{0,0}\| \\ \mathbf{h}_{0,1}\frac{\mathbf{h}_{0,0}^*}{\|\mathbf{h}_{0,0}\|} \end{bmatrix} + \begin{bmatrix} \|\mathbf{h}_{1,0}\| \\ \mathbf{h}_{1,1}\frac{\mathbf{h}_{1,0}^*}{\|\mathbf{h}_{1,0}\|} \end{bmatrix} \tag{3.9}$$

and $\mathbf{K}_z$ is the covariance matrix of the noise plus interference given by

$$\mathbf{K}_z = \mathbf{H}_k\mathbf{H}_k^* + N_0\mathbf{I}. \tag{3.10}$$

3.4 New ICI Mitigation Strategy in m-WiMAX

3.4.1 Three Steps for ICI Mitigation

Most IA techniques only consider the amount of ICI. Considering the ICI environment (e.g., dominant inter-sector interference in its own cell and dominant ICI from other cells), they can further improve the performance. The previous section has briefly discussed how various ICI mitigation techniques can be combined for further improvement of m-WiMAX performance. In this section, we consider combined use of ICI mitigation techniques to maximize the user capacity in the cell boundary region.

The user capacity can be represented as [13]

$$C_i = \rho_i \log_2(1 + \eta \cdot \gamma_i) \tag{3.11}$$

where i is the index of frequency reuse set, η is a parameter related to the implementation loss, ρ_i is the reuse factor for index i and γ_i denotes the CINR of user i. Since the CINR is mainly affected by the ICI in the cell boundary region, we consider the improvement of the capacity with and without the use of IA techniques.

We first determine the frequency reuse set that maximizes the user capacity with the use of IA. Then, we consider the inter-sector cooperation according to the ICI condition, which additionally considers the use of frequency resource. When the receiver is equipped with multiple receive antennas, we also consider the cancellation of dominant interference from neighboring cells to further utilize the frequency resource by IA.

The four frequency reuse sets φ_0, φ_1, φ_2 and φ_3 defined in Section 3.3.1 correspond to frequency reuse factor $\rho_0 = 1$, $\rho_1 = 2/3$, $\rho_2 = 2/3$ and $\rho_3 = 1/3$, respectively. Let P_{3n+k} be the received signal strength (RSS) from sector k of cell n, via parameter *MOB_SCN_REP* [2].

Step 1. Determine the frequency reuse set index j_1 achieving maximum geometry capacity for IA as [13]

$$j_1 = \arg \max_{i \in \{0,1,2,3\}} \rho_i \log_2(1 + \eta G_{1,i}). \tag{3.12}$$

Then, the corresponding maximum geometry capacity is given by

$$C_{1,j_1} = \rho_{j_1} \log_2(1 + \eta G_{1,j_1}). \tag{3.13}$$

Here, $G_{1,i}$ denotes the geometry corresponding to the frequency reuse set φ_i of the 1st STEP, given by [24]

$$G_{1,i} = \frac{g_{array} P_0}{\sum\limits_{k \in \varphi_i} S_k + \sigma_w^2}. \tag{3.14}$$

Here, σ_w^2 is the noise power of the MS, and g_{array} and S_k denote the array gain and the interference from sector k, respectively. g_{array} is equal to 2 when a transmit diversity scheme is used. When a transmit beamforming is employed, it can be shown that [13]

$$g_{array} = E\left\{\mathbf{g}^*\mathbf{g}\right\}$$

$$= E\left\{\left\|\mathbf{h}_{0,0}\right\|^2 + \frac{\left\|\mathbf{h}_{0,1}^*\mathbf{h}_{0,0}\right\|^2}{\left\|\mathbf{h}_{0,0}\right\|^2}\right\} \approx 1.5 \tag{3.15}$$

where each element of $\mathbf{h}_{i,j}$ is assumed to have zero mean and unit variance. The interference term S_k can be represented as

$$S_k = \sum_{n=0}^{N_c-1} P_{3n+k, \neq 0} \tag{3.16}$$

where N_c is the number of neighboring cells, and $k = 0$, 1 and 2 correspond to sector 0, 1 and 2, respectively. Note that the RSS P_0 from the serving sector is excluded in (3.16). After determining an optimum frequency reuse set φ_{j_1}, we consider inter-sector cooperation to increase the user capacity.

Step 2. Determine the frequency reuse set index j_2 achieving the maximum geometry capacity for inter-sector cooperation as [13]

$$j_2 = \arg \max_{i \in D} 0.5 \rho_i \log_2(1 + \eta G_{2,i}) \tag{3.17}$$

where the constant 0.5 is due to simultaneous sharing of the same resource by two sectors.

Then, the corresponding maximum geometry capacity is given by

$$C_{2,j_2} = 0.5\rho_{j_2}\log_2(1 + \eta G_{2,j_2}). \tag{3.18}$$

Here, D denotes the set comprising the frequency reuse set index which is considered to increase the frequency reuse factor determined by the STEP 1, given by

$$D = \begin{cases} \{0\} & j_1 = 1, 2 \\ \{1, 2\} & j_1 = 3. \end{cases} \tag{3.19}$$

The geometry $G_{2,i}$ corresponding to the i frequency reuse set of the STEP 2 is given by

$$G_{2,i} = \frac{g_{array}P_0}{\displaystyle\sum_{k\in\varphi_i} S_k - P_{\mu_i} + \sigma_w^2} \tag{3.20}$$

where μ_i denotes a sector associated with both the inter-sector cooperation and the resource avoided by the STEP 1, given by

$$\mu_i = \begin{cases} 2 & j_1 = 1 \\ 1 & j_1 = 2 \\ i & j_1 = 3. \end{cases} \tag{3.21}$$

It can be known from Equation (3.21) that STEP 2 tries to achieve reuse factor of 1 for reuse factor of 2/3 (i.e., $j_1 = 1$ or $j_1 = 2$) determined by the STEP 1 and reuse factor of 2/3 for reuse factor of 1/3 (i.e., $j_1 = 3$) determined by the STEP 1, respectively. g_{array} denotes the inter-sector cooperation gain which is less than or equal to 4 when an inter-sector diversity is used. When the inter-sector beamforming is employed, it can be shown that [13]

$$g_{array} = E\left\{\mathbf{g}^*\mathbf{g}\right\} \approx E\left\{\left(\|\mathbf{h}_{0,0}\| + \|\mathbf{h}_{1,0}\|\right)^2 + \left\|\frac{\mathbf{h}_{0,1}^*\mathbf{h}_{0,0}}{\|\mathbf{h}_{0,0}\|^2} + \frac{\mathbf{h}_{1,1}^*\mathbf{h}_{1,0}}{\|\mathbf{h}_{1,0}\|^2}\right\|^2\right\}$$

$$\approx 6 + 2\left[\Gamma\left(\frac{5}{2}\right)\right]^2 \tag{3.22}$$

where $\Gamma(\cdot)$ denotes the gamma function [25]. It can be shown that the inter-sector beamforming has an array gain about 2.5 times that of the inter-sector diversity. $G_{2,i}$ corresponding to the case when the BS considers sector μ_i of its own cell in addition to sectors in frequency reuse set φ_{j_1}. Therefore, the target MS can be free from the interference from sector μ_i of its own cell as a consequence of the inter-sector cooperation. When the MS is equipped with multiple receiver antennas, it can suppress dominant interference by a filtering process, enabling the transmission with the increase of reuse factor.

Step 3. Determine the frequency reuse set index j_3 maximizing the geometry capacity represented as

$$j_3 = \arg\max_{i\in D} \rho_i\log_2(1 + \eta G_{3,i}) \tag{3.23}$$

where set D is the same as in (18). The corresponding maximum geometry capacity is given by

$$C_{3,j_3} = \rho_{i_3} \log_2(1 + \eta G_{3,j_3}).\tag{3.24}$$

Here,

$$G_{3,i} = \frac{g_{array} P_0}{\displaystyle\sum_{k \in \varphi_i} S_k - P_{3\hat{m}+\mu_i} + \sigma_w^2}\tag{3.25}$$

where μ_i denotes the candidate sector to cancel out according to frequency reuse set index i and is the same definition as in (3.21). By using ICI cancellation via μ_i, STEP 3 tries to achieve reuse factor of 1 for reuse factor of 2/3 determined by STEP 1 and reuse factor of 2/3 for reuse factor of 1/3 determined by STEP 1, respectively.

$\hat{m}$ is given by

$$\hat{m} = \arg \max_{0 \leq m \leq N_c-1} P_{3m+\mu_i}.\tag{3.26}$$

Note that $G_{3,i}$ corresponds to the case when the BS considers sector μ_i to suppress in addition to sectors of φ_{j_i} and excludes sector μ_i of cell $\hat{m}$ as the most dominant interference source. Thus, the MS does not suffer interference from sector μ_i of cell $\hat{m}$, due to the suppression by the cancellation filter. However, the use of a cancellation filter does not allow an array gain [26] and thus g_{array} is reduced by one half in (3.25). Hence, the optimum frequency reuse set index j_{opt} can be obtained by finding the number of the STEP which yields the maximum geometry capacity among the aforementioned three steps as

$$j_{opt} = j_{\hat{k}}\tag{3.27}$$

where

$$\hat{k} = \arg \max_{k \in \{1,2,3\}} C_{k,j_k}.\tag{3.28}$$

3.4.2 Performance Evaluation

The performance of the proposed ICI mitigation techniques is verified in the downlink of WiBro by computer simulation. The simulation parameters are summarized in Table 3.1. We assume that the MS can obtain the CSI of the two strongest interfering cells using a pilot signal encoded by cell-specific scrambling codes. The PUSC permutation is performed to obtain both the IA and FH. For ease of verification, we assume that the BS allocates the resource to the MS at every frame time. For reference, the actual user capacity is calculated by averaging the capacity obtained from link level simulation at every frame. The simulation is performed in two geographical positions to investigate poor interference environments as illustrated in Figure 3.8, where MS 1 located in a direction of 60 degrees experiences strong ICI from sector 2 of cell 3 and simultaneously inter-sector interference from sector 1 of its own cell.

Table 3.1 Simulation parameters

Parameters	Values
Number of 3-sector cells	19
Carrier frequency	2.3 GHz
Duplex	TDD
Frame duration	5 ms
Channel bandwidth	8.75 MHz
Cell radius	1 km
BS EIRP	57 dBm
MS noise figure	7 dB
# of BS TX/RX antenna	TX: 2, RX: 2
Transmit antenna scheme	Beamforming
Max number of retransmission	3
MCS	Profile for code type CTC [2]
Subcarrier allocation	PUSC
Path loss model	COST 231 - Hata Suburban [27]
BS antenna pattern	65°(-3dB) with 20dB front-to-back ratio
Cell loading factor	1
Channel model	ITU-R Pedestrian A 3km/h
Channel estimation	Perfect
Receiver algorithm	Linear MMSE

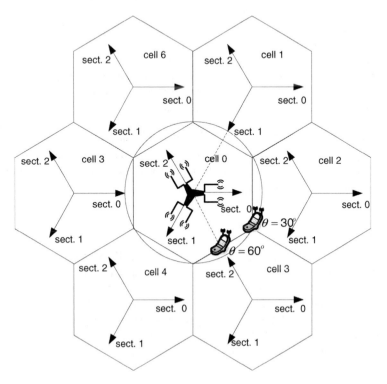

Figure 3.8 Location of MSs in sector 0 of cell 0

It can also be seen that MS 2 in a direction of 30 degrees experiences strong ICI from sector 1 of cell 2 and sector 2 of cell 3 simultaneously. Figure 3.9(a) depicts the geometry capacity associated with the frequency reuse set when MS 1 is located at a direction of 60 degrees. Here, *Step1_FRS0*, *Step1_FRS1*, *Step1_FRS2*, *Step1_FRS3*, *Step2_FRS1* and *Step3_FRS2* denote the geometry capacity calculated by $G_{1,0}$, $G_{1,1}$, $G_{1,2}$, $G_{1,3}$, $G_{2,1}$ and $G_{3,2}$, respectively.*FRS0*, *FRS1*, *FRS2* and *FRS3* denote frequency reuse set of index 0, 1, 2 and 3, respectively. It can be seen that *Step1_FRS3* is chosen first since frequency reuse set φ_3 has the maximum geometry capacity by STEP 1, and then *Step2_FRS1* is chosen for the inter-sector beamforming with frequency reuse set index 1 as the optimum frequency reuse set through STEP 2. It can also be seen that the cancellation of interference from sector 2 of cell 3 with the use of receive filtering (i.e., *Step3_FRS2*) yields a geometry capacity similar to *Step1_FRS3*. Figure 3.9(b) depicts the user capacity, where legends IC and FRF denote the interference cancellation and the frequency reuse factor, respectively. It can be seen that the user capacity has a tendency similar to the geometry capacity. This indicates that the proposed steps can easily be applied to real ICI environments, determining the frequency reuse set that maximizes the user capacity. It can be seen that the proposed scheme that combines IA and inter-sector beamforming outperforms sole use of IA [8] and ICI cancellation [12] which do not consider combining other ICI mitigation techniques.

It can be seen from Figure 3.9(b) that combined use of IA and ICI cancellation (i.e., IA with FRF of 2/3+IC) can be an alternative solution when the inter-sector beamforming is not available.

Figure 3.10 (a) depicts the geometry capacity according to the proposed step when MS 2 is located at a direction of 30 degrees. It can be seen that *Step1_FRS0* is chosen at a distance of 0.7 km, *Step1_FRS1* at 0.8 km and *Step1_FRS3* at a distance between 0.9 km and 1.1 km by the STEP 1, respectively. However, STEP 2 is not chosen because MS 2 is not near the sector boundary. It is shown that *Step1_FRS3* corresponding to frequency reuse set index φ_3 is chosen by STEP 1 from 1.0 km to 1.1 km since MS 2 experiences strong interference from sectors 1 and 2. *Step3_FRS2* is chosen by the STEP 3 at 0.9 km and has a geometry capacity similar to *Step1_FRS3* since MS 2 can cancel out the interference from sector 2 of cell 3 and can avoid ICI from sector 1 with IA. Combined use of interference cancellation and IA (i.e., IA with FRF of 2/3+IC) can be an alternative solution when the reuse factor 1/3 is not available. It can be seen from Figure 3.10(b) that the proposed scheme outperforms the use of single ICI mitigation technique in terms of the user capacity. It can also be seen that the user capacity has a tendency similar to the geometry capacity.

3.5 Conclusion

This chapter has described ICI mitigation techniques for OFDMA systems. We have proposed a new ICI mitigation strategy that maximizes the user capacity near the cell boundary. The new strategy combines ICI mitigation techniques according to the ICI condition. In particular, combined use of IA and FH in the m-WiMAX system was considered using multiple reuse factors in the PUSC mode. Inter-sector beamforming was introduced to avoid inter-sector interference and to obtain beamforming gain. Simulation results show that the new strategy increases the user capacity near the cell boundary and that the scheduler can allocate the resources to users with flexibility.

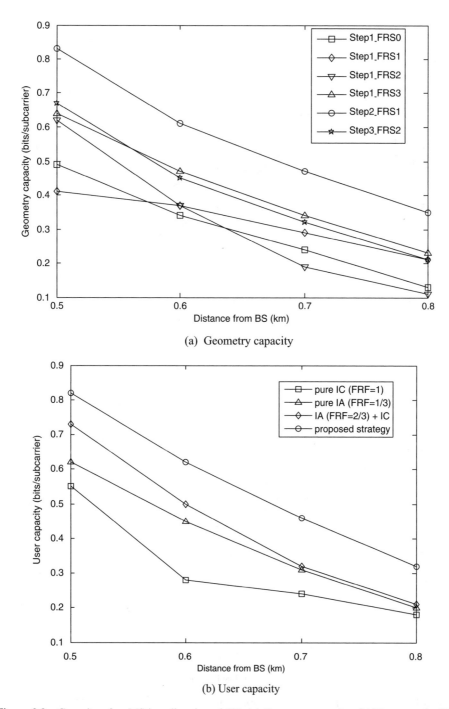

(a) Geometry capacity

(b) User capacity

Figure 3.9   Capacity of an MS in a direction of 60°: (a) Geometry capacity; (b) User capacity [22]

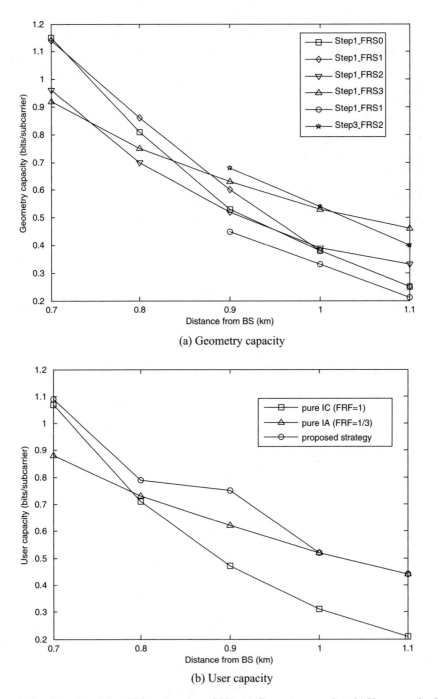

(a) Geometry capacity

(b) User capacity

Figure 3.10   Capacity of the MS in a direction of 30°: (a) Geometry capacity; (b) User capacity [22]

References

[1] G. Lawton, 'What Lies Ahead for Cellular Technology?', *IEEE Computer*, **38**, June 2005, 14–17.

[2] IEEE Std 802.16e, 'Part 16: Air Interface for Fixed and Mobile Broadband Wireless Access Systems', Dec. 2005.

[3] T.S. Rappaport, *Wireless Communications*, Englewood Cliffs, NJ: Prentice Hall, 1996.

[4] A.J. Viterbi, *CDMA: Principles of Spread Spectrum Communication*, Reading, MA: Addison-Wesley Publishing, 1995.

[5] 3GPP TR 25.814, 'Physical Layer Aspects for Evolved UTRA, section 71.2.6', V7.1.0, Sept. 2006.

[6] I. Katzela and M. Naghshineh, 'Channel Assignment Schemes for Cellular Mobile Telecommunication Systems: A Comprehensive Survey', *IEEE Wireless Commun.*, **3**(3), June 1996, 10–31.

[7] J. Zander, 'Generalized Reuse Partitioning in Cellular Mobile Radio', in *Proc. IEEE VTC'93*, May 1993, 181–184.

[8] 3GPP TSG-RAN R1-050896, 'Description and Simulations of Interference Management Technique for OFDMA Based E-UTRA Downlink Evaluation', Qualcomm Europe, Sept. 2005.

[9] 3GPP TSG-RAN R1-060368, 'Performance of Inter-cell Interference Mitigation with Semi-static Frequency Planning for EUTRA Downlink', Texas Instruments, Feb. 2006.

[10] 3GPP TSG-RAN R1-050764, 'Inter-cell Interference Handling for E-UTRA', Ericsson, Aug. 2005.

[11] 3GPP TSG-RAN R1-050608, 'Inter-cell Interference Mitigation Based on IDMA', RITT, June 2005.

[12] 3GPP TSG-RAN R1-061233, 'Performance Evaluation of STTD and Cyclic Shift Diversity in the Presence of Inter-cell/Sector Interference in Downlink MIMO System for LTE', Nortel, May 2006.

[13] D. Tse and P. Viswanath, *Fundamentals of Wireless Communication*, Cambridge: Cambridge University Press, 2005.

[14] H. Dai, A.F. Molisch and H. Vincent, 'Downlink Capacity of Interference-limited MIMO Systems with Joint Detection', *IEEE Trans. Wireless Commun*, **3**, Mar. 2004, 442–453.

[15] 3GPP TSG-RAN R1-050615, 'Investigations on Inter-sector Diversity in Evolved UTRA Downlink', NTT DoCoMo, June 2005.

[16] A. Morimoto, K. Higuchi and M. Sawahashi, 'Performance Comparison between Fast Sector Selection and Simultaneous Transmission with Soft-combining for Intra-node B Macro Diversity in Downlink OFDM Radio Access', in *Proc. IEEE VTC'06*, Mar. 2006, 157–161.

[17] J. Tan and G.L. Stuber, 'Multicarrier Delay Diversity Modulation for MIMO Systems', *IEEE Trans. Wireless Commun.*, **3**(5), Sept. 2004, 1756–1763.

[18] S.A. Jafar, G.J. Foschini and A. Goldsmith, 'PhantomNet: Exploring Optimal Multicellular Multiple Antenna Systems', in *Proc. IEEE VTC'02*, 1, Sept. 2002, 24–28.

[19] A. Goldsmith, S.A. Jafar, N. Jindal and S. Vishwanath, 'Capacity Limits of MIMO Channels', *IEEE J. Select. Areas Commun.*, **21**, June 2003, 684–702.

[20] L. Shao and S. Roy, 'Downlink Multicell MIMO-OFDM: An Architecture for Next Generation Wireless Networks', in *Proc. IEEE WCNC'05*, **2**, Mar. 2005, 13–17.

[21] A. Ghosh and D.R. Wolter, 'Broadband Wireless Access with WiMAX/802.16: Current Performance Benchmarks and Future Potential', *IEEE Comm. Magn.*, **43**(2), Feb. 2005, 129–136.

[22] J. Yeom and Y. Lee, 'Mitigation of Inter-cell Interference in the WiMAX System', in *Proc. IEEE Mobile WiMAX Symp'07*, pp. 26–31, Mar. 2007.

[23] WiMAX Forum, 'WiMAX Forum Mobile System Profile v1.0.0', May 2006.

[24] 3GPP2 TSG-C C30-20021021-011, '1xEV-DV Evaluation Methodology-Addendum (Proposed V7)', Nokia, Oct. 2002.

[25] J.G. Proakis, *Digital Communications*, 4th edn, Maidenhead: McGraw-Hill Higher Education, 2001.

[26] A. Paulraj, R. Nabar and D. Gore, *Introduction to Space-Time Wireless Communications*, Cambridge: Cambridge University Press, 2003.

[27] V.S. Abhayawardhana and I.J. Wassell, 'Comparison of Empirical Propagation Path Loss Models for Fixed Wireless Access Systems', in *Proc. IEEE VTC'05*, **1**, June 2005, 73–77.

4

Overview of Rate Adaptation Algorithms and Simulation Environment Based on MIMO Technology in WiMAX Networks

Tsz Ho Chan, Chui Ying Cheung, Maode Ma and Mounir Hamdi

4.1 Introduction

Broadband wireless communication networks have undergone rapid development for many years. From the 2.5G GPRS technology to the 3.5G HSDPA/HSUPA technologies, the data rates have increased by around a hundred times theoretically. However, the data rate is still low compared to the traditional wired networks such as IEEE 802.3 Ethernet. The IEEE 802.11n standard was developed to provide data rates of over 240 Mbps theoretically for local area network. In order to extend the coverage to a metropolitan scale, the IEEE 802.16 (WiMAX) standard was developed and ratified as the IEEE 802.16d [1] standard in June 2004. In December 2005, enhancements were made to the WMANs mobility support to WiMAX and it was officially approved as the IEEE 802.16e [2] standard. The WiMAX standard identifies two major groups of frequency bands: the 10–66 GHz licensed bands and the frequency bands below 11GHz. The 10–66 GHz licensed band has typical channel bandwidths of 25 MHz or 28 MHz and a raw data rate over 120 Mbps. Since the wavelength is shorter, line-of-sight (LOS) is required and it is suitable for point-to-multipoint (PMP) access serving applications in office environment. The frequency bands below 11 GHz can further be classified as licensed or license-exempt bands. Because of its longer wavelength, LOS is not necessary, but multipath effects in NLOS operation can degrade its performance significantly. Another problem with the license-exempt band comes from the fact that it allows other primary devices to utilize the same radio spectrum (primarily 5–6 GHz), which introduces additional interference and

Mobile WiMAX   Edited by Kwang-Cheng Chen and J. Roberto B. de Marca.
© 2008 John Wiley & Sons, Ltd

co-existence issues. The PHY and MAC layers have to incorporate additional mechanisms such as dynamic frequency selection (DFS) to cope with them. The IEEE 802.16e standard defines the mobile WMANs for combined fixed and mobile broad bandwidth access, supporting subscriber stations moving at vehicular speeds operating in licensed bands below 6 GHz. With the enhancements on mobility support, WiMAX systems can provide broadband wireless access with higher data rates (around 70 Mbps) and high mobility. It can compete with existing wired systems such as coaxial systems using cable modems and digital subscriber line (DSL) links.

Regarding data rates support, one important issue in the WiMAX networks is to develop efficient algorithms to support data rate adaptation. The WiMAX standard has specified different modulation schemes and message formats for the systems to deliver broadband service. However, the policy on how and which modulation scheme should be used under various link conditions has not been specified in the standard. It has provided a wide field for researchers to develop and evaluate different data rate adaptation algorithms to enhance the performance of the WiMAX systems. In this chapter, we will investigate the research issues in deriving rate adaptation algorithms in WiMAX. We will also construct a framework in which the PHY layer metrics can be passed into the MAC layer in a practical simulation environment, which will facilitate future development on rate adaptation algorithms. The following section gives some background knowledge on the WiMAX physical layer and the MAC layer related to the PHY frame structure and message formats supporting rate adaptation algorithms. In Section 4.3, we discuss various research issues on the MIMO-based rate adaptation algorithms. Section 4.4 presents a simulation framework for rate adaptation algorithms and evaluates the simulation results on the MIMO-based WiMAX PHY layer. Finally, Section 4.5 presents a conclusion.

4.2 WiMAX Physical and MAC Layer Description

The WiMAX physical (PHY) layer protocol specifies the electrical specification, collision control and timing dynamics of hardware such as antennas. It defines the conversion between the data bits from upper layer and the corresponding electrical signals transmitting over the air. The standard identifies two entities in a WiMAX network, the base stations (BSs) and the subscriber stations (SSs). The standard specifies five air interfaces for the PHY layer allowing flexibility for the service providers to optimize system deployments:

- *WirelessMAN-SC*: A single-carrier modulated air interface for 10-66 GHz frequency bands.
- *WirelessMAN-SCa*: A single-carrier modulated air interface for licensed bands below 11 GHz.
- *WirelessMAN-OFDM*: An orthogonal-frequency division multiplexing scheme which consists of 256 orthogonal carriers for licensed bands below 11 GHz.
- *WirelessMAN-OFDMA*: An OFDM scheme with 2048 carriers for licensed bands below 11 GHz. Multiple access for SSs is achieved by assigning each SS a subset of the 2048 carriers.
- *WirelessHUMAN*: Wireless High-speed Unlicensed Metropolitan Area Networks. It is used for license-exempt bands below 11 GHz. The SCa, OFDM and OFDMA modulation schemes can be applied with an additional DFS mechanism to ensure there is no harmful interference to the primary devices identified by the regulation.

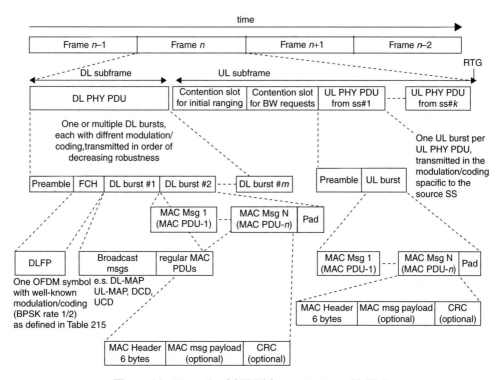

Figure 4.1 Example of OFDM frame structure with TDD

The first four air interfaces support both the time-division duplexing (TDD) and frequency-division duplexing (FDD) while the WirelessHUMAN air interface supports only TDD. Both TDD and FDD configurations adopt an adaptive burst profiling framing mechanism where transmission parameters such as modulation and coding schemes (MCS) can be adjusted individually to each SS on a frame-by-frame basis. Figure 4.1 shows an example of an OFDM frame structure with TDD [1].

The uplink (UL) PHY is based on a combination of time-division-multiple-access (TDMA) and demand-assigned multiple-access (DAMA) where the channel is divided into a number of time slots. The downlink (DL) channel is TDM at the BS where the MAC protocol data unit (PDU) for each SS is multiplexed onto a single stream of data and is received by all SSs within the coverage sector of the BS.

The WiMAX medium access control (MAC) layer protocol describes and specifies the issues of message composition and transmission, service provisioning, resource allocation and connection maintenance. The standard has defined frameworks supporting both the point to multipoint (PMP) and the Mesh topologies. In the PMP mode operation, within a given frequency channel and coverage of the BS sector, all SSs in the downlink receive the same transmission, or parts of it. The BS is the only transmitter operating in this direction and it transmits without having to coordinate with other stations. The downlink is used for information broadcasting. In cases where the message DL-MAP does not explicitly indicate that a portion of the downlink subframe is for a specific SS, all SSs are able to listen to that portion. The

SSs check the connection identifiers (CIDs) in the received protocol data units (PDUs) and retain only those PDUs addressed to them. SSs share the uplink to the BS on a demand basis. Depending on the class of service at the SSs, the SSs may be issued continuing rights to transmit or the transmission rights granted by the BS after receipt of requests from SSs. In addition to individually addressed messages, messages may also be sent by multicast to group of selected SSs and broadcast to all SSs. In each sector, SSs are controlled by the transmission protocol at the MAC layer and they are enabled to receive services to be tailored to the delay and bandwidth requirements of each application. It is accomplished by five types of uplink sharing schemes, which are unsolicited bandwidth grants, polling, and bandwidth requests contention. The IEEE 802.16e standard has also defined a set of schemes to support mobility for the ranging, handover process, power control and sleeping of mobile WiMAX devices. When SS registers to a BS, one transport connection is associated with one scheduling service. The IEEE 802.16e standard has defined five scheduling services:

- Unsolicited Grant Service (UGS)
- Real-time Polling Service (rtPS)
- Extended rtPS (ertPS)
- Non-real-time Polling Service (nrtPS)
- Best Effort (BE).

The transmission scheme at the MAC layer is connection-oriented. All data communications are defined in the context of a connection. Service flows can be provisioned at an SS and connections are associated with these service flows, each of which is to provide transmission service with requested bandwidth to a connection. The service flow defines the QoS parameters for the PDUs that are exchanged on the connection. The concept of a service flow on a connection is the key issue to the operation of the MAC protocol. Service flows provide a mechanism for uplink and downlink QoS management as bandwidth allocation processes. An SS requests uplink bandwidth for each connection. Bandwidth is granted by the BS to an SS as an aggregate of grants in response to each connection request from the SS. The BS and each SS have agreed on using a particular burst profile when the SS registered to the BS and the SSs have to measure the link conditions and to report statistics such as Carrier to Interference-plus-Noise Ratio (CINR) to the BS periodically. If the CINR is beyond the agreed operation threshold of that burst profile, the SS sends a downlink-burst-profile-change request (DBPC-REQ) to the BS if it has uplink bandwidth granted, or sends a RNG-REQ message during initial ranging interval, to signify a change in the burst profile. Figures 4.2 and 4.3 show the transitions to a more robust and a less robust burst profiles respectively [1].

The standard does not specify the exact algorithm for SSs to determine its optimal burst profile. It is up to individuals to implement their own algorithm and it is this flexibility which leads to research on different kinds of rate adaptation algorithms in the WiMAX systems.

4.3 Research Issues on the MIMO-based Rate Adaptation Algorithms

Limited frequency bandwidth and multipath fading have long been problems for wireless communication. Larger bandwidth can provide higher data transmission rates while multipath fading over the air channel causes signal quality degradation and hence lowers the overall

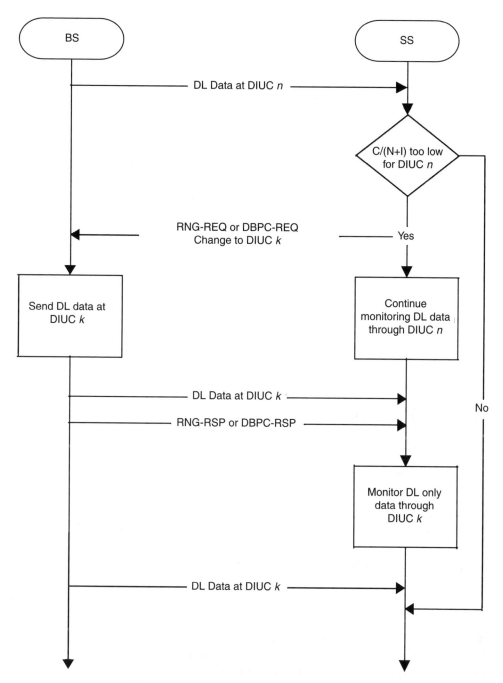

Figure 4.2　Transition to a more robust burst profile

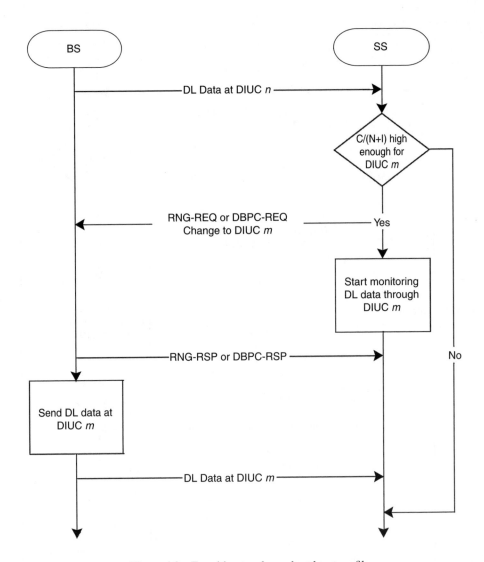

Figure 4.3   Transition to a less robust burst profile

achievable throughput. To mitigate the signal quality degradation issue, increasing transmitted power level may be needed. However, the transmitted power is regulated and it has to comply with various standards in different countries. Therefore, the data transmission rates cannot easily be increased using that approach. With the development of wireless communication technology, the emerging Multiple-Input-Multiple-Output (MIMO) technology provides a promising solution to improve the data transmission rates. This section discusses various research issues on MIMO-based rate adaptation algorithms.

4.3.1 Physical Layer Enhancement by MIMO: Spatial Diversity vs Spatial Multiplexing

MIMO has been a hot research topic in the field of wireless communication since the first literature on this topic was released, around twenty years ago. Hundreds of papers were published regarding different aspects of MIMO such as coding schemes and performance analysis. However, it was not until the past two to three years that MIMO-based routers have appeared on the consumer market. MIMO systems exploit spatial diversity by using multiple antennas at the transmitter side or the receive side of the communication link. The multipath property of wireless signals transmission becomes an important basis in constructing theories in MIMO. The WiMAX standard has specified the adoption of MIMO techniques as an optional feature for WiMAX systems, where at most four antennas are supported on a WiMAX device. The gain in spatial diversity provided by the MIMO technique improves robustness to channel fading and hence WiMAX BSs and SSs can communicate at higher data rates as compared with a WiMAX system with a single antenna only.

In a MIMO-based WiMAX system, each antenna can be treated as one spatial stream basically, in which a higher level MCS requires a higher SNR to maintain a low Bit Error Rate (BER). MIMO supports two modes of operation: spatial diversity for better signal quality and spatial multiplexing for higher throughput. These two MIMO modes can be configured by sending data over variable number of spatial streams. Since each WiMAX device supports up to four antennas, when the number of spatial streams is fewer than the number of transmit antennas, the remaining antennas can be exploited for diversity gain. In general, Space Time Code (STC) is the technique to exploit spatial diversity. It consists of a set of coding algorithms which adjust and optimize the joint encoding across the transmit antennas, aiming to increase the reliability of the wireless link. Alamouti [3] proposed the Space Time Block Code (STBC) which can be implemented in the DL of the WiMAX systems with 2×1 or 2×2 transmit-receive antennas at a coding rate of 1. STBC can be used in low delay spread environment and it enhances the PER performance. STBC was later generalized to support arbitrary number of transmit antennas. Research has been conducted in the area of MIMO coding algorithms over the years. For example, Space Time Trellis codes (STTC) is another STC developed with improvements in coding, diversity gain and BER. The complexity of STTC lies on the non-linear decoding procedure at the receiver where a Viterbi decoder is used. This contrasts with the linear Maximum Likelihood (ML) scheme for the decoding of STBC. Spatial multiplexing (SM) is the technique for increasing throughput at a given SNR in the wireless channel. The WiMAX BS transmits independent data streams from each transmit antenna to SSs. Source data streams are multiplexed into N transmit antennas and they are sent to the WiMAX SSs with N receive antennas resulting in a N times higher rate. The Bell Labs Layered Space-Time (BLAST) scheme, such as VBLAST or D-BLAST, sends parallel symbols streams from the transmitter to the receiver. The main disadvantage with SM schemes is that the channel has to be in good condition with high SNR for it to be feasible. The decoding procedure at the WiMAX SS is simple and all the received signals can be recombined linearly by methods, such as ML, Zero Forcing (ZF) or Minimum Mean Square Error (MMSE). A WiMAX BS can either use all its antennas to exploit spatial diversity or spatial diversity. Reference [4] shows that it may not be an optimal case in terms of maximizing throughput if a WLAN system, which has more than two antennas, operates at the two extremes. Reference [4] proposes a hybrid 4 x 4 scheme which combines spatial diversity and spatial multiplexing to provide both increased throughput and diversity to the system. This chapter gives insights to the design of rate adaptation algorithms in the PHY layer.

The WiMAX standard only specifies MIMO feedback messages framework and does not define the exact implementation of different spatial diversity or spatial multiplexing techniques. Research issues such as MIMO encoding and decoding using information theory, beamforming and channel capacity analysis are still continuing in this vibrant and active research area. Advances in either of these aspects definitely influence the design, the choice of MCS and rate adaptation algorithms in the MAC layer.

4.3.2 Closed-loop and Open-loop Link Adaptations in WiMAX

The availability of the channel state information (CSI) categorizes MIMO systems into two types and it deeply affects the performance of the rate adaptation algorithms. The WiMAX standard has defined a PMP frame structure for DL and UL of SSs. It supports both the open-loop and closed-loop link adaptive feedback from the SSs to the BS. The closed loop operation assumes perfect CSI is available at the transmitter, either through explicit feedback from the receiver using specific control frames or through channel sounding and calculation between the transmitter and the receiver. However, the computation complexity of the precoding matrix [5] and the communication overhead are large and thus difficult to implement. In closed-loop rate adaptation designs, the MIMO PHY scheme maximizes the data rate for a target BER. It selects a subset of the transmit antennas and chooses the best constellation which can be supported on each of the selected antennas. The selected rate setting is then sent back from the receiver to the transmitter in the MAC design. In open-loop rate adaptation designs, there is no feedback from the receiver to the transmitter and hence the transmitter does not have any channel knowledge. It uses the same MCS and the same TX power for each spatial stream for data transmission. The open-loop multi-antenna approach is easier in implementation, at the expense of the potential channel capacity utilization.

Research papers published specifically on this aspect of WiMAX systems have been extremely rare. This may be due to the absence of a suitable WiMAX simulator to evaluate performance in this area. References [5] and [6] focus on the theoretical analysis of the closed-loop feedback and limited feedback mechanism in non-WiMAX context. It is expected that more research work will be done in this area on the availability of suitable simulation tools. The research trend in this area has been towards deriving rate adaptation algorithms with limited channel feedback as suboptimal solutions to closed-loop feedback adaptation.

4.3.3 Channel Quality Measurement and Channel Characterization

The WiMAX standard provided a framework for BSs and SSs to obtain signal quality statistics during their communications. It defined fields for the Received Signal Strength Indication (RSSI) and Carrier to Interference-plus-Noise Ratio (CINR) in the feedback response message body. The RSSI and CINR statistics, which are mandatory, but vendor-specific in implementation, can aid the BS assignment and burst adaptive profile selection. SSs can obtain RSSI measurement statistics from the OFDM DL preambles and then SSs report the updated mean and the standard deviation of the RSSI and CINR, in units of dBm and dB respectively, back to the BS via REP-RSP messages. With the RSSI and CINR statistics, the antenna at the BS can

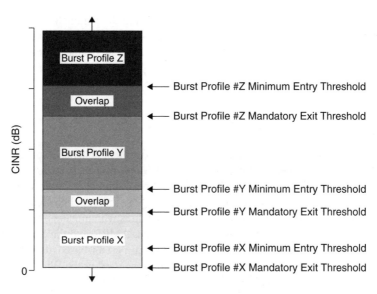

Figure 4.4 Burst profile usage

evaluate the channel quality and determine the transmission rate to be used in the DL to SSs. There is little research in this area. Practically no paper has been published on rate adaptation algorithms using RSSI and CINR statistics in the WiMAX systems. This may be due to the fact that no commercial product has been released on the consumer market and individual researchers could not evaluate the performance of their algorithms. Research on this area is limited to companies that develop WiMAX products.

An accurate channel model which captures the characteristics of the propagation environment allows the transmitter at the BS or SSs a better knowledge of the channel condition. It also helps in channel capacity analysis and the design of MIMO coding techniques and rate adaptation algorithms which match the channel statistics. In the WiMAX specification [1], only the usage of the burst profile threshold is defined, as shown in Figure 4.4 and there is no definition on when each of the MIMO techniques will be used.

An active and open research area is how to use the statistical data and characterize the MIMO channel so as to facilitate the choice of spatial multiplexing or spatial diversity techniques. Reference [7] points out that most papers on the topic of MIMO use a statistical channel model which is an idealized abstraction of spatial propagation characteristics and assumes independent and identically distributed (i.i.d.) fading between different transmit–receive antenna pairs. The use of an idealized channel model helps in STC design since it is tractable in capacity analysis. However, the realistic capacity of MIMO channels can be substantially lower as the channel coefficients between different transmit-receive antenna pairs exhibit correlation due to clustered scattering in realistic environments. Parametric physical models, which explicitly model signals arriving from different directions, are another heuristic in the channel modeling research. Nevertheless, physical channel parameters, such as the angles of signals direction, introduce nonlinear dependence on the models and this increases the difficulty of calculating the channel capacity and designing STC. Research on this topic aims to characterize the channel accurately while preserving the linear properties in the parameters of the models for analysis. References [7] and [8] propose a virtual representation of MIMO fading channel by introducing pairwise error probability (PEP), rank and eigenvalue characterization. It is expected such characterization could provide simpler parameters set to the upper layer and hence provide clues to the rate adaptation algorithms in cross-layer design.

4.3.4 Automatic Request (ARQ) at the MAC Layer

The WiMAX standard does not mandate an acknowledgement scheme in the MAC layer to ensure packets have been received correctly in the BS and SSs. It is an optional feature to implement the Automatic Request (ARQ) or the Hybrid Automatic Request (hARQ) in the MAC layer to provide such capability. The lack of ARQ or similar acknowledgement scheme creates difficulties when devising rate adaptation algorithms which make use of the feedback acknowledgment statistics from the other stations. Onoe, AMRR and the Sample Rate algorithms used in IEEE 802.11 WiFi environments, which are credit-based algorithms counting the number of successful or failed transmissions, cannot be used in the WiMAX systems. An open research issue is how to devise a rate adaptation algorithm when ARQ is absent. The WiMAX standard only defines a channel quality indicator (CQI) channel to report channel state information (CSI), which includes physical and effective CINR and RSSI mean and standard deviation values and MIMO mode selection data, to the BS to aid the process of adaptive modulation and coding (AMC) selection.

On the other hand, if the WiMAX devices incorporate ARQ or HARQ schemes into the MAC layer, rate adaptation schemes can make use of the feedback data to realize AMC. However, the lack of a suitable comprehensive simulator in this stage is the main issue when devising and verifying such algorithms. Simulators such as NS-2 and OPNET are the most widely adopted tools in the wireless research field, [9] works on the performance evaluation of different ARQ algorithms in WiMAX systems, but there are no comprehensive WiMAX MAC packages released so far. As mentioned in previous subsections, the accuracy of link quality measurement and characterization constitutes a main factor in verifying the performance of rate adaptation algorithms. Currently, only the WiMAX devices vendors are able to carry out performance tests on their products. It is still at an early stage for individual researchers to develop simulation tools in this area.

4.4 Constructing a Practical Rate Adaptation Simulation Model for Mimo-Based WiMAX Systems

In this section, we construct a WiMAX rate adaptation simulation model and evaluate the results on the MIMO-based WiMAX PHY layer. Research on WiMAX PHY and MAC layers has been conducted for many years. However, the focus of most published work has been either PHY enhancement without MIMO or MAC layer quality of service (QoS). Some researchers used OPNET to carry out WiMAX simulation, but the simulation model does not take MIMO techniques into account. Nevertheless, corporations, such as Intel and Nokia, etc, are actively engaged in the development and deployment of MIMO-based WiMAX systems. The WiBro network in Korea is one example. It is expected that rate adaptation algorithms could be evaluated more easily when more commercial products become available.

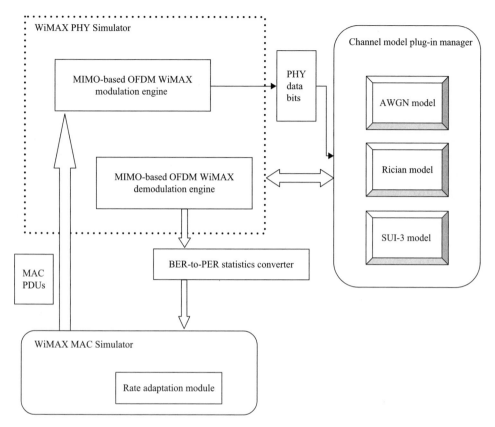

Figure 4.5   Structure of the simulation framework

4.4.1 Simulation Model Structure and Features

Since the introduction of MIMO technology into the WiMAX systems, how to select the best combination of antennas which could maximize the data transmission rates has been an issue. This leads to research topics in cross-layer optimization where the MAC design has to take account of the PHY layer parameters. The evaluation methodology would be a combination of the signaling level simulation in the PHY layer and the protocol level simulation in the MAC layer. However, the absence of a suitable WiMAX simulator leads to little evaluation of the research in this area. In order to evaluate rate adaptation algorithms in WiMAX for practical simulations, we construct a MIMO-based WiMAX simulation model on the PHY layer, based on the work of [10] on MIMO channel characterization. Figure 4.5 shows the structure of the simulation framework. The PHY model adopts the WirelessMAN-OFDM air interface specification. It is based on OFDM-256 modulation and designed for NLOS application. In OFDM-256, the waveform is composed of 256 orthogonal carriers and so selective fading can be confined to a number of carriers which can be equalized easily. With the use of OFDM symbol time and a cyclic prefix, the inter-symbol interference (ISI) problems and the adaptive equalization problems can be handled.

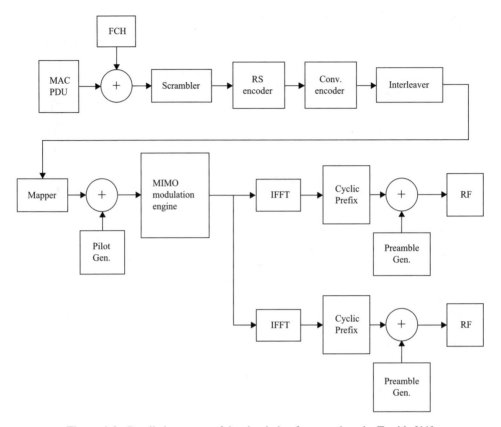

Figure 4.6 Detailed structure of the simulation framework at the Tx side [11]

A basic requirement of a practical simulation environment to evaluate rate adaptation algorithms is the ability to provide support and switching of different modulation schemes. Adaptive modulation allows a WiMAX system to adopt different MCSs depending on the signal to noise ratio (SNR) of the radio link. The simulation model is capable of performing binary phase shift keying (BPSK), quaternary phase shift keying (QPSK), 16-quadrature amplitude modulation (16-QAM) and 64-QAM which comply with the four mandatory modulation coding schemes specified in the WiMAX standard for uplink (UL) and downlink (DL) connections. Figure 4.6 depicts the detailed structure of our simulation model on PHY layer at the transmitter side.

- *Frame Control Header (FCH)*: This is added to the data frame according to the WiMAX specification.
- *Scrambler*: Data randomization is performed on each burst of data with padding.
- *Reed-Solomon encoding*: This is derived from a systematic RS code using GF(28) and other parameters specified in the WiMAX standard. The code is shortened and punctured to enable variable block sizes and variable error-correction capability.

Table 4.1 Supported rate combinations

Rate_ID	Modulation RS-CC rate
0	BPSK 1/2
1	QPSK 1/2
2	QPSK 3/4
3	16-QAM 1/2
4	16-QAM 3/4
5	64-QAM 2/3
6	64-QAM 3/4
7–15	*reserved*

- *Convolutional code*: Each RS block is encoded by the binary convolutional encoder which has native rate of 1/2.
- *Interleaver*: This is a two-stage permutation block interleaver with a block size corresponding to the number of coded bits per allocated subchannels per OFDM symbol.
- *Modulation coding schemes (MCS)*: BPSK, gray-mapped QPSK, 16-QAM and 64-QAM are supported for the constellation mapper.
- *Pilot insertion*: Pilot subcarriers are inserted into each data burst in order to constitute the symbol and they are modulated according to their carrier location within the OFDM symbol.
- *MIMO modulation engine*: Symbols are processed to exploit spatial diversity or spatial multiplexing.

Table 4.1 shows the supported rate combinations with the RSCC rates in the simulator. The WiMAX standard does not specify a channel modeling for practical simulation. It is up to researchers to decide which channel model should be adopted for their simulations. The simulation model, with a plug-in interface, supports additive white Gaussian Noise (AWGN) channel model, Rician fading channel model and the SUI-3 channel model.

4.4.2 Simulation Results and Discussion

As discussed in the previous section, there are various aspects of research issues which have to be considered when deriving rate adaptation algorithms. For example, methods such as the Frobenius norm can be used to characterize a channel and its capacity at a given SNR. Reference [10] proposes another method based on the Euclidean distance calculations to characterize channel H with the use of the Demmel condition number. The paper proposes a fixed rate adaptation, where the spatial modulation scheme is varied between the Alamouti MIMO diversity scheme with 16-QAM modulation and spatial multiplexing with 4-QAM transmit constellations. Based on the idea of using the Demmel condition number proposed in [10], we have extended and built the WiMAX PHY simulator. Three groups of channel matrix H capturing the essence of different channel conditions are simulated using the MCSs defined by the IEEE 802.16e standard with MIMO technology incorporated into the simulation.

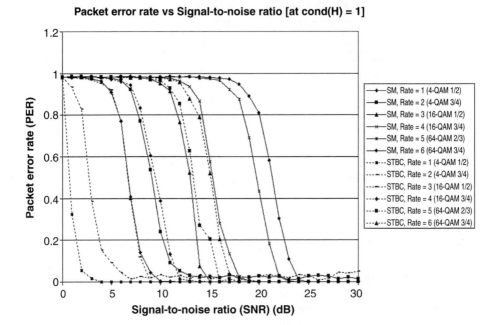

Figure 4.7   PER vs SNR at cond(H) = 1

Alamouti STBC is used for spatial diversity while V-BLAST is used for spatial multiplexing as the MIMO technique.

Based on the results in [11], the PER-SNR curves are obtained from the BER-to-PER statistics converter. In each figure, under the same MIMO technique, a higher MCS will shift the PER-SNR curve to the right. This corresponds with the fact that under a given target PER, a higher MCS would require a higher SNR to achieve the same performance. Figure 4.7 shows the simulation result where cond$D$(H) = 1. The value is small and it implies the channel coefficients in the channel matrix H are more independent of each other. Either STBC or SM schemes can result in similar PER. Therefore, spatial multiplexing should be adopted in such cases since the data rate can be boosted up. Figure 4.8 and Figure 4.9 show the simulation results where the cond$D$(H) values become larger and larger (5 and 20).

It is worthy of note that as the value goes up, the set of STBC curves and the set of SM curves start to separate from each other. The set of STBC curves goes towards the left while the set of SM curves moves towards the right. This suggests that when the SNR is high, SM with a higher MCS should be adopted and when the SNR is low, it is better to be more conservative to stay with the STBC scheme. A general observation is to use spatial multiplexing to achieve higher data throughput at high SNR regions as long as the packet error rates (PER) are small. When the channel quality is detected to be poorer, data transmission should be switched to exploit the spatial diversity so that the transmission reliability can be increased at low SNR regions.

After the abstract PHY model is built, PHY layer parameters such as the channel matrix, the Demmel condition numbers and SNR can be passed into the MAC layer simulator. In terms of

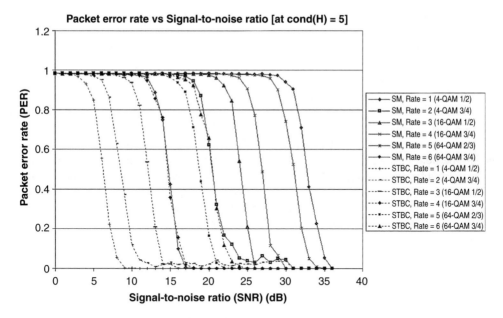

Figure 4.8   PER vs SNR at cond(H) = 5

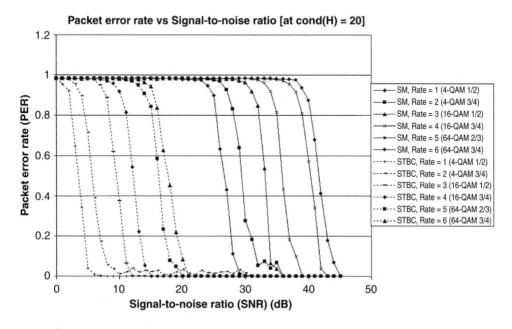

Figure 4.9   PER vs SNR at cond(H) = 20

practical rate adaptation algorithms, the complexity of the parameters passing from the PHY layer into the MAC layer should be small so that the computation in the WiMAX systems hardware is affordable without performance depreciation. The simulation model makes use of the Demmel condition number to capture the channel characteristic, which satisfies such a requirement. From the simulation results, we can select some effective SNR thresholds to switch between different MCSs and MIMO coding techniques under different condD(H) values in the WiMAX PMP mode. The MAC layer of WiMAX adopts a request-grant approach and the rate information is defined in the Uplink Channel Descriptor (UCD) burst profile and the Downlink Channel Descriptor (DCD) burst profile. The exact MCS is specified inside the FEC code type field of the profiles. Rate adaptation algorithms in MAC layer simulator can be implemented based on the recently released WiMAX MAC layer in ns-2 [12]. The direction of devising rate adaptation algorithms could be on deriving a credit-based algorithm with reference to the RSSI or SNR statistics and the channel feedback data from the PHY layer. These algorithms, together with QoS research solutions, can be evaluated thoroughly by using the constructed WiMAX simulation model.

4.5 Conclusion

In this chapter, we have given a comprehensive overview on the IEEE 802.16e PHY and the MAC layers and the background knowledge related to devising rate adaptation algorithms in WiMAX. We have also discussed the various research issues in devising rate adaptation algorithms with the incorporation of MIMO techniques. The major issue in the research of rate adaptation algorithms in WiMAX nowadays is the lack of comprehensive WiMAX simulation tools and framework in the open research domain. The majority of the work is limited to WiMAX vendors internally. We have demonstrated a practical simulation model to evaluate IEEE 802.16e rate adaptation algorithms with Alamouti STBC and V-BLAST as the MIMO techniques at the WiMAX PHY layer. The model has made use of the Demmel condition values to address the channel characterization issue and has the advantage of a reduced parameter set passed to the WiMAX MAC layer for practical and representative simulation. More contributions from academia to the construction of a publicly available WiMAX simulator are expected in the near future.

References

[1] IEEE 802.16-2004, IEEE Standard for Local and Metropolitan Area Networks – Part 16: Air Interface for Fixed Broadband Wireless Access Systems, October 2004.

[2] IEEE 802.16e-2005 and IEEE 802.16-2004/Cor1-2005, IEEE Standard for Local and Metropolitan Area Networks – Part 16: Air interface for Fixed Broadband Wireless Access Systems, Amendment 2: Physical and Medium Access Control Layers for Combined Fixed and Mobile Operation in Licensed Bands and Corrigendum 1, February 2006.

[3] S.M. Alamouti, 'A Simple Transmit Diversity Technique for Wireless Communications', *IEEE J. Sel. Areas Commun.*, **16**(8), October 1998, 1451–1458.

[4] A. Doufexi, A. Nix, and M. Beach, 'Combined Spatial Multiplexing and STBC to Provide Throughput Enhancements to Next Generation Wlans', in *IST Mobile and Wireless Communications Summit*, Dresden, 2005.

[5] J. Zhang, T. Reid, K. Kuchi, N.V. Waes, and V. Stolpman, 'Closed-Loop MIMO Precoding with CQICH Feedbacks', *IEEE 802.16 Broadband Wireless Access Working Group*, 2005.

[6] W. Santipach and M.L. Honig, 'Achievable Rates for MIMO Fading Channels with Limited Feedback and Linear Receivers', in *IEEE Eighth International Symposium on Spread Spectrum Techniques and Applications*, 2004.

[7] Z. Hong, K. Liu, R.W. Heath, and A.M. Sayeed, 'Spatial Multiplexing in Correlated Fading Via the Virtual Channel Representation', *IEEE Journal Selected. Areas Commun.*, **21**(5), June 2003.

[8] A.M. Sayeed, 'Deconstructing Multiantenna Fading Channels', *IEEE Trans. Signal Process.*, **50**(10), October 2002, 2563–2579.

[9] M.S. Kang and J. Jang, 'Performance Evaluation of IEEE 802.16d Arq Algorithms with Ns-2 Simulator', paper presented at 2006 Asia-Pacific Conference on Communications.

[10] R.W. Heath and A.J. Paulraj, 'Switching Between Diversity and Multiplexing in MIMO Systems', *IEEE Trans. Commun.*, **53**(6), 2005.

[11] T.H. Chan, C.Y. Cheung, M. Ma, and M. Hamdi, 'Overview of Rate Adaptation Algorithms Based on MIMO Technology in WiMAX Networks', *IEEE Mobile WiMAX Symposium, 2007*, 98–103.

[12] The National Institute of Standards and Technology, 'IEEE 802.16 module for NS-2', http://www.antd.nist.gov/seamlessandsecure/doc.html

5

Phase Noise Estimation in OFDMA Uplink Communications

Yi-Ching Liao, Chung-Kei Yu, I-Hsueh Lin and Kwang-Cheng Chen

5.1 Introduction

OFDM transmission technique has been adopted in several wireless communication standards for its ability to combat channel multipath fading with relatively low complexity while providing high spectral efficiency in comparison to single carrier transmission. An OFDMA system divides the available subcarriers into groups, called subchannels, and assigns one or multiple subchannels to multiple users for simultaneous transmission. Signals from different users overlap in frequency domain but occupy different subcarriers, and the orthogonality among subcarriers prevents multiple access interference (MAI) among users.

On the other hand, OFDM is much more sensitive to carrier frequency offset and phase noise than single carrier systems [1] because of loss of orthogonality among OFDM subcarriers result in the appearance of common phase error (CPE) and inter-carrier interference (ICI). OFDMA inherits from OFDM the fact that is more sensitive to both these problems than single carrier multiple access systems. It is shown that if all users' signals are received with equal power and all transceivers have the same phase noise spectrum, the degradation for OFDMA uplink is the same as for OFDM [2]. However, the users have distinct phase noise spectrum due to different oscillators and the base station may suffer from the near-far effect when the received signal power levels among subcarriers from different users are different. Users with higher power can consequently interfere severely with those users with lower power. On the other hand, poor phase noise spectrum (usually due to unsatisfactory oscillators) can ruin the overall system performance [19]. Therefore, phase noise is more detrimental to uplink OFDMA systems if not carefully compensated.

Various methods to suppress phase noise in OFDM systems have been proposed in the literature [3]–[5]. However, they are specifically suitable for dealing with single phase noise. To mitigate multiple phase noise in OFDMA uplink, the adopted subcarrier assignment scheme

---

Mobile WiMAX   Edited by Kwang-Cheng Chen and J. Roberto B. de Marca.
© 2008 John Wiley & Sons, Ltd

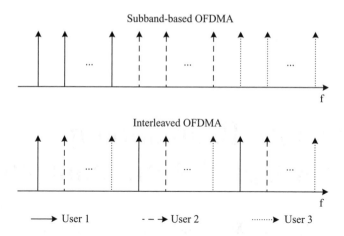

Figure 5.1   Illustration of subband-based and interleaved sub-carrier assignment schemes. Source: 16

needs to be taken into account since it affects the amount of MAI in the system. In this chapter, we study two major subcarrier assignment schemes: sub-band-based and interleaved [6]. The former divides the whole bandwidth into small continuous sub-bands, each user is assigned to one or several sub-bands. In the latter, subcarriers assigned to different users are interleaved over the whole bandwidth. An example of both schemes is illustrated in Figure 5.1.

The rest of this chapter is organized as follows. Section 5.2 introduces the modeling of phase noise. We discuss the effect of phase noise in OFDM systems and several methods to correct phase noise in Section 5.3. With some modification of the correction schemes in OFDM systems, two algorithms based on least-square and maximum-likelihood criterion for estimation of Wiener phase noise in uplink OFDMA communications are introduced in Section 5.4.

5.2 Modeling of Phase Noise

To effectively analyze the distortion and interference caused by phase noise, accurate modeling of these factors is quite important. There are many factors that would generate phase noise. Here, high PAPR with nonlinear amplifier, IQ imbalance and oscillator phase noise are our major interests in OFDM(A) systems.

High PAPR with nonlinear amplifier. The power amplifier is usually operated in the linear region and as close to the saturation region as possible to make its operation power efficient, as shown in Figure 5.2. When the dynamic range of input signal power is too large for the high PAPR, the operation might not stay in the linear region. If the input power is below the minimum detectable signal (MDS), the noise dominates. On the other hand, if it goes into the saturation range, the output power may start to saturate and phase noise then appear due to nonlinearity [7]. Even if the dynamic range falls into the linear region, since the amplifier could not be made ideally, the operation is still not perfectly linear.

A widely accepted power amplifier model [8] is a nonlinear memoryless transformation between the complex envelopes of the input and output signals. The expression is written as

$$A(r) = \frac{vr}{[1 + (vr/A_0)^{2p}]^{\frac{1}{2p}}} \tag{5.1}$$

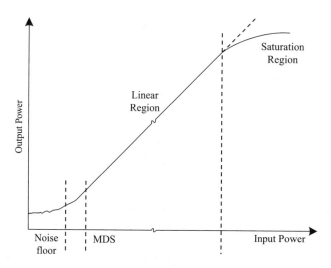

Figure 5.2   Power amplifier transfer function

where r is input amplitude, v is the small signal gain, A_0 is the saturating amplitude and p is a parameter which controls the smoothness of the transition from the linear region to the saturation region.

IQ imbalance. In practical analog electronic component manufacture, the amplitude and phase imbalance between I and Q branches will occur. A solution to IQ imbalance is to implement the IQ modulation in digital circuits. In this way, however, it requires a sufficiently high sampling rate to generate an appropriate intermediate frequency (IF).

With IQ gain imbalance ε and phase imbalance θ, the signals of I and Q branches become

$$x_I(t) = s_I \cdot (1 + \varepsilon) \cos(\omega t)$$

$$x_Q(t) = s_Q \cdot \sin(\omega t + \theta) \tag{5.2}$$

as shown in Figure 5.3.

Oscillator phase noise. Ideally the spectrum of the oscillator is an impulse at center frequency as shown in Figure 5.4(a) while the spectrum of a practical oscillator cannot be a perfectly

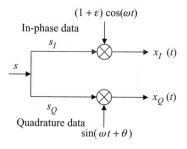

Figure 5.3   IQ modulator model

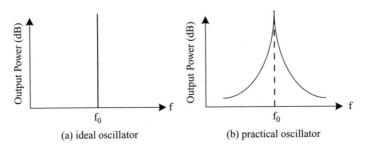

Figure 5.4   Spectrum of the oscillator

impulse as Figure 5.4(b). Therefore, the oscillator frequency is not stable with center frequency, which will cause phase noise when the signals are mixed with such a practical oscillator. Phase noise caused by the instability of the oscillator can be classified into two categories. When the system is only frequency-locked, the resulting phase noise is slowly varying but not limited, and it is modeled as a zero-mean, nonstationary, infinite-power Wiener process. When the system is further phase-locked, the resulting phase noise is with compared light effect and modeled as a zero-mean, stationary, finite-power random process [9].

In the frequency-locked only system, phase noise $\phi(t)$ is conventionally modeled as a Wiener process for which

$$E\left[\phi(t)\right] = 0$$

$$E\left[\phi(t + t_0) - \phi(t_0)\right]^2 = 4\pi\beta\left|t\right| \tag{5.3}$$

where β(Hz) denotes the one-sided 3 dB linewidth of the Lorentzian power density spectrum where we usually use the word 'linewidth' to describe the 3 dB bandwidth of phase noise power spectrum [1]. The Lorentzian spectrum is the squared magnitude of a first order lowpass filter transfer function [10]. The single-sided spectrum $S_\Phi(f)$ is given by

$$S_\Phi(f) = \frac{2/\pi\beta}{1 + f^2/\beta^2} \tag{5.4}$$

The Lorentzian spectrum with different one-sided 3 dB linewidth is shown in Figure 5.5.

On the other hand, for the classical phase noise modeling of the phase-locked system $\phi(t)$ can be modeled as a stationary Gaussian process with zero mean and the specified power spectrum density (PSD). Two common types of phase noise PSD are given in the literatures and they differ in the decay of the transition band [14]. The well-known model proposed by [11] which has a linear decay is given by

$$S_\Phi(f) = 10^{-c} + \begin{cases} 10^{-a} & |f| \le f_l \\ 10^{-\dfrac{(f - f_l)b}{f_h - f_l} - a} & f_l < f \\ 10^{-\dfrac{(f + f_l)b}{f_h - f_l} - a} & f < -f_l \end{cases} \tag{5.5}$$

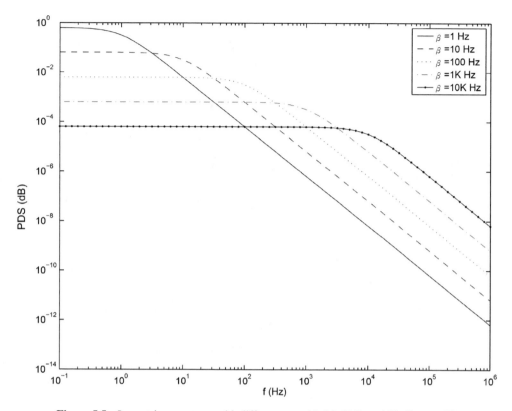

Figure 5.5   Lorentzian spectrum with different one-sided 3 dB linewidth. Source: [5]

where c determines the white phase noise floor of the oscillator. The parameter a gives the phase noise level near the center frequency up to $\pm f_l$ and b is the steepness of noise reduction with increasing frequency distance up to $\pm f_h$ where the noise floor becomes dominant.

Another important characterization of stationary phase noise is its autocorrelation function (ACF). The ACF of a phase noise process can be obtained by performing the inverse Fourier transform to its PSD. In [12], the ACF of (5.5) is given by

$$
R_\Phi(\tau) = 10^{-c}\delta(\tau)
$$
$$
+ 2 \cdot 10^{-a}\left[f_l \cdot \mathrm{sinc}\, 2f_l\tau + \frac{p\cos 2\pi\tau\, f_l - (2\pi\tau)\sin 2\pi\tau\, f_l}{p^2 + (2\pi\tau)^2} \right] \qquad (5.6)
$$

where

$$
p = \frac{b\ln 10}{f_h - f_l}.
$$

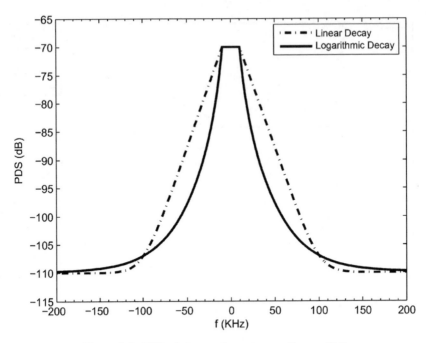

Figure 5.6   PSD of phase-noise processes. Source: [14]

An alternative PSD model [13] which has a logarithmic decay is given by

$$S_\Psi(f) = 10^{-c} + \begin{cases} 10^{-a} & |f| \leq f_l \\ 10^{-a}(f/f_l)^{-b} & f_l < f \\ 10^{-a}(-f/f_l)^{-b} & f < -f_l \end{cases} \tag{5.7}$$

And its ACF is

$$R_\Psi(\tau) = 10^{-c}\delta(\tau) + 2 \cdot 10^{-a} \cdot f_l \cdot \operatorname{sinc} 2f_l\tau$$
$$+ 2f_l^b \cdot 10^{-a} \int_{f_l}^{\infty} f^{-b} \cos 2\pi f\tau df \tag{5.8}$$

An example of the two-phase noise PSD model where $a = 7, b = 4$ and $c = 11$ is illustrated in Figure 5.6.

5.3 Phase Noise in OFDM

Consider a general OFDM system using N-point inverse fast Fourier transform (IFFT) for modulation. Assume the frequency domain subcarrier index set is composed of three mutually exclusive subsets defined by

$$D \overset{\Delta}{=} \{d_1, d_2, \ldots, d_{N_d}\}$$

$$P \overset{\Delta}{=} \{p_1, p_2, \ldots, p_{N_p}\}$$

$$V \overset{\Delta}{=} \{v_1, v_2, \ldots, v_{N_v}\} \tag{5.9}$$

where D denotes the set of indices for N_d data-conveying subcarriers, P is the set of indices for N_p pilot subcarriers and V stands for N_v virtual subcarriers. Then, the set of indices for N_u useful subcarriers U can be defined as

$$U \overset{\Delta}{=} \{u_1, u_2, \ldots, u_{N_u}\} = D \cup P \tag{5.10}$$

where $N_u = N_d + N_p$. Let $X_m(l)$ be the modulated symbol on the lth subcarrier of the mth OFDM symbol. For $l \in U$, $X_m(l)$ is taken from some constellation with zero mean and average power $\sigma_X^2 = E\{|X_m(l)|^2\}$. The output of the IFFT has a duration of T seconds which is equivalent to N samples. We denote $x_m(n)$ the mth OFDM symbol time domain signal. Consequently, the baseband time domain discrete-time transmitted signal can be represented as

$$x_m(n) = \frac{1}{N} \sum_{l \in U} X_m(l) e^{j2\pi \frac{ln}{N}}, \quad 0 \leq n \leq N - 1 \tag{5.11}$$

Before being transmitted over the channel, this signal is preceded by a cyclic prefix (CP) of length N_g longer than the channel impulse response to eliminate inter-symbol interference (ISI).

High PAPR with nonlinear amplifier. After the nonlinear amplifier with the transfer function $A(r)$ expressed in (1), the time domain signal becomes

$$\begin{aligned} x_m'(n) &= x_m(n) \frac{A(|x_m(n)|)}{x_m(n)|} \\ &= x_m(n) + \left[x_m(n) \frac{A(|x_m(n)|)}{x_m(n)|} - x_m(n) \right] \\ &= x_m(n) + p_m(n) \end{aligned} \tag{5.12}$$

where $p_m(n)$ denotes the mth OFDM symbol distortion caused by the nonlinear amplifier.

Focusing on the part of phase noise, the channel is assumed ideal without AWGN noise. After FFT at the receiver, the frequency domain signal is

$$\begin{aligned} \hat{X}_m(l) &= \sum_{n=0}^{N-1} x_m'(n) e^{-j2\pi \frac{ln}{N}} \\ &= \sum_{n=0}^{N-1} [x_m(n) + P_m(n)] e^{-j2\pi \frac{ln}{N}} \\ &= X_m(l) + P_m(l) \end{aligned} \tag{5.13}$$

where $P_m(l)$ is the error term which includes phase noise in the frequency domain.

IQ imbalance. Now we consider the OFDM system with IQ modulation implemented in analog and there exists an IQ imbalance as mentioned in Equation (5.2):

$$x'_m(n) = \frac{1}{N} \sum_{l \in U} \left[X_m(l)(1 + \varepsilon) \cos\left(2\pi \frac{ln}{N}\right) + X_m(l)j \sin\left(2\pi \frac{ln}{N} + \theta\right) \right]$$

$$= \frac{1}{N} \sum_{l \in U} \left\{ X_m(l)e^{j2\pi \frac{ln}{N}} + X_m(l)\frac{\varepsilon}{2}(e^{j2\pi \frac{ln}{N}} + e^{-j2\pi \frac{ln}{N}}) \right.$$

$$\left. + X_m(l) \cdot j \left[\sin\left(2\pi \frac{ln}{N}\right) \cos\theta + \cos\left(2\pi \frac{ln}{N}\right) \sin\theta - \sin\left(2\pi \frac{ln}{N}\right) \right] \right.$$

(assume θ is small)

$$= \frac{1}{N} \sum_{l \in U} \left[X_m(l)e^{j2\pi \frac{ln}{N}} + X_m(l)\left(\frac{\varepsilon + j\theta}{2}\right)(e^{j2\pi \frac{ln}{N}} + e^{-j2\pi \frac{ln}{N}}) \right] \qquad (5.14)$$

Again we assume the channel is ideal without AWGN noise. The received signal in the frequency domain is

$$\hat{X}_m(k) = \sum_{n=0}^{N-1} x'_m(n)e^{-j2\pi \frac{kn}{N}}$$

$$= \frac{1}{N} \sum_{n=0}^{N-1} \sum_{l \in U} \left[X_m(l)e^{j2\pi \frac{ln}{N}} + X_m(l)\left(\frac{\varepsilon + j\theta}{2}\right)(e^{j2\pi \frac{ln}{N}} + e^{-j2\pi \frac{ln}{N}}) \right] e^{j2\pi \frac{kn}{N}}$$

$$= \frac{1}{N} \sum_{n=0}^{N-1} \sum_{l \in U} \left[X_m(l)e^{j2\pi \frac{n(l-k)}{N}} + X_m(l)\left(\frac{\varepsilon + j\theta}{2}\right)(e^{j2\pi \frac{n(l-k)}{N}} + e^{-j2\pi \frac{n(l-k)}{N}}) \right]$$

$$= X_m(k) \cdot \left(1 + \frac{\varepsilon + j\theta}{2}\right) + X_m(N - k) \cdot \frac{\varepsilon + j\theta}{2} \qquad (5.15)$$

In (5.15), the term $1 + (\varepsilon + j\theta)/2$ can be regarded as a common phase error since this term is all the same for all the subcarriers of a symbol. The last term, $X_m(N - k) \cdot (\varepsilon + j\theta)/2$ can be treated as a random noise since $X_m(N - k)$ is the data on the $(N - k)$-th subcarrier.

Oscillator phase noise. The impact of oscillator phase noise on OFDM systems is that the orthogonality among OFDM subcarriers may be destroyed because of the appearance of common phase error (CPE) and inter-carrier interference (ICI) [1]. Considering the multiplicative phase noise caused by the instability of the transceiver oscillator and the additive white noise, the received nth sample of the mth OFDM symbol can be written as

$$r_m(n) = x_m(n) \otimes h_m(n)e^{j(\phi_m(n)+\theta)} + w_m(n) \qquad (5.16)$$

and

$$\phi_m(n) = \phi\left(m(N + N_g) + N_g + n\right) \qquad (5.17)$$

where $h_m(n)$ and $\phi(t)$ represent the channel impulse response and the phase noise respectively, while $w_m(n)$ denotes the AWGN noise and θ is the initial phase of the phase noise process. After removing the cyclic prefix and performing the FFT, the frequency domain symbol can be expressed by

$$R_m(l) = \Phi_m(0)H_m(l)X_m(l) + \underbrace{\sum_{\substack{q \in U \\ q \neq l}} \Phi_m(l-q)H_m(q)X_m(q)}_{I_m(l)} + W_m(l) \tag{5.18}$$

where $H_m(l)$ is the channel frequency response and $W_m(l)$ denotes the frequency domain expression of $w_m(n)$. $\Phi_m(i)$ is the discrete Fourier transform of the phase noise process given by

$$\Phi_m(i) = \frac{1}{N} \sum_{n=0}^{N-1} e^{j(\phi_m(n)+\theta)} e^{-j2\pi \frac{ln}{N}} \tag{5.19}$$

and it can be viewed as a weighting function on the transmitted frequency domain symbols. In particular, when $i = 0$, $\Phi_m(0)$ is actually the time average of the phase noise process within one OFDM symbol duration. This term is usually known as the common phase error (CPE) which causes the same phase rotation and amplitude distortion to each transmitted frequency domain symbol. On the other hand, when $i \neq 0$, the second term in (5.18) is the inter-carrier interference (ICI) resulting from contributions of other subcarriers by the weighting of $\Phi_m(i)$ due to the loss of orthogonality. We call $\Phi_m(i)$ the ICI weighting function.

Based on Equation (5.18), the received frequency domain vector can be given by

$$r_m = \Phi_m(0)H_m x_m + i_m + z_m$$
$$= \Phi_m(0)H_m x_m + e_m \tag{5.20}$$

where

$$H_m \triangleq diag(H_m(0), H_m(1), \ldots, H_m(N-1))$$
$$x_m \triangleq [X_m(0)X_m(1)\ldots X_m(N-1)]^T$$
$$l_m \triangleq [I_m(0)I_m(1)\ldots I_m(N-1)]^T$$
$$\zeta_m \triangleq [\zeta_m(0)\zeta_m(1)\ldots \zeta_m(N-1)]^T \tag{5.21}$$

and $diag(\cdot)$ is a diagonal matrix. From now on, we shall add a second subscript to one of the vector or matrix variables defined in Equations (5.20) and (5.21) to indicate its sub-vector or sub-matrix which is taken according to one of the subcarrier index sets in Equations (5.9) and (5.10). The second subscript may be chosen from $\{p, d, v, u\}$ which relates to $\{P, D, V, U\}$ respectively. For example,

$$r_{m,p} = [R_m(p_1) \ R_m(p_2) \ \cdots \ R_m(p_{N_p})]^T \tag{5.22}$$

stands for the received pilot vector.

Conventionally, $r_{m,p}$ is utilized to obtain the channel response and common phase error to carry out equalization on $r_{m,p}$ and finally send the equalized results to the detection block to get the decisions. Since accurate channel estimation in OFDM systems can be obtained by either preambles or pilot symbols [10], we assume that the channel frequency response is acquired by the receiver in subsequent sections.

Most OFDM systems employ pilots to facilitate receiver synchronization since data-aided estimation gives better and steadier estimate. However, since pilots cost system utilization, the number of pilots should be kept as small as possible which confines the performance of pilot-aided CPE correction algorithms. Comparatively, decision-directed approaches enjoy a larger observation space. However, to ensure the acceptable correctness of the decision, they can only be operated when phase noise is small. Therefore, use pilots to acquire the CPE and provide an initial compensation, then enlarge the observation space by including the tentative decisions as the sufficient statistics to perform the final estimate can benefit from the advantages of both approaches. This method is called pilot-aided decision-directed (PADD) approach [5].

Considering the pilot-aided decision-directed approach, the sufficient statistics is $r_{m,u}$ and can be modeled as a complex Gaussian random vector with mean vector $\Phi_m(0)H_m x_m$ and covariance matrix $C_{r_{m,u}}$. Assume that we have acquired the covariance matrix, the log-likelihood function of $\Phi_m(0)$ can be given by

$$
\Lambda(\Phi_m(0)) = 2\Re\left\{x_{m,u}^H H_{m,u}^H C_{r_{m,u}}^{-1} r_{m,u} \Phi_m^*(0)\right\}
$$
$$
-x_{m,u}^H H_{m,u}^H C_{r_{m,u}}^{-1} H_{m,u} x_{m,u} |\Phi_m(0)|^2 \tag{5.23}
$$

Differentiate (23) with respect to $\Phi_m(0)$, the ML estimation of the common phase error

$$
\hat{\Phi}_m(0) = \frac{x_{m,u}^H H_{m,u}^H C_{r_{m,u}}^{-1} r_{m,u}}{x_{m,u}^H H_{m,u}^H C_{r_{m,u}}^{-1} H_{m,u} x_{m,u}} \tag{5.24}
$$

By assuming the inter-carrier interference on each subcarrier to be independently identically distributed, the ML estimator can be further degenerated to

$$
\hat{\Phi}_m(0) = \frac{\sum_{k\in P} R_m(k)X_m^*(k)H_m^*(k)}{\sum_{k\in P} |X_m(k)H_m(k)|^2} \tag{5.25}
$$

which is the least-square (LS) estimator proposed in [4].

Now we need to derive the covariance matrix $C_{r_{m,u}}$. It can be expressed as

$$
C_{r_{m,u}} = C_{\iota_{m,u}} + \sigma_z^2 I \tag{5.26}
$$

Since σ_z^2 can be obtained by preamble signals as proposed in the literature [15], we can assume that it is known by the receiver henceforth.

For Wiener phase noise, ICI can be modeled as a random variable which is independent of $W_m(l)$, since $X_m(l)$ and $H_m(l)$ are assumed to be known in PADD circumstance. The mean of

ICI depends on the mean of the ICI weighting function $\Phi_m(i)$ which can be expressed as

$$E\left[\Phi_m(i)\right] = \frac{1}{N} \sum_{n=0}^{N-1} E\left[e^{j(\phi_m(n)+\theta)}\right] e^{-j2\pi \frac{in}{N}}$$

$$= \frac{1}{N} \sum_{n=0}^{N-1} E\left[e^{j\phi_m(n)}\right] E\left[e^{j\theta}\right] e^{-j2\pi \frac{in}{N}} \qquad (5.27)$$

The second equality is because since θ is independent with $\phi_m(n)$, the expectation in Equation (5.27) can be decomposed as the product of $E[e^{j\phi_m(n)}]$ and $E[e^{j\theta}]$, and by the characteristic function of uniformly distributed random variable $\Psi_\theta(\omega)$, we have

$$E[e^{j\theta}] = \Psi_\theta(\omega)|_{\omega=1}$$

$$= \frac{2}{2\pi\omega} \sin(\frac{2\pi\omega}{2}) e^{j\pi\omega}|_{\omega=1}$$

$$= 0 \qquad (5.28)$$

Therefore, the mean of ICI becomes zero.

On the other hand, for stationary phase noise, again since $X_m(l)$ and $H_m(l)$ are assumed to be known in PADD circumstance, ICI can be modeled as a random variable which is independent of $W_m(l)$. The mean of ICI depends on the mean of the ICI weighting function $\Phi_m(i)$ which can be expressed as

$$E\left[\Phi_m(i)\right] = \frac{1}{N} \sum_{n=0}^{N-1} E\left[e^{j\phi_m(n)}\right] e^{-j2\pi \frac{in}{N}} \qquad (5.29)$$

Since $\Phi_m(i)$ is stationary, $E[\Phi_m(i)]$ will be a constant and can be taken out of the summation. And since

$$\sum_{n=0}^{N-1} e^{-j2\pi \frac{in}{N}} = 0, \forall i \neq 0 \qquad (5.30)$$

therefore, the mean of ICI becomes zero, too.

Let the element of $C_{r_{m,u}}$ be denoted by $\sigma_i(k1, k2)$, it can be expressed as

$$\sigma_i(k1, k2) = \sum_{\substack{l_1 \in U \\ l_1 \neq u_{k1}}} \sum_{\substack{l_2 \in U \\ l_2 \neq u_{k2}}} H(l_1) H^*(l_2) \tilde{X}(l_1) \tilde{X}^*(l_2) R_\Phi(u_{k1} - l_1, u_{k2} - l_2) \qquad (5.31)$$

The autocorrelation function of the ICI weighting function $\Phi_m(i)$ for Wiener phase noise is

$$R_\Phi(p, q) = E\left[\Phi_m^{(k)}(p)\Phi_m^{(k)*}(q)\right]$$

$$= \frac{\delta_{(p-q)N}}{N^2}\left[N + \frac{1 - N + Ne^{z_1} - e^{Nu}}{2 - e^{z_1} - e^{-z_1}} + \frac{1 - N + Ne^{z_2} - e^{Nu}}{2 - e^{z_2} - e^{-z_2}}\right]$$

$$+ \frac{(1 - \delta_{(p-q)N})}{N^2} \frac{1 - e^{Nu}}{1 - e^{(z_1-z_2)}}\left(\frac{1}{1 - e^{z_1}} - \frac{1}{1 - e^{\bar{z}_1}} + \frac{1}{1 - e^{z_2}} - \frac{1}{1 - e^{\bar{z}_2}}\right) \qquad (5.32)$$

where

$$z_1 \equiv u + jv_1 \equiv -\frac{2\pi}{N}[\beta T + jp]$$

$$z_2 \equiv u + jv_2 \equiv -\frac{2\pi}{N}[\beta T + jq] \tag{5.33}$$

and $(\cdot)_N$ denotes the modulo by N, $\bar{z}$ represents the complex conjugate of z. The proof is shown in [5].

The autocorrelation function of the ICI weighting function $\Phi_m(i)$ for stationary phase noise is

$$R_\Phi(p, q) = \sum_{n_1=0}^{N-1} \sum_{n_2=0}^{N-1} \frac{E[e^{j(\phi_m(n_1) - \phi_m(n_2))}]}{N^2} e^{-j2\pi \frac{n_1 p - n_2 q}{N}}$$

$$= \sum_{n_1=0}^{N-1} \sum_{n_2=0}^{N-1} \frac{e^{R_\phi(n_1 - n_2) - R_\phi(0)}}{N^2} e^{-j2\pi \frac{n_1 p - n_2 q}{N}} \tag{5.34}$$

Because $R_\phi(0)$ is already available from Equations (5.6) and (5.8), the ML CPE estimator (5.24) can be performed [14].

5.4 Phase Noise in OFDMA

OFDM(A) systems are sensitive to phase noise/jitter. This challenge becomes more critical when the users have distinct phase noise spectrum due to their own distinct oscillators and the power levels of all subcarriers received at the base station are different due to the near-far effect and frequency selective channels, which is common in OFDMA uplink communications. The subcarriers with higher power interfere severely on others with lower power. Moreover, the subcarriers with poor phase noise spectrum (usually due to unsatisfactory oscillators) may ruin the overall system performance [19]. We therefore should take more care when compensating for phase noise in OFDMA uplink communications.

In this section, we consider the uplink of an OFDMA system consisting of N subcarriers and K users. The frequency domain subcarrier index set is composed of three mutually exclusive subsets: D, P and V, which correspond to data, pilot and virtual subcarriers with size Nd, Np and Nv respectively. In addition, U, which is the union of D and P, represents the index set of useful subcarriers with size U. Among the N subcarriers, the kth user is assigned to a subset of U_k subcarriers with the index set

$$U_k = \left\{ u_1^{(k)}, u_2^{(k)}, \cdots, u_k^{(k)} \right\}, \tag{5.35}$$

where the superscript $(\cdot)^{(k)}$ denotes the kth user.

We denote $x_m^{(k)}$ the mth block frequency domain symbols sent by the kth user. The lth entry of $x_m^{(k)}$, say $X_{m,l}^{(k)}$ is nonzero if and only if $l \in U_k$. Consequently, the corresponding discrete-time baseband signal can be represented as

$$s_m^{(k)}(n) = \frac{1}{N} \sum_{l \in U_k} X_{m,l}^{(k)} e^{j2\pi \frac{ln}{N}}, \quad 0 \le n \le N - 1. \tag{5.36}$$

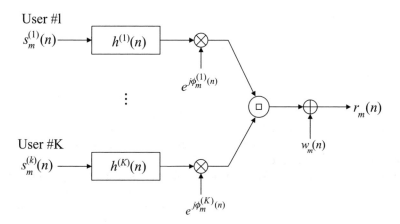

Figure 5.7   Simplified uplink OFDMA signal model with phase noise. Source: [16]

Prior to transmission over the channel, this signal is preceded by a cyclic prefix (CP) of length N_g longer than the channel impulse response to eliminate interblock interference (IBI).

Let the discrete-time composite channel impulse response with order L_k between the kth user and the uplink receiver be denoted by $h^{(k)}(n)$ and the channel frequency response on the lth subcarrier of kth user's channel be denoted by $H_l^{(k)}$, we have

$$H_l^{(k)} = \sum_{n=0}^{L_k-1} h^{(k)}(n)e^{-j2\pi \frac{nl}{N}}. \tag{5.37}$$

Denoting the discrete-time phase noise process and AWGN impairing the kth user by $\phi_m^{(k)}(n)$ and $w_m(n)$ respectively, after the removal of CP, as depicted in Figure 5.7, we have

$$r_m(n) = \sum_{k=1}^{K} \left[s_m^{(k)}(n) \otimes h^{(k)}(n) \right] e^{j\phi_m^{(k)}(n)} + w_m(n) \tag{5.38}$$

where $\otimes$ is the circular convolution and $n \in \{0, 1, \cdots, N-1\}$. Note that here we assume the phase noise only exists in mobile stations, that is, transmitter side for uplink OFDMA communications. This is because we can usually use high cost oscillators with sufficiently low phase noise at base stations.

After taking the FFT, the frequency domain received symbol on the lth subcarrier of the mth block is

$$R_{m,l} = \Phi_m^{(k)}(0)H_l^{(k)}X_{m,l}^{(k)} + \underbrace{\sum_{i=1}^{K}\sum_{q\in U_i} \Phi_m^{(i)}(l-q)H_q^{(i)}X_{m,q}^{(i)}}_{I_{m,l}} + W_{m,l}, \; l \in U_k \tag{5.39}$$

where $W_{m,l}$ is the AWGN noise in frequency domain and

$$\Phi_m^{(k)}(q) = \frac{1}{N}\sum_{n=0}^{N-1} e^{j\phi_m^{(k)}(n)} e^{-j\frac{2\pi nq}{N}}.$$ (5.40)

From Equation (5.39), we find the effect of phase noise in multiuser OFDM to be different from that of single user OFDM. First of all, the CPE term $\Phi_m^{(k)}(0)$ varies according to the index k and means that each user suffers from different CPE and they need to be acquired separately for each user. Secondly, the ICI from a certain user not only affects himself, but all the other users due to the simultaneous access.

To characterize any subcarrier assignment scheme and facilitate further signal processing, it is essential to express the frequency domain received symbols in matrix form. This unified frequency domain signal model [16] can be obtained by including the frequency domain dummy symbols transmitted by each active user in Equation (5.39) as follows

$$R_{m,l} = \sum_{k=1}^{K}\sum_{q=0}^{N-1} \Phi_m^{(k)}(l-q) H_q^{(k)} X_{m,q}^{(k)} + W_{m,l}.$$ (5.41)

To simplify the illustration, we can define $Y_{m,l}^{(k)} \triangleq H_l^{(k)} X_{m,l}^k$. Then the mth block frequency domain received vector which comprises the frequency domain received symbols from the N subcarriers can be expressed as

$$r_m = \sum_{k=1}^{K} \mathbf{y}_m^{(k)} \Phi_m^{(k)}(0) + \sum_{k=1}^{K} \mathbf{Y}_m^{(k)} \varphi_m^{(k)}(0) + \zeta_m$$ (5.42)

where

$$\mathbf{y}_m^{(k)} = [\, Y_{m,0}^{(k)} \; Y_{m,1}^{(k)} \; \cdots \; Y_{m,N-1}^{(k)} \,]^T$$ (5.43)

$$\mathbf{Y}_m^{(k)} = \begin{bmatrix} Y_{m,N-1}^{(k)} & Y_{m,N-2}^{(k)} & \cdots & Y_{m,1}^{(k)} \\ Y_{m,0}^{(k)} & Y_{m,N-1}^{(k)} & \cdots & Y_{m,2}^{(k)} \\ \vdots & \vdots & \ddots & \vdots \\ Y_{m,N-2}^{(k)} & Y_{m,N-3}^{(k)} & \cdots & Y_{m,0}^{(k)} \end{bmatrix}_{N\times(N-1)}$$ (5.44)

and

$$\varphi_m^{(k)} = [\, \Phi_m^{(k)}(1) \quad \Phi_m^{(k)}(2) \quad \cdots \quad \Phi_m^{(k)}(N-1) \,]^T$$ (5.45)

$$\zeta_m = [\, W_{m,0} \quad W_{m,1} \quad \cdots \quad W_{m,N-1} \,]^T.$$ (5.46)

We can define $l_m^{(k)} = \mathbf{Y}_m^{(k)} \varphi_m^k$ which represents the ICI vector contributed from the kth user at the mth block. In addition, the sum of the ICI vectors from each active user is defined as

$$\xi_m = \sum_{k=1}^{K} l_m^{(k)}.$$ (5.47)

The overall interference plus noise vector can be defined by $\varepsilon_m \overset{\Delta}{=} \zeta_m + \zeta_m$ and the overall frequency domain received vector can be rearranged as

$$r_m = \sum_{k=1}^{K} \mathbf{y}_m^{(k)} \Phi_m^{(k)}(0) + \varepsilon_m. \tag{5.48}$$

We further define $\tilde{\varphi}_m^{(k)} \overset{\Delta}{=} [\Phi \ \tilde{\varphi}_m^{(k)}(0) \quad \varphi_m^{(k)}]^T$ and $\tilde{\mathbf{Y}}_m^k \overset{\Delta}{=} [\mathbf{y}_m^{(k)} \quad \mathbf{Y}_m^{(k)}]_{N \times N}$, then, the overall frequency domain received vector can be arranged as

$$r_m = \sum_{k=1}^{K} \tilde{\mathbf{Y}}_m^{(k)} \tilde{\varphi}_m^{(k)}(0) + \zeta_m. \tag{5.49}$$

To obtain a more concise expression and facilitate the parameter estimation, we proceed to define

$$\boldsymbol{\theta}_m = [\ \Phi_m^{(1)}(0) \ \Phi_m^{(2)}(0) \cdots \Phi_m^{(K)}(0)\ ]^T \tag{5.50}$$

and

$$\mathbf{B}_m = [\ \mathbf{y}_m^{(1)} \ \mathbf{y}_m^{(2)} \cdots \mathbf{y}_m^{(K)}\ ]_{N \times K} \tag{5.51}$$

The matrix representation of the frequency domain received symbols can be summarized as

$$r_m = \mathbf{B}_m \boldsymbol{\theta}_m + \varepsilon_m. \tag{5.52}$$

To represent the frequency domain received vector according to the subcarrier index sets defined previously, we add a second subscript to one of the vector or matrix variables defined in Equations (5.42)–(5.49) and (5.51)–(5.52) to indicate its sub-vector or row-wise sub-matrix which is taken according to P, D, U and one of the subcarrier index sets defined in Equation (5.35). The second subscript may be chosen from $\{p, d, v, u_k\}$ which relate to $\{P, D, V, U_k\}$ respectively. In particular, the row-wise submatrix is taken from the rows corresponding to the subcarriers indicated by the second subscript.

Since preamble is used to assist channel estimation [17] and accurate channel estimation algorithms are available [10], in the following sections we assume that the channel frequency response is acquired when doing a CPE estimation. Please note that this channel estimation includes the CPE observed on the preamble because they are not distinguishable from each other. Although the CPE estimation generated from this channel estimation will deviate from the exact CPE, the overall combination of CPE and channel estimation shall cancel out the deviation and the frequency domain equalization can still be performed correctly.

Pilot-Aided CPE Estimation: Least-Square Approach [16]. In the pilot-aided scenario, the sufficient statistics are the P received pilot symbols collected from each active user

$$r_{m,p} = \mathbf{B}_{m,p} \boldsymbol{\theta}_m + \varepsilon_{m,p}. \tag{5.53}$$

There exists a linear model between the CPE vector and the sufficient statistics. Since the precise statistical characteristics of ε_m is not available, the least-square approach is favorable since it only requires the linear signal model which is valid in our case. Therefore, the least-square criterion [4] can be applied to yield

$$\tilde{\theta}_{m,LS} = (\mathbf{B}_{m,p}^T \mathbf{B}_{m,p})^{-1} \mathbf{B}_{m,p}^T \mathbf{r}_{m,p} \tag{5.54}$$

Note that the choice of subcarrier assignment scheme does not alter the form of the proposed estimator, thus making the proposed algorithm applicable to any subcarrier assignment schemes.

Pilot-Aided Decision-Directed CPE Estimation: Maximum Likelihood Approach [16]. Pilot-aided decision-directed approach implies using pilots to acquire the CPE to provide an initial compensation, then enlarging the observation space by including the tentative decisions in the sufficient statistics to perform the final estimate. In this case, the first two order statistics of the ICI can be used to enhance the CPE estimation.

(1)*Maximum Likelihood CPE Estimator:* When pilot symbols and tentative-decisions from each active user are available, the sufficient statistics can be given by

$$\mathbf{r}_{m,u} = \mathbf{B}_{m,u}\boldsymbol{\theta}_m + \boldsymbol{\varepsilon}_{m,u}. \tag{5.55}$$

In [5], it is shown that ξ_m can be approximated as a complex Gaussian vector which makes $\varepsilon_{m,u}$ complex Gaussian distributed too. Then, the maximum likelihood estimator [5] of θ can be given by

$$\tilde{\theta}_{m,ML} = (\mathbf{B}_{m,u}^T \mathbf{C}_{\varepsilon_{m,u}}^{-1} \mathbf{B}_{m,u})^{-1} \mathbf{B}_{m,u}^T \mathbf{r}_{m,u} \tag{5.56}$$

(2)*Statistical Characteristics of the ICI:* Since the ξ_m and ζ_m are independent, we have

$$\mathbf{C}_{\varepsilon_{m,u}} = \mathbf{C}_{\xi_{m,u}} + \mathbf{C}_{\zeta_{m,u}} \tag{5.57}$$

where $\mathbf{C}_{\zeta_{m,u}}$ is a diagonal matrix with each diagonal element equals to σ_w^2. Since σ_w^2 can be obtained by preamble signal and has been proposed in the literature [15], we can assume that σ_w^2 is known by the receiver. To acquire the covariance matrix of $\xi_{m,u}$, we need to check its mean first, and the mean of $\xi_{m,u}$ can be expressed as

$$E[\xi_m] = \sum_{k=1}^{K} \mathbf{Y}_m^{(k)} E[\varphi_m^{(k)}]. \tag{5.58}$$

In [5], it is shown that $E[\varphi_m^{(k)}]$ is two orders of magnitude smaller than both r_m and $\varphi_m^{(k)}$, therefore, we can ignore $E[\varphi_m^{(k)}]$ and treat $\varphi_m^{(k)}$ as a zero-mean vector when finding the covariance matrix of $\varphi_m^{(k)}$.

The covariance matrix of $\xi_{m,u}$ can be given by

$$\begin{aligned}
\mathbf{C}_{\varepsilon_{m,u}} &= E[\xi_m \xi_m^H] \\
&= E\left[\sum_{k1=1}^{K} \boldsymbol{\iota}_m^{(k1)} \sum_{k2=1}^{K} \boldsymbol{\iota}_m^{(k2)H} \right] \\
&= \sum_{k1=1}^{K} \sum_{k2=1}^{K} \mathbf{Y}_m^{(k1)} E[\varphi_m^{(k1)} \varphi_m^{(k2)}] \mathbf{Y}_m^{(k2)H}.
\end{aligned} \tag{5.59}$$

Since the phase noise processes corrupting different users are independent [2], the covariance matrix of $\varphi_m^{(k)}$ can be arranged as

$$C_{\varepsilon_{m,u}} = \sum_{k=1}^{K} \mathbf{Y}_m^{(k)} \mathbf{R}_{\varphi_m^{(k)}} \mathbf{Y}_m^{(k)H} \qquad (5.60)$$

where $\mathbf{R}_{\varphi_m^{(k)}}$ is the auto-correlation matrix of $\varphi_m^{(k)}$

$$\mathbf{R}_{\varphi_m^{(k)}} = \begin{bmatrix} R_\Phi(1,1) & R_\Phi(1,2) & \cdots & R_\Phi(1, N-1) \\ R_\Phi(2,1) & R_\Phi(2,2) & \cdots & R_\Phi(2, N-1) \\ \vdots & \vdots & \ddots & \vdots \\ R_\Phi(N-1, 1) & R_\Phi(N-1, 2) & \cdots & R_\Phi(N-1, N-1) \end{bmatrix}_{(N-1) \times (N-1)} \qquad (5.61)$$

and $R_\Phi(p, q)$ is shown in Equation (5.32). Therefore, by Equation (5.60), $C_{\varepsilon_{m,u}}$ can be obtained.

We can find the performance of these two CPE estimations by simulation [16]. Consider an OFDMA system with 64 subcarriers in the 5 GHz frequency band. The signal bandwidth is 20 MHz. There are four sub-channels in the system, each contains 13 subcarriers. Each active user uses one sub-channel and the configuration and frequency domain structure of each subchannel are identical. We denote N_p as the number of pilot subcarriers in a sub-channel and it varies from 1 to 4 in our experiments.

The channel response of each user is generated according the IEEE 802.11a channel model with root-mean-square delay spread equals to 50 ns. The channel coefficients are modeled as independent and complex-valued Gaussian random variables with zero-mean and an exponential power delay profile

$$E\left\{|h_k(l)^2|\right\} = \lambda \cdot e^{-l}, \quad l = 0, 1, \cdots, 10. \qquad (5.62)$$

The constant λ is chosen such that the signal power of each user is normalized to unity. The phase noise is generated by the Lorentzian model with β equals to 1 kHz. Two typical subcarrier assignment schemes: sub-band-based subcarrier assignment and interleaved subcarrier assignment as illustrated in Figure 5.1 are used. Each simulation point is conducted using $3 \cdot 10^5$ frames, each frame consists of 16 OFDM symbols.

Figure 5.7 shows the symbol error rate (SER) performance of the two proposed CPE estimators in comparison with both no-phase-noise and no-phase-noise-correction cases with QPSK. Since the number of pilot subcarriers affects the spectrum efficiency and the capacity of an OFDMA system, N_p is set to 1 in the simulation generating these two figures. Figure 5.8(a) refers to sub-band-based subcarrier assignment while Figure 5.8(b) corresponds to interleaved subcarrier assignment.

First of all, the maximum likelihood (ML) approaches always have more improvement than least square (LS) ones, which is not surprising because the statistics of ICI term are taken into consideration. We can observe that when the number of active users increases, interleaved subcarrier assignment suffers more from the multiple-access interference because the other active users' signal is at nearer subcarriers.

Figure 5.8 illustrates how the performance of the proposed schemes changes with phase noise levels. The number of pilot symbols N_p is set to 1. The aim of the proposed schemes is to correct medium to small phase noise, i.e., for phase noise variance βT_s less than 10^{-4}. Figure 5.9 shows that when phase noise variance is greater than 10^{-5}, the OFDMA system

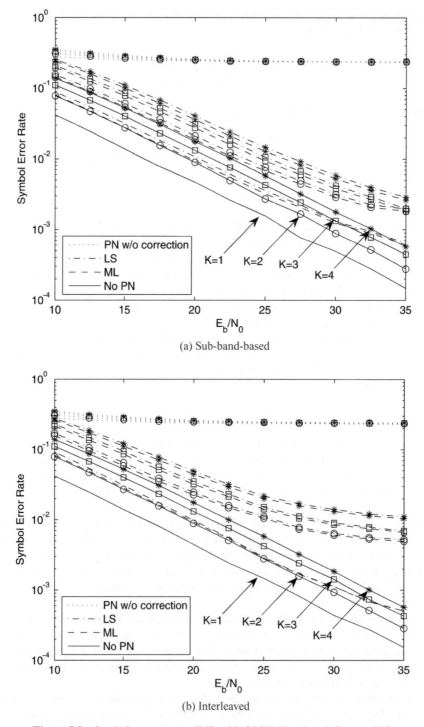

(a) Sub-band-based

(b) Interleaved

Figure 5.8   Symbol error rate vs. SNR with QPSK, K = 1 to 4. Source: [16]

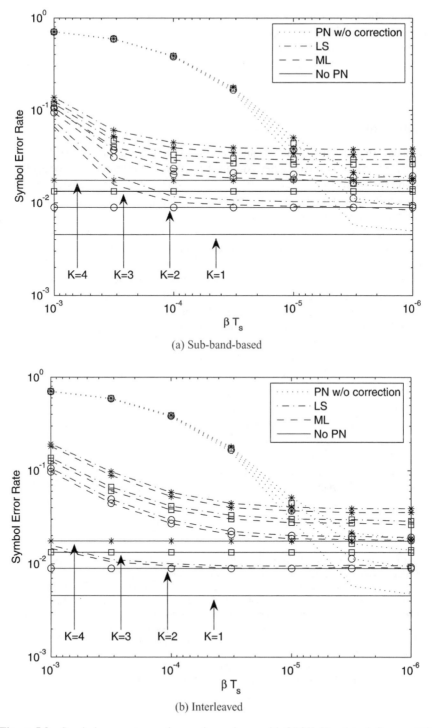

(a) Sub-band-based

(b) Interleaved

Figure 5.9   Symbol error rate vs. phase noise variance with QPSK, K = 1 to 4. Source: [16]

suffers remarkable performance degradation. However, the proposed CPE estimation schemes provide significant performance improvement over no-phase-noise-correction case.

When phase noise variance is less than 10^{-5} for a OFDMA system employing QPSK, we can see the error floor of the proposed schemes. For this phase noise variance range, it is not necessary to make the CPE correction to correct multiple phase noise.

5.5 Conclusion

Several models of phase noise sources and the corresponding effects in OFDM and OFDMA systems were introduced in this chapter. Among them, usually the primary source of phase noise is the instability of oscillators. Regarding the oscillator phase noise, we discussed the estimation of Wiener phase noise with the system which is only frequency-locked, and stationary phase noise with the system which is further phase-locked.

Pilots are usually employed in OFDM systems to facilitate receiver synchronization, while pilots also cost system utilization, which means that the number of pilots should be kept as small as possible. The result is the confinement of pilot-aided CPE correction algorithms performance. On the other hand, decision-directed approaches enjoy a larger observation space, while they can only be operated when phase noise is small to ensure the acceptable correctness of the decision. The method called pilot-aided decision-directed (PADD) approach which combines pilots and decision feedback as the sufficient statistics is discussed in this chapter and shown to benefit from the advantages of both.

Then, two algorithms, LS approach and ML approach, were proposed to mitigate the effects of multiple phase noises in uplink OFDMA systems. The LS approach which only uses the pilots to estimate phase noise provides acceptable performance with low complexity while the ML approach which combines pilot-aided and decision-directed approaches considers the second-order statistics of the ICI to enhance the performance. The proposed schemes aim to compensate for CPE, the major effect of phase noise for medium to low phase noise levels where phase noise correction is applicable. Moreover, it is shown that these two multiuser phase noise correction schemes are stable within a wide range of phase noise levels and applicable to any subcarrier assignment scheme, which exhibits the potential in practical applications.

References

[1] T. Pollet, M. Van Bladel, and M. Moeneclaey, 'BER Sensitivity of Ofdm Systems to Carrier Frequency Offset and Wiener Phase Noise', *IEEE Trans. on Comm.*, **43**, Feb./March/April 1995, 191–193.

[2] H. Steendam, M. Moeneclaey, and H. Sari, 'The Effect of Carrier Phase Jitter on the Performance of Orthogonal Frequency-Division Multiple-Access Systems', *IEEE Trans. on Comm.*, **46**(4), April 1998, 456–459.

[3] V. Abhayawardhana and I. Wassell, 'Common Phase Error Correction for OFDM in Wireless Communication', in *Proc. IEEE Global Telecommun. Conf (Globecom'02)*, **1**, Taipei, ROC, Nov. 2002, 17–21.

[4] S. Wu and Y. Bar-Ness, 'A Phase Noise Suppression Algorithm for OFDM-based WLANs', *IEEE Commun. Lett.*, **6**, Dec. 2002, 535–537.

[5] Yi-Ching Liao, and Kwang-Cheng Chen, 'Estimation of Wiener Phase Noise by the Autocorrelation of the ICI Weighting Function in OFDM Systems', in *Proc. Personal, Indoor and Mobile Radio Communications (PIMRC 2005)*, Berlin, Germany, Sept. 2005.

[6] Z. Cao, U. Tureli, and Y.-D. Yao, 'Deterministic Multiuser Carrier Frequency Offset Estimation for Interleaved OFDMA Uplink', *IEEE Trans. on Comm.*, **52**(9), Sept. 2004, 1585–1594.

[7] K. Chang, *RF and Microwave Wireless Systems*, Chichester: John Wiley & Sons, Ltd., 2000.

[8] Elena Costa and Silvano Pupolin, 'M-QAM-OFDM System Performance in the Presence of a Nonlinear Amplifier and Phase Noise', *IEEE Trans. on Comm.*, **50**(3), March 2002, 462–472.

[9] L. Piazzo and P. Mandarini, 'Analysis of Phase Noise Effects in OFDM Modems', *IEEE Trans. on Comm.*, **50**(10), Oct. 2002, 1696–1705.

[10] R. Van Nee and R. Prasad, *OFDM for Multimedia Wireless Communications*. Boston: Artech House, 2000.

[11] P. Robertson and S. Kaiser, 'Analysis of the Effects of Phase Noise in Orthogonal Frequency Division Multiplexing (OFDM) Systems', in *Proc. ICC'95*, 1652–1657.

[12] B. Stantchev and G. Fettweis, 'Time-Variant Distortion in OFDM', *IEEE Commun. Letters*, **4**(9), Sept. 2000, 312–314.

[13] Z. Jianhua, H. Rohling, and Z. Ping, 'Analysis of ICI Cancellation Scheme in OFDM Systems with Phase Noise', *IEEE Trans. on Broadcasting*, **50**(2), June 2004, 97–106.

[14] Yi-Ching Liao, and Kwang-Cheng Chen, 'Estimation of Stationary Phase Noise by the Autocorrelation of the ICI Weighting Function in OFDM Systems', *IEEE Trans. on Wireless Comm.*, **5**(12), Dec. 2006: 3370–3374.

[15] S. He and M. Torkelson, 'Effective SNR Estimation in OFDM System Simulation', in *Proc. IEEE Global Telecommun. Conf. (Globecom'98)*, **2**, Nov. 1998, 945–950.

[16] Yi-Ching Liao, and Kwang-Cheng Chen, 'Multiuser Common Phase Error Estimation for Uplink OFDMA Communications', in *IEEE Wireless Communications and Networking Conference (WCNC 2007)*, Mar. 2007.

[17] IEEE Std. 802.16a, Air Interface for Fixed Broadband Wireless Access Systems-Amendment 2: Medium Access Control Modifications and Additional Physical Layer Specifications for 2–11 GHz, 2003.

[18] P.H. Moose, 'A Technique for Orthogonal Frequency Division Multiplexing Frequency Offset Correction', *IEEE Trans. on Comm.*, **42**, Oct. 1994, 2908–2914.

[19] A.G. Armada, V. P.G. Jimenez, and J. Darriba, 'Analysis of Phase Noise Effects in Multi-User OFDM', in *IEEE Intl. Symposium on Control, Communications and Signal Processing (ISCCSP'04)*, Hammamet, Tunis, Mar. 2004, 63–66.

Part Two

Medium Access Control and Network Architecture

6

Optimizing WiMAX MAC Layer Operations to Enhance Application End-to-End Performance

Xiangying Yang, Muthaiah Venkatachalam, and Mohanty Shantidev

6.1 Introduction

Application end-to-end performance in wireless networks is critical to user experience. WiMAX is designed to support a large set of existing and emerging applications, such as Voice-over-IP (VoIP), media streaming, multi-cast broadcast services, online gaming, interactive conferencing, web browsing and instant text messaging. All these applications have different Quality of Service (QoS) requirements. Different techniques at the Media Access Control (MAC) layer are used to handle different service classes. We here take one example to understand how to optimize MAC layer design to enhance the performance of TCP-based applications.

TCP-based applications such as web browsing, email, and FTP are among the most popular Internet applications and should be supported by WiMAX with good performance. Although such applications are typically classified as 'best effort', TCP's performance is indeed very sensitive to delay, jitter and packet loss. Therefore, in a WiMAX network, where heterogeneous services and applications co-exist, resource allocation for 'best-effort' applications may be limited and TCP performance optimization is particularly interesting. There has been a myriad of research work on optimizing TCP performance over wireless networks. To name only a few relevant to the scope of our discussions, [2], [3], and [4] studied the stochastic modeling of TCP performance over lossy channel. TCP performance over cellular network and its interaction with radio link protocol such as ARQ and MAC layer are investigated in [5], [6], [7] and [8]. In a

Mobile WiMAX   Edited by Kwang-Cheng Chen and J. Roberto B. de Marca.
© 2008 John Wiley & Sons, Ltd

WiMAX network with centralized scheduling, MAC layer scheduling is critical for application performance. Proportional Fair (PF) scheduling is geared at exploiting multiuser diversity to enhance system throughput while maintaining fairness. 'Throughput optimal' scheduling policies such as Modified Largest Weighted Delay First (MLWDF) and Exponential Rule (ER) add delay or queue length as weight in metric calculation to optimally ensure queue stability. In addition, a proportional fair scheduling variant designed for TCP is studied in.

The main focus of this chapter is to show that the flexible MAC framework of WiMAX is the key to optimizing system-level application performance. Limited work has been done on utilizing WiMAX MAC features for TCP performance enhancement. In addition, most existing work focuses on the performance of individual TCP connection, instead of the aggregate network performance, which is the key interest for network operators. We present the benefits of asymmetric link adaptation for the TCP data channel and the ACK channel, which is enabled by the connection-based WiMAX MAC layer and achieves a good tradeoff between overall system capacity and individual TCP performance. In addition, we show that different schedulers have subtle impacts on TCP performance in terms of throughput and fairness, a fact that has not been fully appreciated in the literature. Based on WiMAX's MAC QoS framework, we propose that schedulers specifically designed for service classes are indeed desirable.

This chapter is organized as follows. In Section 6.2, we briefly review the relevant features of the WiMAX MAC layer. In Section 6.3, we propose enhancing TCP performance by coupling asymmetric link adaptation with ARQ. In Section 6.4, we investigate the benefit of service-class specific scheduling. We present simulation results in Section 6.5 and briefly discuss other MAC optimization techniques in Section 6.6.

6.2 Overview of WiMAX MAC features

6.2.1 Connection-Based Service Differentiation

In WiMAX, QoS is maintained at the granularity of 'connection', which is identified by a Connection ID (CID). The flexibility of WiMAX MAC framework does not specify how CIDs should be associated with mobile subscriber station and its applications. In a typical implementation, each CID is characterized by a Mobile Subscriber Station (MSS) ID, link direction and QoS service class. For example, as shown in Figure 6.1, each voice call can establish one CID in one direction for its stringent QoS requirement on delay. Multiple TCP flows from the same MSS may choose to share only one CID to reduce overheads, given their best-effort QoS requirement.

6.2.2 Scheduling Types and Opportunistic Scheduler

Mobile WiMAX standard [13] specifies the following scheduling types: Unsolicited Grant Service (UGS), real-time Polling Service (rtPS), non-real-time Polling Service (nrtPS) and Best Effort (BE). As their names suggest, these scheduling types are designed for CBR real-time applications, VBR real-time applications, VBR non-real-time applications and best-effort applications, respectively. In this chapter, we focus on BE scheduling service for TCP-based best-effort applications such as FTP or HTTP. If not specified, we assume a download scenario where a MSS client downloads data from an Internet-attached server. However, our main results are not limited to download scenarios.

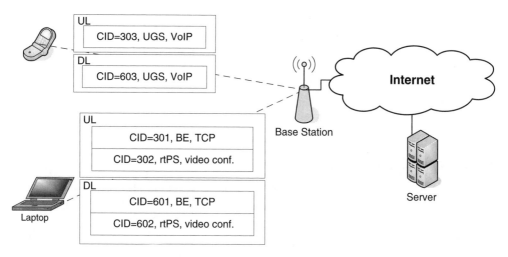

Figure 6.1   Connection (CID)-based WiMAX MAC layer [19]

A connection-based QoS framework enables the scheduler to perform resource allocation at a fine granularity. In the following, we show that it is desirable to have scheduling algorithms specific to QoS classes to achieve good performance. A 'TCP-friendly' scheduler can naturally fit in the WiMAX MAC framework for TCP-based best-effort applications.

6.2.3 Best-Effort Service Class in WiMAX

BE service provides neither minimal rate nor minimal delay guarantee for flows. This means BE-class connections usually have the lowest scheduling priority at the BS. In addition, BE service is designed to consume minimal resources for uplink grant request. BE service allows MSS to use contention-based bandwidth request procedure as shown in Figure 6.2 in the

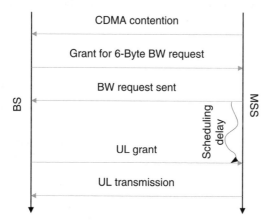

Figure 6.2   BE service class uplink contention-based bandwidth request

uplink. Contention-based bandwidth request is slow but incurs less cost compared to polling services. A MSS with data to transmit contends with a randomly selected CDMA code in the UL contention channel. Upon successful contention, i.e., a successful code detection by the base station (BS), the BS allocates a 6-bytes grant for this MSS to transmit a BW_REQ message containing the amount of bandwidth requested. Then the BS performs uplink scheduling and MSS transmits its data with the allocated uplink grants. The current BW_REQ in WiMAX is sufficient for most best-effort applications that are delay-tolerant.

6.2.4 Link Adaptation and ARQ

Link adaptation and Automatic-Repeat-reQuest (ARQ) are all closely connected with QoS guarantee and application performance. They are all managed at each CID individually.

Link adaptation dictates the Modulation Coding Schemes (MCS) that should be used given the current channel condition. With adaptive MCS, spectral efficiency can be significantly improved. Link adaptation is associated with some target Packet Error Rate (PER) or block error rate (BER), i.e., by properly defining thresholds for MCS switch, the system can manage the average PER in channel fluctuation. As shown in Figure 6.3, given the average channel condition, the spectral efficiency of the system measured at physical layer (PHY) layer increases in PER initially but drops quickly as PER approaches 1. Therefore, given the WiMAX frame structure, the PHY capacity also will first increase and then decrease in the target PER. Depending on the higher layer protocol design, application performance may not align with PHY capacity, e.g., TCP performance typically drops more quickly in PER but the performance drop may be mitigated by error-correction mechanisms. WiMAX MAC framework allows per CID link adaptation, which helps tweak link adaptation strategies for individual QoS service class or even individual application needs. This is one of the main ideas we want to highlight.

ARQ is a retransmission mechanism at the upper MAC layer, sometimes also referred to as the Radio Link Control (RLC) layer. An example of ARQ operation is shown in Figure 6.4, in which each packet is a MAC layer packet containing one or more ARQ blocks. ARQ also ensures in-order delivery of packets to the IP layer. MAC layer ARQ alone does not improve the spectral efficiency. However, with its retransmissions mechanism to correct packet errors at the cost of extra delays, ARQ provides a more reliable link layer as seen by applications.

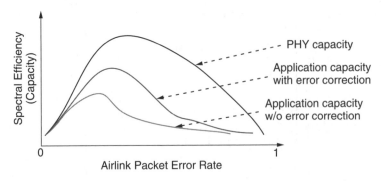

Figure 6.3   Spectral efficiency (capacity) measured at different layers changes in target PER [19]

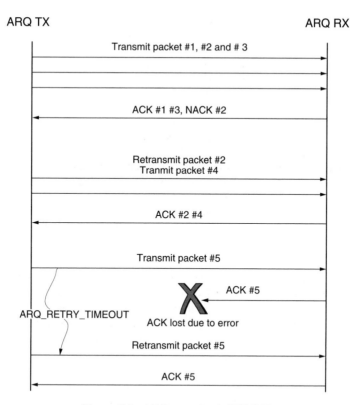

Figure 6.4   ARQ operation in WiMAX

ARQ also permits aggressive link adaptation for higher spectral efficiency. As we shall discuss in more detail later, this is critical for the performance of TCP.

Hybrid ARQ (HARQ) [14], [15], and [16] is another retransmission mechanism operating at the physical layer. Different from ARQ operating at the MAC layer, HARQ allows the receiver to perform soft-combining of retransmitted packets and therefore typically provides 10–20% improvement on the spectral efficiency. For simplicity, here we consider a system with only MAC ARQ functionality. When both ARQ and HARQ are implemented in the system, ARQ and HARQ interaction need to be addressed to avoid redundant retransmissions from different protocol layers, which we will briefly discuss in Section 0. Since HARQ operates differently from ARQ, the operation optimization associated with HARQ will be different as well.

6.3 Asymmetric Link Adaptation for TCP

6.3.1 TCP Performance on Wireless Network

The AIM of this section is to provide some intuition of TCP performance on a WiMAX network, instead of deriving accurate analytical results for delay and throughput performance. We assume for each downlink or uplink CID, packet size is constant, e.g., TCP segments in

Table 6.1   Description of notations

B	TCP end-to-end throughput
b	the number of TCP packets acknowledged by one ACK
R	TCP average round-trip time
T_o	TCP timeout value
p	wire-link loss rate
p_{PHY}	down-link physical layer packet error rate
q_{PHY}	up-link physical layer packet error rate
m	maximal ARQ retransmissions
p_{ARQ}	residual down-link packet error rate after ARQ
q_{ARQ}	residual up-link packet error rate after ARQ
D_d	typical down-link wireless delay
D_u	typical up-link wireless delay
K	average number of ARQ (re)transmissions per TCP packet

the downlink and TCP ACKs in the uplink. We also assume packets are transmitted in the granularity of TCP packets instead of MAC PDUs. Therefore, we capture the packet errors with independent events. We also ignore the detailed impact on delays from network load, scheduling, uplink bandwidth request and ARQ feedback mechanism. Instead, we assume an average delay per transmission, i.e., D_d and D_u for downlink and uplink respectively. We want to use this simplified model to motivate the proposed scheme. In Table 6.1 we list the notations that will be used in the chapter.

TCP is designed to ensure reliable end-to-end transmission. TCP uses congestion control to avoid overloading the network. We here focus on the end-to-end performance of TCP, i.e., throughput. In [2], an analytical result for TCP throughput B without window constraint is obtained as follows.

$$B_{wire} = \frac{1}{R_{wire}\sqrt{\frac{2bp}{3}} + T_o \min\left(1, 3\sqrt{\frac{2bp}{3}}\right) p(1 + 32p^2)} \qquad (6.1)$$

According to Equation (6.1), since T_o is generally conservatively selected to include delay variance, TCP performance is mainly determined by Round Trip Time (RTT) and loss indication probability p. If one wants to achieve a very high TCP throughput on a single TCP session, the only way is to ensure an extremely reliable link with small p because it is hard to manage *RTT* across the network, which is true for both wireless and wired links. On the other hand, it is rare to need a high throughput from a single TCP connection and therefore a reasonably good performance can be obtained by balancing *RTT* and p, as we shall discuss below.

In a wireless network without ARQ, packet errors are independent, which happens with probability p_{PHY} and q_{PHY} for uplink and downlink respectively. A lossy wireless access link degrades TCP performance significantly as it contributes not only increased RTT, but also higher overall packet loss rate. In this scenario, we obtain TCP throughput as

$$B_{noARQ} = \frac{1}{\hat{R}\sqrt{\frac{2b\hat{p}}{3}} + T_o \min\left(1, 3\sqrt{\frac{2b\hat{p}}{3}}\right) \hat{p}(1 + 32\hat{p}^2)}, \qquad (6.2)$$

where $\hat{R} \approx R_{wire} + D_d + D_u$ and $\hat{p} \approx p + p_{PHY} + q_{PHY}$, with the first order approximation. This means the delay measured by TCP protocol includes both wired link delay plus the wireless link delay. In addition, any packet error/drop occurred due to either router buffer overflow or unreliable wireless transmission will affect the TCP performance. Therefore, without ARQ, a WiMAX network has to ensure a small enough physical layer PER, e.g., about the same order of magnitude as wired link loss rate p, so that TCP performance does not suffer significantly. While this is possible, low PER requires a very conservative link adaptation, i.e., choosing lower MCS order for better reliability, which leads to low spectral efficiency and limits the network capacity.

When ARQ is enabled on the wireless link, it achieves a tradeoff between PER and delay. On one hand, ARQ retransmission mechanism is able to correct packet errors and thus significantly reduces the loss rate seen by the TCP layer. On the other hand, retransmissions incur extra delay that contributes to the overall RTT. It is difficult to obtain analytical results for TCP performance over a WiMAX-specific ARQ implementation. Yet with some simplifications, the analysis will be possible.

Assume T_o is much greater than the average ARQ layer delay, supported by empirical studies in 0, and thus we can ignore the case where the jitter caused by ARQ retransmissions leads to TCP timeout. Ignore the ARQ block structure and assume ARQ transmits TCP packets back to back. Instead of modeling WiMAX's timer-based ARQ retransmission management, we simply consider a maximal retransmission count m. With this first-order approximation, we indeed hide the complexity of ARQ within the definition of D_d, D_u and m, and we use the high-level TCP packet error rate p_{ARQ} and q_{ARQ} to capture the aggregation of MAC layer PDU loss rate after ARQ. Note they are functions of maximal ARQ retransmission count m as well as packet error rate before ARQ, i.e., p_{PHY} and q_{PHY}. Assume ARQ can transmit up to m times for each block. One can rewrite Equation (6.1) for TCP over ARQ as follows:

$$B_{ARQ} = \frac{1}{\tilde{R}\sqrt{\frac{2b\tilde{p}}{3}} + T_O \min\left(1, 3\sqrt{\frac{2b\tilde{p}}{3}}\right)\tilde{p}(1 + 32\tilde{p}^2)},$$

$$\tilde{R} = R_{wire} + k(p_{PHY})D_d + k(q_{PHY})D_u, \qquad (6.3)$$

$$\tilde{p} \approx p_{ARQ} + q_{ARQ} + p.$$

With the simplifications mentioned above, we approximately calculate $p_{ARQ} \approx p_{PHY}^m$, $q_{ARQ} \approx q_{PHY}^m$ and $k(p) \approx \frac{\sum_{i=1}^{m} p^{i-1}(1-p)i}{1-p^m}$. Note that residual PER, p_{ARQ} and q_{ARQ} decreases almost geometrically in m, while ARQ associated delay scales only sub-linearly and will be mitigated by R_{wire}, it is obvious that enabling ARQ significantly improves TCP performance over lossy wireless links.

6.3.2 TCP Usage Model in Broadband Wireless Networks

It is important to highlight that a broadband access network supports many different flows from different users. The usage model we are interested in is the 'TCP capacity' of the network, which represents the aggregated throughput one can achieve with many concurrent TCP flows. From the point of view of network operators, aggregated TCP capacity measured in the total throughput that can be supported is more important than what the maximal throughput a single TCP flow can achieve. In a broadband wireless network with a relatively heavy load, these two objectives are unlikely to be achieved at the same time because concurrent flows are competing for limited radio resources. Therefore, it is critical for the network to efficiently allocate resources while maintaining good throughput and reasonable fairness among individual flows. With an understanding of TCP performance over ARQ, we can now study how the link adaptation can be coupled in the optimization for the best TCP capacity.

6.3.3 Asymmetric Link Adaptation for TCP-Based Applications

A TCP connection is highly asymmetric in the sense that its data channel, defined as from source to destination, carries the bulk of the transmissions, while its ACK channel, defined as from destination to source, only carries small TCP signaling packets such as ACK and SYN. This TCP asymmetry motivates us to consider an asymmetric link adaptation strategy to optimize the aggregate TCP performance at the network level. With the assumed download scenario, the TCP data channel is on a downlink CID and ACK channel is on an uplink CID for each MSS.

6.3.3.1 TCP Data Channel Resource Allocation

For the TCP data channel, we propose to aggressively use more efficient MCS, at the cost of higher PHY PER, to increase the bandwidth of the data pipe and improve the aggregate TCP throughput from all users.

This is similar to the case of pushing BER for better spectral efficiency, as shown in Figure 6.3. However, aggressive MCS selection must be coupled with ARQ. Otherwise, due to the sensitivity of TCP performance to packet loss rate, when the packet loss on the wireless link becomes comparable to or larger than the loss rate on wired link, TCP performance will drop much earlier before the PHY spectral efficiency approaches its maximum.

Combined with proper ARQ settings, increased PHY PER will not significantly worsen the link error rate observed by TCP and only marginally increase the RTT. Therefore individual TCP performance remains largely uncompromised. For example, according to link-level simulation, pushing target PER from 1% to 10% is usually sufficient for switching to a higher-order MCS in the profile, which can offer 1.5–4 times higher spectral efficiency. We will prove this in our simulations. On the other hand, with maximal m retransmissions, the residual PER will be $p_{ARQ} = 10^{-m}$, which will be sufficiently small if m is large enough.

6.3.3.2 TCP ACK Channel Resource Allocation

In the ACK channel, TCP flows only carry application requests and TCP signaling messages such as ACK and SYN most of the time, which are generally small compared to the packets in the data channel. Although one can apply the same link adaptation strategy as for the data channel, the small traffic volume in the ACK channel will give marginal rewards. Instead, an ACK channel having a large pipe is not as critical as robustness and response time. For example, each ACK packet roughly has the same contribution to the packet loss rate and RTT, even much smaller in size than a data packet. In addition, one ACK carries acknowledgement for b TCP segments, which makes ACK loss highly undesirable. Finally, in a download scenario, uplink transmissions of ACK typically require bandwidth requests which incur an extra delay of more than 4 frames, which likely makes ACK retransmissions contribute more to the overall RTT, i.e., $D_u > D_d$. Therefore, we propose choosing conservative MCS to achieve low loss rate and low delay, which in turn benefits the end-to-end performance of TCP.

6.3.4 Optimizing ARQ Setting

The idea of asymmetric link adaptation for TCP relies on ARQ, particularly for the data channel, to mitigate the increased PER due to aggressive MCS selection for TCP data channel. ARQ layer parameters must be properly set up to offer the desired reliability.

The WiMAX standard defines timer-based ARQ operations. In particular, retransmissions are triggered by either ARQ_RETRY_TIMEOUT at the transmitter or an ARQ NACK feedback from the receiver. Each ARQ block has a maximum lifetime defined by ARQ_BLOCK_LIFETIME. Therefore, neither the retransmission interval nor the maximal retransmission count is explicitly defined.

ARQ_RETRY_TIMEOUT should be selected conservatively such that retransmissions do not happen too fast. The reason is that when aggressive link adaptation is applied, packet errors happen mostly due to link adaptation error, i.e., current MCS does not reflect a recent change in channel condition, such as a deep fade. Depending on the channel coherence time, such link adaptation error may last for a period until the downlink channel condition is updated by a Channel Quality Indicator (CQI).[1] Therefore, a very fast ARQ retransmission may likely incur packet error again, which causes wasted radio resource. Therefore, the rule of thumb for setting ARQ_RETRY_TIMEOUT (in the downlink) is this parameter should not be faster than the CQI update interval.

For ARQ_BLOCK_LIFETIME, which implicitly controls the estimated maximal retransmission count, should be enough to maintain a low residual PER seen by TCP. For example, if the data channel target PER is 20%, loss indication rate p from the wired network is typically around 1%, ARQ needs at least total 4 (re)transmissions to lower the wireless link loss rate p_{ARQ} to be an order of magnitude smaller than p, e.g., 0.16% in this case assuming i.i.d. packet error events. Therefore, conservatively setting ARQ_BLOCK_LIFETIME to a large value is generally safe as long as the buffer allows and this will be unlikely to cause extremely long delays as consecutive packet errors happen with geometrically decreasing probability.

6.4 Service-Class Specific Scheduling

Yet another important factor determining application performance is scheduling. So far, we have largely ignored its subtle impact on system performance. WiMAX MAC framework allows service differentiation, which implies different scheduling policies can be used to different service classes. There are two questions we try to answer in this section. Do we really need separate scheduling algorithms for different applications or service classes? If so, what kind of benefits does this bring to the overall performance? We shall revisit these questions from the TCP performance aspect.

---

[1] Note that CQI may be configured to be only once for a few frames. Otherwise, there will be a significant feedback overhead.

6.4.1 Relevant Scheduling Policies

An opportunistic scheduler can generally be written as

$$\text{schedule } i^*, \text{ s.t. } i^* = \arg\max_i (W_i), \tag{6.4}$$

where W_i is some metrics calculated for each user or connection. PF scheduler schedules users based on $W_i = a_i C_i / \bar{C}_i$, where a_i is some weighting constant, C_i is the instantaneous data rate (channel capacity) and $\bar{C}_i$ is the average data rate allocated for user i. PF efficiently exploits multi-user diversity for throughput enhancement and maintains much better fairness than a greedy scheduler that simply considers $W_i = C_i$.

Queue-length weighted schedulers like queue-length-based MLWDF and ER give higher priority to queues with longer queue length. In queue-length-based MLWEF, $W_i = a_i \times q_i \times C_i / \bar{C}_i$, while in EP, $W_i = a_i C_i \exp(\frac{q_i - \bar{q}}{1+\sqrt{\bar{q}}})$, where q_i is the queue length of user i and $\bar{q}$ is the average queue length among all users. These 'throughput optimal' schedulers work well both in achieving optimal throughput and maintaining queue stability when traffic flow is independent of queue states. Therefore, they are good candidates for VoIP like traffic.

A variant of a queue-length-based scheduler is a delay-based scheduler, e.g., in delay-based MLWDF, $W_i = a_i \times d_i \times C_i / \bar{C}_i$, where d_i is the head of line packet delay for user i. In the scenario with independent stationary traffic sources, it is essentially the same as queue-length-based scheduling by considering Little's theorem [17]: E[queue length] = E[delay] × arrival_rate. However, when traffic sources are not independent of queue states, e.g. due to TCP congestion control, delay-based schedulers totally diverge from queue-length-based schedulers in performance.

6.4.2 Scheduling Impacts End-to-End TCP Performance

TCP's congestion control mechanism reacts to the network condition it measures on an end-to-end basis. Assume the wireless link quality of TCP flow-1 is better than that of TCP flow-2. As shown in Figure 6.5, for flow-1, both the TCP window size and the queue length at BS station

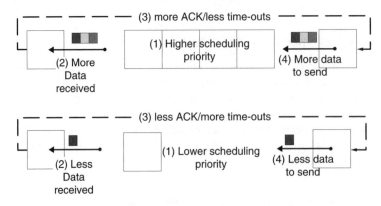

Figure 6.5   A queue-length based scheduler leads to poor fairness among TCP flows [19]

tend to be large. Giving priority to flow-1's larger queue length during scheduling encourages it to further increase its window size and potentially take more resource allocation. On the other hand, flow-2 does not have a large window size and thus its queue length is small. Getting less resource allocation causes flow-2 to further reduce its window. Such an ill-behaved close loop can lead to extreme unfairness when the network is congested. In the worst case, flows with insufficient resource allocation can only periodically send 1-byte segments in order to re-open the TCP window. But again a queue-length based scheduler will make these small probes hard to go through. Queue-length-based schedulers thus become greedy for TCP applications and can lead to lack of fairness among users. On the other hand, delay-based schedulers try to maintain a proportional delay. When the wireless network is congested, wireless link delay plays an important role in the overall RTT and this causes all TCP flows to experience similar delays, assuming their R_{wire} values are similar. Thus delay-based MLWDF forces all TCP flows to have a similar throughput. Such fairness comes at the cost of low aggregate throughput because the scheduler does not exploit the 'diversity' of multiple TCP flows, i.e., different queue lengths corresponding to different TCP throughputs.

Therefore, we argue that in a WiMAX network BE service class should have its own scheduler, different from those for real-time traffic. Service-class specific scheduling is easy to implement in the WiMAX MAC framework. To avoid pure priority-based scheduling among different service classes, multiple metrics from different schedulers can be considered jointly with proper weight assignment and admission control [18]. We propose the following TCP-friendly scheduler.

$$\text{Schedule } i^*, \text{ s.t. } i^* = \arg\max_i(\gamma_i \times C_i/\bar{C}_i \times d_i \times q_i) \qquad (6.5)$$

where γ_i is a weighting constant, d_i is the head-of-line delay and q_i is the number of packets in queue. Intuitively, it achieves a tradeoff between exploiting diversity and maintaining fairness among TCP flows. Making q_i a packet count instead of byte count offers higher priority for small signaling messages, if traffic flows happen in both downlink and uplink. We will verify these observations through simulations.

6.5 Simulations

6.5.1 Simulation Setup

We use an OPNET Modeler to construct a WiMAX network. We measure the performance in a typical *sector*, typically 3 sectors per cell, with radius 1Km where MSSs are uniformly located in the sector. Each MSS is running a heavy HTTP download from an HTTP server connected to the Internet. The wired delay has a shifted-gamma distribution with mean 100ms and standard deviation 20ms. TCP throughput is measured in KByte/s.

We model the wireless channel with ITU path-loss model with path-loss exponent $\alpha = 4$. We model the fast fading using Jake's model with ITU PED-B parameters, i.e., 3Km/hr low speed mobility scenario. The packet error events are triggered by PER calculated from instantaneous channel quality, where the mapping between PER and channel quality is derived from our link-level simulations. We estimate the interference from nearby cells, with average loading factor 75% on the downlink and 50% on the uplink.

The WiMAX frame structure has 5ms frame length, 10MHz bandwidth with 1024 FFT. Downlink/uplink ratio is 1:1. We consider frequency reuse 1:3:3 to enhance sector edge coverage. The MCS profile is CTC-based and includes the following code rates: QPSK(1/2, 3/4), 16QAM(1/2, 3/4) and 64QAM(1/2, 2/3, 3/4). Link adaptation thresholds for a target PER are obtained by separate empirical studies. The downlink CQI update interval is 4 frames for each MSS, i.e., 20ms.

6.5.2 Optimizing ARQ Parameter Setting

We first verify the ARQ parameter setting proposed in Section 0 by changing the parameter at one MSS client. We set the uplink PER to be 1% and the downlink PER to be 20%. We vary downlink ARQ_RETRY_TIMEOUT and adjust ARQ_BLOCK_LIFETIME to be three times larger. As shown in Figure 6.6, TCP performance suffers from a very small ARQ_RETRY_TIMEOUT, even more significantly than selecting a large value. Recall our CQI update is 20ms. When retransmission is too fast, link adaptation error may likely remain and cause consecutive packet error. Since the system has limited maximal retransmission count, this reduces the effectiveness of ARQ. On the other hand, if ARQ_RETRY_TIMEOUT is too large, it will contribute to larger RTT and we again observe throughput drop.

Similarly, we investigate the impact of ARQ_BLOCK_LIFETIME timer on TCP performance. We set ARQ_RETRY_TIMEOUT to be 30ms and we vary $\frac{ARQ\_BLOCK\_LIFETIME}{ARQ\_RETRY\_TIMEOUT}$ from 3 to 10. As shown in Figure 6.7, we observe that there is a clear improvement after four estimated retransmissions, which corresponds to a ARQ residual PER negligible for TCP layer. Choosing an even larger ARQ_BLOCK_LIFETIME does benefit the performance, albeit marginally.

6.5.3 Capacity Improvement with Asymmetric Link Adaptation

We fix the ARQ setting so that ARQ_RETRY_TIMEOUT = 30ms and ARQ_BLOCK_LIFETIME = 300ms. We then verify how well the asymmetric link adaptation improves

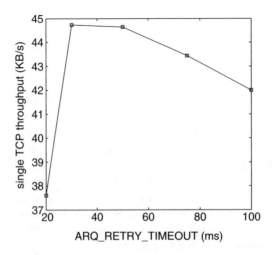

Figure 6.6   TCP throughput versus ARQ_RETRY_TIMEOUT setting [19]

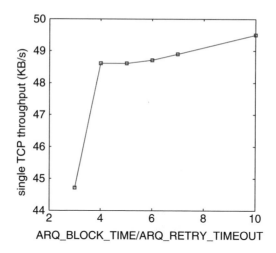

Figure 6.7 TCP throughput versus ARQ maximal retransmission count setting [19]

the system aggregated throughput. We start with the case that only a single TCP session exists in the network. We switch the link adaptation thresholds to obtain different target PER before ARQ. As shown in, the single TCP throughput remains largely unchanged until the PER is larger than 20%. This matches the conclusion we drew from (6.3) very well, i.e., TCP only sees $\tilde{R}$ and $\tilde{p}$ after ARQ as long as ARQ can manage to maintain a low delay and residual PER. However, beyond a certain point, TCP performance drops quickly as a combined result of failed ARQ functionality and reduced overall PHY spectral efficiency.

Meanwhile, as shown in Figure 6.9, the resource used by this single TCP connection reduces almost linearly in target PER. Such efficiency improvement is slower when the target PER is higher because ARQ retransmissions consume extra resources. Therefore we observe that there is a regime, e.g., target PER below 20%, where we can improve the efficiency of the

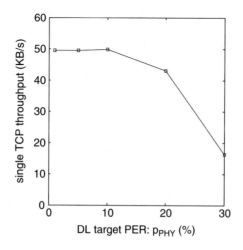

Figure 6.8 Single TCP throughput versus DL target PER (before ARQ) [19]

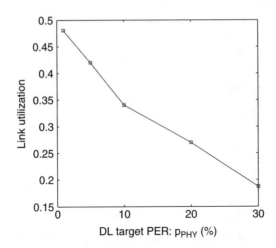

Figure 6.9   Single TCP resource utilization versus DL target PER (before ARQ) [19]

link without significantly compromising the individual TCP performance. This translates to a larger aggregated TCP throughput achievable by multiple TCP sessions in a network.

On the other hand, consider the ACK channel in the uplink. As shown in Figure 6.10, TCP performance exhibits similar trends as with downlink PER, but appears to be marginally more sensitive in uplink PER. The reason is that the uplink WiMAX may take more time for transmission due to bandwidth request procedure. When we consider a single TCP flow, the network is not congested. Therefore, such extra delay in uplink becomes significant, i.e., $D_u >$ D_d. This causes the increased sensitivity of TCP throughput in uplink PER q_{PHY}.

We are now ready to examine the improvement on the aggregated TCP throughput in the network. We consider a network with 10 MSS clients, each is running a TCP session. We fix the ARQ setting as before and further fix the uplink target PER $q_{PHY} = 5\%$. We then vary the

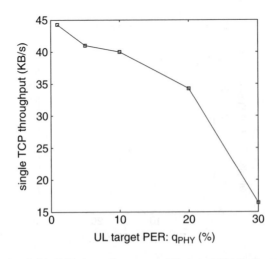

Figure 6.10   Single TCP throughput versus UL target PER (before ARQ) [19]

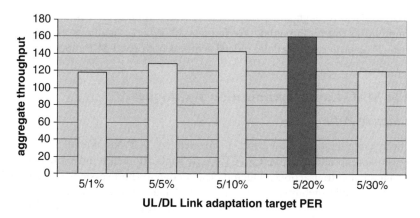

Figure 6.11   Aggregate network TCP throughput versus target PER before ARQ [19]

downlink target PER p_{PHY} from 1% to 30%. As shown in Figure 6.11, when we measure the aggregated throughput in the network, indeed it is maximized at the target PER around 20%. The gain compared to the most conservative link adaptation with target PER = 1% is almost 33%.

6.5.4 Performance of TCP-Aware Scheduler

We consider a network with 20 MSS clients. We compared the performance of different schedulers in terms of overall throughput and fairness, i.e., the standard deviation among individual throughput. As shown in Figure 6.12, simulation results confirm that queue-length-based

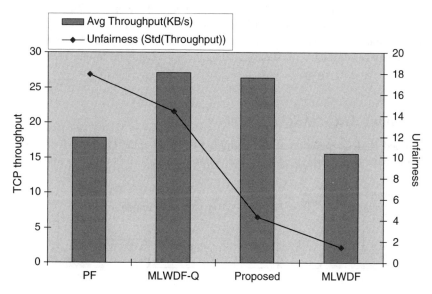

Figure 6.12   Average TCP performance (aggregate throughput normalized by the number of MSS) and fairness measure with different scheduling policies [19]

MLWDF is greedy in throughput by sacrificing fairness while delay-based MLWDF significantly compromises overall throughput for fairness. PF achieves neither good throughput nor good fairness. The proposed scheduler indeed achieves both good TCP capacity and fairness.

6.6 Other MAC Layer Optimization Techniques

6.6.1 Adaptive Polling

We have been using the term 'best effort' to characterize TCP-based applications. This may not always be appropriate since many TCP-based applications are indeed interactive, such as online banking and transactions, which typically contain short sessions that demand short delays as well as good reliability. An adaptive polling mechanism can be used to serve these applications, without compromising the overheads. For example, a minimal polling interval T_{min} and maximal interval T_{max} are defined based on maximal/minimal QoS requirements, respectively. During session ongoing period, T_{min} is used for polling, while the polling interval is automatically adjusted to T_{max} if no traffic is detected over idle periods. This provides a much better service than the existing BE service class, and yet consumes less overhead than the existing polling protocols. Since all control and management are done at the BS, there is no extra complexity required on the MSS to implement adaptive polling.

6.6.2 Enhance Contention-Based Bandwidth Request

The contention-based bandwidth request procedure for BE service class, as shown in Figure 6.2, can also be optimized. Recall that we have multiple CDMA codes available. One can use a subset of these codes to designate a particular bandwidth request value, predefined based on applications. The bandwidth request procedure can be further simplified to only two signaling messages on the radio link as shown in Figure 6.14, because the bandwidth request value has been implicitly indicated by the selected CDMA code. This is particularly useful for TCP ACK and ARQ accumulative ACK, whose sizes are known in advance. With this approach, uplink scheduling delay is minimized for short signaling messages and end-to-end TCP performance is improved.

6.6.3 Coupling ARQ-HARQ Operations

We mentioned HARQ operation at the beginning of the chapter. One wonders whether it is redundant to have two overlapping error-correction protocols. It is true that most of the time

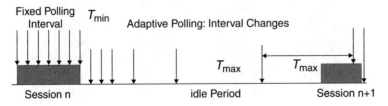

Figure 6.13 Adaptive polling mechanism

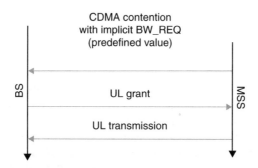

Figure 6.14   Enhanced contention-based bandwidth request

this is redundant and HARQ indeed provides faster error-recovery because of shorter feedback delays at the physical layer. However, HARQ has two drawbacks. Firstly, HARQ does not provide in-order packet delivery due to its simplicity in buffer management. Secondly, HARQ feedbacks are not CRC protected because it is simple signaling at the physical layer instead of a MAC layer, and thus HARQ feedbacks are more prone to errors. Therefore, one would still want to keep ARQ for high reliability and packet reordering. On the other hand, HARQ can handle most of the retransmissions when packets are in error, without triggering relatively slow MAC ARQ retransmissions. An example is using HARQ feedbacks to generate ARQ NACK messages, as shown in Figure 6.15, which eliminates the ARQ feedback overheads on the air link. Note that additional ARQ feedbacks may still be triggered when the receiver side detects HARQ feedback errors via the reception of unexpected transmissions. Such HARQ-ARQ interaction requires the MAC design to be HARQ-aware and can significantly improve delays associated with error correction while reducing air-link signaling overheads. In addition, with HARQ-ARQ interaction, the optimization results for ARQ-only operations mentioned in this chapter may need to be adjusted accordingly.

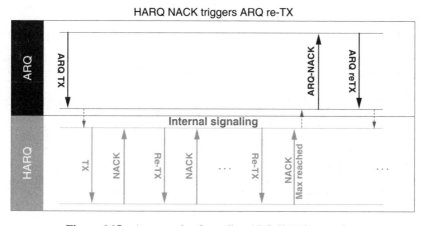

Figure 6.15   An example of coupling ARQ-HARQ operations

6.7 Conclusion

We have shown that the flexibility of WiMAX MAC framework offers new dimensions for cross-layer optimization. By jointly optimizing link adaptation, ARQ and scheduling at a fine granularity, overall TCP throughput is significantly enhanced, without compromising the performance of individual flows or fairness among users. We also introduce other promising approaches such as adaptive polling, enhanced contention-based bandwidth request and coupling HARQ-ARQ operations, geared at achieving better performance and system capacity.

We want to emphasize that although we mainly focus on TCP performance optimization, WiMAX MAC framework can be exploited to optimize the performance of other applications as well. For example, the adaptive polling technique is actually suitable for any traffic source that exhibits on/off or bursty traffic pattern; and the enhanced bandwidth request approach will also help to speed up the network entry by reducing the ranging delay.

MAC layer enhancement alone is not enough to improve application end-to-end performance, particularly in mobile WiMAX networks. Optimization associated with the network operations is also very important, in order to handle the mobility and battery power constraint of MSS. Optimized hard handover as well as related sleep/idle mode operations should be carefully studied to guarantee seamless mobile computing experience.

References

[1] X. Yang, M. Venkatachalam and S. Monhanty, 'Exploiting the MAC Layer Flexibility of WiMAX to Systematically Enhance TCP Performance', in *Proc. IEEE Mobile WiMAX Symposium*, March 2007.

[2] J. Padhye, V. Firoiu, D. Towsley, and J. Kurose, 'Modeling TCP Throughput: A Simple Model and its Empirical Validation', in *Proc. ACM SIGCOMM*, Sept. 1998.

[3] A.A. Abouzeid, S. Roy, and M. Azizoglu, 'Stochastic Modeling of TCP over Lossy Links', in *Proc. IEEE INFOCOM*, March 2000.

[4] E. Altman, K. Avrachenkov, and C. Barakat, 'A Stochastic Model of TCP/IP with Stationary Random Losses', in *Proc. ACM SIGCOMM*, Sept. 2000.

[5] Gang Bao, 'Performance Evaluation of TCP/RLP Protocol Stack over CDMA Wireless Link', *Wireless Networks*, 2(3), Sept.1996, 229–237.

[6] H. Lin and S.K. Das, 'TCP Performance Analysis of CDMA Systems with RLP and MAC Layer Retransmissions', in *Proc. IEEE MASCOTS*, Oct. 2002.

[7] A. Simonsson, J. Peisa, and J. Pettersson, 'Analytic Study of TCP Performance over a Soft Rate Switching WCDMA Bearer', in *Proc. IEEE PIMRC*, Sept. 2005.

[8] A-F. Canton and T. Chahed, 'End-to-End Reliability in UMTS: TCP over ARQ', in *Proc. IEEE GLOBECOM*, Nov. 2001.

[9] P. Bender, P. Black, M. Grob, R. Padovani, N. Sindhushayana, and A. Viterbi, 'CDMA/HDR: A Bandwidth-Efficient High-Speed Wireless Data Service for Nomadic Users', *IEEE Communication Magazine*, 38(7), July 2000, 70–77.

[10] M. Andrews, K. Kumaran, K. Ramanan, A. Stolyar, P. Whiting, and R. Vijayakumar, 'Providing Quality of Service over a Shared Wireless Link', *IEEE Communication Magazine*, 39(2), Feb. 2001, 150–154.

[11] S. Shakkottai and A.L. Stolyar, 'Scheduling for Multiple Flows Sharing a Time-Varying Channel: The Exponential Rule', *American Mathematical Society Translations, Series 2*, ed. Y.M. Suhov, 207, Feb. 2000, 150–154.

[12] H. Zheng T.E. Klein and K.K. Leung, 'Enhanced Scheduling Algorithms for Improved TCP Performance in Wireless IP Networks', in *Proc. IEEE GLOBECOM*, Nov. 2004.

[13] IEEE 802.16e, 'IEEE Standard for Local and Metropolitan Area Networks – Part 16: Air Interface for Fixed and Mobile Broadband Wireless Access Systems, Amendment 2 and Corrigendum 1', IEEE Standards, Feb. 2006.

[14] S. Lin and P. Yu, 'A Hybrid ARQ Scheme with Parity Retransmission for Error Control of Satellite Channels', *IEEE Trans. on Communications*, 30(72), July 1982, 1701–1719.

[15] S. Kallel and D. Haccoun, 'Generalized Type II Hybrid ARQ Scheme Using Punctured Convolutional Coding', *IEEE Trans. on Communications*, **38**(11), Nov. 1990, 1938–1946.

[16] J.F. Cheng, 'Coding Performance of Hybrid ARQ Schemes', *IEEE Trans. on Communications*, **54**(6), June 2006, 1017–1029.

[17] D. Bertsekas and R. Gallager, *Data Networks*, 2nd edn, Englewood Cliffs, NJ: Prentice Hall, 1992, 152–162.

[18] H. Kim and G. de Veciana, 'Losing Opportunism: Evaluating Service Integration in an Opportunistic Wireless System', in *Proc. IEEE INFOCOM*, May 2007.

[19] X. Yang, M. Venkatachalam, and S. Mohanty, 'Exploiting the MAC Layer Flexibility of WiMAX to Systematically Enhance TCP Performance', *Proceeding IEEE Mobile WiMAX Symposium*, Orlando, 2007.

7

A Novel Algorithm for Efficient Paging in Mobile WiMAX

Mohanty Shantidev, Muthaiah Venkatachalam, and Xiangying Yang

7.1 Introduction

Mobile WiMAX [4] based on IEEE 802.16e standard [1] enables high speed data communications anywhere and any time. For significant time durations, mobile stations (MSs) are powered in mobile WiMAX networks but are not in active call sessions. To use these periods as battery-conserving opportunities, idle mode, location update, and paging operations are specified in IEEE 802.16e standard [1]. Per these procedures, an MS enters a low-power mode called 'Idle mode'. Upon entering idle mode, the MS relinquishes all of its connections and states associated with the base station (BS) it was last registered with. The idle MS is tracked by the network at the granularity of a group of BSes (commonly called a paging group) as opposed to a non-idle MS which is tracked at the granularity of a BS. While in idle mode the MS periodically listens to the radio transmissions for paging messages, in a deterministic fashion that is decided a priori between the network and itself. The period for which the MS listens to paging messages is known as 'paging listen interval' (PLI) and the period for which the MS powers off its radio interface is known as the 'paging unavailable interval' (PUI). The amount of power saving an MS can achieve in idle mode is tightly coupled to the duty cycle of the MS, which is the ratio of paging listen to the paging unavailable interval. One paging unavailable interval and one paging listening interval constitute a paging cycle (PAGING_CYCLE) as shown in Figure 7.1. Therefore, once in every PAGING_CYCLE interval the idle-mode MS wakes up and listens for paging messages. When traffic arrives for the idle-mode MS the network performs paging to locate the MS and to bring it back to active mode.

The operation of idle mode and paging, in mobile WiMAX networks, can be summarized as follows.

Maintaining the location information of an idle-mode MS: This is achieved by logically dividing the network coverage area into different paging groups. A paging group (PG) refers

Mobile WiMAX   Edited by Kwang-Cheng Chen and J. Roberto B. de Marca.
© 2008 John Wiley & Sons, Ltd

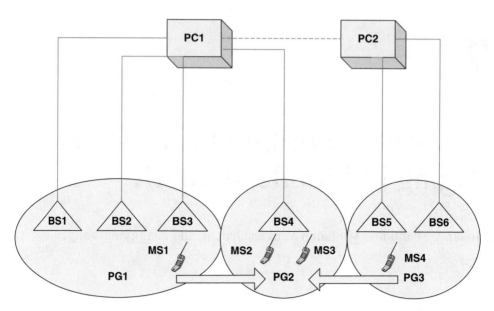

Figure 7.1   Network reference model [6]

to the coverage area of one or more base stations (BSes). A Paging Controller (PC) administers
one or more paging group(s). There could be one or more PCs in the network. When an MS
goes into idle mode, a PC, referred to as anchor PC, creates an entry in its database noting the
PG where the MS is initially located. When the MS moves from one PG to another, it updates
the location with the anchor PC. Therefore, while in idle mode the location of an MS is known
up to the granularity of one PG.

Paging an idle-mode MS: When the network wants to locate an idle-mode MS, or has
incoming data buffered for it, or for administrative purposes, the PC initiates paging the MS
by a broadcasting *mobile paging advertisement* (MOB-PAG-ADV) message to all the BSes
in the PG; the BSes in turn broadcast this message on the airlink. This is because when the
location information stored at a PC is correct; the MS is expected to reside in the coverage area
of at least one of these BSes. If the paging advertisement happens during the PLI of the MS,
then the MS is expected to receive the page and perform network re-entry or location update
in response to the page if it is alerted to do so.

This chapter provides detailed description of an algorithm that is developed to implement
efficient paging in IEEE 802.16e-based Mobile WiMAX networks. The main contribution of
this work is the design, development, and performance evaluation of a simple yet efficient
paging algorithm for Mobile WiMAX networks. The performance of paging mechanism in
Mobile WiMAX networks can be specified by two parameters: paging signaling overhead and
paging latency.

- *Definition 1: Paging signaling overhead (χ)* is defined as the number of bits per second used
  for paging one idle-mode MS.
- *Definition 2: Paging latency (τ)* is defined as the time delay between the initiation of paging
  operation by the network and the completion of MS's response to the paging operation.

Most of the applications specify an upper bound for paging latency. Thus, the objective of our research is to minimize paging signaling overhead without increasing the paging latency beyond what is required by different applications. A paging architecture is proposed for mobile WiMAX networks in [2]. However, it does not analyze the tradeoff between paging signaling overhead and paging latency. To the best of our knowledge there is no existing work that investigates the tradeoff between these two parameters of paging procedures in mobile WiMAX networks.

In our proposed paging algorithm, idle-mode MSs are grouped into different sets, referred to as paging sets, in such a way that the paging information of these MSs can be aggregated into one MOB-PAG-ADV message. This reduces the overhead associated with paging operation. However, aggregation of paging information of MSs may result in too many MSs transitioning from idle mode to normal/active mode at the same time causing high contention load during network re-entry, timeouts and potentially resulting in further retransmissions of paging broadcasts. This may increase the paging latency beyond what is required by the applications. To eliminate this, the proposed paging algorithm determines the maximum cardinality of each paging set that guarantees an upper bound for paging latency. Thus, the proposed paging algorithm achieves a good tradeoff between the paging latency and paging signaling overhead.

The remainder of the chapter is organized as follows. In Section 7.2 we describe idle mode and paging operation in mobile WiMAX networks. We present the proposed paging algorithm in Section 7.3. We carry out performance analysis of the proposed paging algorithm in Section 7.4. Finally, we summarize the advantages of the proposed paging algorithm and draw conclusions in Section 7.5.

7.2 Overview of Idle Mode and Paging Operation in Mobile WiMAX Networks

7.2.1 Paging Architecture

Figure 7.1 depicts a representative network reference model used to describe the idle mode operation in WiMAX networks. It consists of the three PGs (PG1, PG2, and PG3) and two PCs (PC1 and PC2). PC1 manages PG1 and PG2. PC2 manages PG3. PG1 comprises three BSs, PG2 comprises one BS, and PG3 comprises two BSs. Each PC maintains a location database that keeps information about all the MSs that have gone into idle mode in the PG(s) managed by that PC. shows a snapshot at time t, for four representative MSs (MS1, MS2, MS3, and MS4). At time t, all four MSs are in coverage area of BS4, and in PG2. Prior to t, MS1 was in coverage area of BS3 (i.e. in PG1). Prior to t, MS4 was in coverage area of BS5 (i.e. in PG3). While only four idle-mode MSs are shown in this scenario, there may be several more MSs (both idle mode and active mode) in real deployments in BS4 coverage area.

The entire idle mode operation in IEEE 802.16e based mobile WiMAX networks can be divided into following stages: idle mode initiation, idle mode entry, operation during idle mode, and idle mode exit [3]. NDISP, we provide a brief description of these states.

- *Idle mode initiation*: The idle mode can be initiated either by the BS or MS when the MS does not have any ongoing traffic. In case of MS initiated idle mode, the MS sends a *deregistration request* (DREG-REQ) message to the BS [1], [2]. Similarly, in the case of a BS-initiated idle mode, the BS sends a *deregistration command* (DREG-CMD) message to the MS. When the MS receives the DREG-CMD message, it sends the DREG-REQ message to the BS [1], [2]. In each case, the BS receives a DREG-REQ message from the MS.

- *Idle mode entry*: When a BS receives the DREG-REQ message from one of its MSs, it sends a message to the anchor PC (how the anchor PC is chosen is a matter of the network configuration and is beyond the scope of this chapter). This message contains certain MS service and operation information referred to as idle mode retain information (IMRI) [1]. IMRI can be used to expedite the MS's network re-entry from idle mode. PC stores the MS IMRI and transmits a backbone message to BS that includes numerical values for PAGING_CYCLE, PAGING_OFFSET, and MS Paging Listening Interval (PLI) for the MS [1]. Note that PC may use its own algorithm or negotiate with the BS and/or MS to decide the numerical values of PAGING_CYCLE and PLI. On the other hand, it can determine the PAGING_OFFSET using its own algorithm. Once BS receives the backbone message from the PC, it sends the DREG-CMD message to the MS that includes the idle mode entry time (IMET), PAGING_CYCLE, PAGING_OFFSET, and PLI values. The MS enters into idle mode at IMET.

- *Idle mode operation*: While in idle mode the MS alternates between PUI and PLI. The idle mode operation of two MSs (MS1 and MS2) is illustrated in Figure 7.2. In this case both MS1 and MS2 have the same PAGING_CYCLE and PLI (which is the case in most network deployments). However, MS1 and MS2 have different PAGING_OFFSETs of T1 and T2, respectively. Therefore, when the network wants to page MS1 and MS2, it does so through two different MOB-PAG-ADV messages at different times. It may be noted that the network needs to send two different MOB-PAG-ADV messages although it wants to page these two idle-mode MSs at the same time because the MSs have non-overlapping paging listening intervals.

- *Idle mode exit*: An MS in idle mode exits from idle mode if it has data to send to the BS or it receives a paging message indicating that its traffic is waiting at the BS. In this case, the MS terminates idle mode operation and carries out network re-entry procedures as specified in IEEE 802.16e [1]. As a part of network re-entry the idle-mode MS may perform contention based initial ranging. In another instance the BS may assign dedicated ranging region to the MS for initial ranging [1].

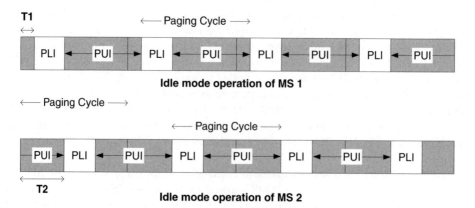

Figure 7.2   Network reference model [6]

Table 7.1 Format of MOB-PAG-ADV message

Syntax	Size (bits)	Note
Generic MAC header	48 (L1)	The generic MAC header
MOB-PAG-ADV_Format () {		
Management Message Type = 62	8 (L2)	
Num_Paging_Group_IDs (N_PG)	16 (L3)	Number of paging group IDs in this message
For (i = 0; i < N_PG; i++) {		
Paging Group ID	16 (L4)	
}		
Num_MACs	8 (L5)	Number of MS MAC addresses
For (j = 0; j < N_MACs; j++) {		Depends on the PHY specification
MS MAC Address hash	24 (L6)	The hash is obtained by computing a CRC24 on MS 48 bit MAC address
Action Code	2 (L7)	Paging action instruction to MS
Reserved	6 (L8)	
}		
Padding	Variable (L9)	Padding bits to ensure octet aligned
TLV encoded information	Variable (L10)	TLV specific
}		

Source: [1].

7.2.2 Paging Overhead

The format of MOB-PAG-ADV message is shown in Table 7.1. This message is appended with a 48 bit MAC header and transmitted over the air link. Out of all the bits in a MOB-PAG-ADV message the number of bits that carry information specific to a particular MS is

$$l_s = \sum_{i=6}^{i=8} L_i \tag{7.1}$$

The number of bits in a MOB-PAG-ADV message that does not carry MS specific information is

$$l_1 = \sum_{i=1}^{i=5} L_i + \sum_{L=9}^{L=10} L_i \tag{7.2}$$

Therefore, using Equations (7.1) and (7.2), when n numbers of MAC addresses are present in a MOB-PAG-ADV message, the total length of the MOB-PAG-ADV message is

$$l = l_1 + n l_s \tag{7.3}$$

Thus, when the paging information of n number of idle-mode MSs is aggregated into one MOB-PAG-ADV message, the number of bits used for paging one idle-mode MS is given by

$$\psi = \frac{l_1 + n l_s}{n} \tag{7.4}$$

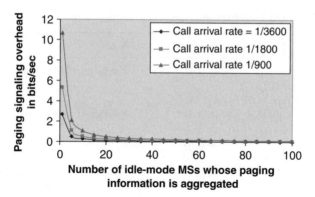

Figure 7.3 Paging signaling overhead for different amount of paging information aggregation [6]

Therefore, when the paging information of n number of idle-mode MSs is aggregated into one MOB-PAG-ADV message, paging signaling overhead (defined in Section 7.1) is given by

$$\chi = \lambda \psi = \frac{\lambda}{n}(l_1 + nl_s) \tag{7.5}$$

where λ is the average call arrival rate, i.e., number of calls arrived per second, for idle-mode MSs.

Figure 7.3 shows the relationship between paging signaling overhead and the number of MS whose paging information is aggregated into one MOB-PAG-ADV message for different call arrival rate (λ). The results show that significant paging overhead reduction is achieved even with the aggregations of paging information of small number of MSs.

7.2.3 Paging Latency

After successful paging, an idle-mode MS performs network re-entry via initial ranging access [1]. After initial ranging access the MS carries out other network entry procedures such as registration, authentication, and IP address acquisition etc. Paging latency which is defined as the time delay between the initiation of paging operation by the network and the reception of MS's response to the paging operation (i.e., completion of network entry) is given by

$$\tau = D_{pb} + \gamma + D_{ir} + D_r + D_a + \delta \tag{7.6}$$

where D_{pb} is the delay in the backbone network between the PC and BS, γ is processing delay at the BS (including scheduling and queuing delays), and D_{ir} is the initial ranging delay. D_r, D_a, δ are the time required for MS's registration and authentication, and IP address acquisition, respectively. As pointed out earlier, D_{ir} depends on the number of MSs that perform initial ranging at the same time. As our focus is the effect of D_{ir} on paging latency, we assume that other parameters in Equation (7.4) are unaffected when the paging messages of many idle-mode MSs are aggregated. Therefore, the desired D_{ir} to meet the required τ can be calculated using Equation (7.6).

7.3 Proposed Paging Algorithm for Mobile WiMAX Networks

7.3.1 Overview of the proposed paging algorithm

The objective of the proposed paging algorithm is to decide the maximum number of MSs whose paging information can be aggregated in one MOB-PAG-ADV message. Once this number is determined, the idle-mode MSs can be grouped into different sets with each set containing the maximum number of MSs. These sets are referred as paging sets. The proposed paging algorithm also develops methodologies to assign the idle-mode MSs into these paging sets.

7.3.2 Description of the proposed paging algorithm

For the description of the proposed paging algorithm, we assume that the numerical values of PAGING_CYCLE and the PLI for each MS are already known to the PC. Moreover, we consider that the numerical values of PAGING_CYCLE and PLI are same for all idle-mode MSs.

We define the following parameters:

T: The upper limit for paging latency, i.e., the paging latency should be limited to T **N:** average number of idle-mode MSs in a PG **R:** maximum number of idle-mode MSs in a cell that can perform initial ranging simultaneously such that paging latency does not exceed its upper limit T **λ:** call arrival rate (number of calls per second) of each idle-mode MS **K:** number of cells in a PG **L:** maximum cardinality of a paging set **p:** probability that an idle-mode MS is paged in a paging cycle. This is the probability that one or more calls are arrived for the idle-mode MS during a paging cycle. The expression for p is given by

$$p = 1 - e^{-\lambda(PAGING\_CYCLE)} \tag{7.7}$$

Therefore, assuming that idle-mode MSs are uniformly distributed among the cells in a PG, the number of idle-mode MSs of a particular paging set that are paged simultaneously in a cell in one paging cycle is

$$\alpha = \frac{Lp}{K} \tag{7.8}$$

To ensure that the paging latency is limited to T, α should be equal to R. Now the question is how to determine the numerical value of R. We will answer this question in Section 7.4. Therefore, using Equations (7.7) and (8), the expression for L is given by

$$L \leq \frac{KR}{p} = \frac{KR}{1 - e^{-\lambda(PAGING\_CYCLE)}} \tag{7.9}$$

Therefore, the number of paging sets, m, to accommodate N number of idle-mode MSs in a PG is

$$m = \left\lceil \frac{N}{L} \right\rceil \tag{7.10}$$

The expression for N is given by

$$N = \rho(K \pi r^2)\omega \eta \tag{7.11}$$

where ρ is the population density of the area where a WiMAX network is deployed, r is the radius of one WiMAX cell, ω is the fraction of population using WiMAX network, and η is the fraction of WiMAX users in idle mode. We denote the paging sets as $\{S_1, S_2, S_3, \ldots S_m\}$. Once, m and L are determined, the PC uses the following steps to assign idle mode users to individual sets.

1. Upon cold start of the system, when a PC receives the first idle mode initiation message for an MS, it assigns the corresponding MS to the first paging set and assigns the PAGING_OFFSET value corresponding to this paging set to the MS. This set is referred as the current paging set. A current paging set is defined as the paging set to which a new idle mode user can be added. The cardinality of the current paging set is at least one less than L.
2. When during normal operations, another idle mode initiation message arrives at the PC; the PC adds the new MS to the current paging set and assigns the corresponding PAGING_OFFSET to the MS. Then, the PC checks if the number of idle-mode MSs in the current paging set has reached the maximum value L, so that adding one more MS will cause the expected contention latency to exceed T. If not, PC keeps the current paging set unchanged. If yes, then the PC chooses another paging set as the current paging set. Therefore, at any given time there are S_c ($c < m$) number of paging sets that are occupied with idle mode users. These paging sets, $\{S_1, S_2, \ldots, S_c\}$ are referred as active paging sets. One of these active paging set is the current paging set that a new idle-mode MS can be assigned to. The remaining paging sets $\{S_{c+1}, S_{c+2}, \ldots, S_m\}$ are referred as empty paging sets.
3. When an MS in idle mode decides to terminate the idle mode and become active by re-entering the network, it is removed from the corresponding paging set. Therefore, the number of MSs in each active paging set changes when some of the idle-mode MSs terminate their idle mode. These vacancies can be used to accommodate new idle-mode MSs. As pointed out earlier, a new idle-mode MS is assigned to the current available paging set. When the current paging set reaches its optimum capacity, the following procedures are used to update the current paging set.

Case I: When the current paging set attains it optimum capacity, i.e., the number of users in this set becomes equal to L, all other active paging sets are checked for any vacancies (Note that these vacancies are created when some of the idle-mode MSs terminate their idle state.) If an active paging set with vacancies is found then that become the current paging set.
Case II: If all the active sets are full, then an empty set is added to the active paging set and this set becomes the current paging set.

When the need to page idle-mode MSs arises, the PC aggregates the paging information of all the MSs that need to be paged and that share the common PAGING_OFFSET value into one paging announcement message (Note that all these MSs belong to one paging set.) Then, the PC broadcasts this paging announcement message to all the BSs in it PG. Each of these BSs sends the MOB-PAG-ADV message over the air link during the corresponding

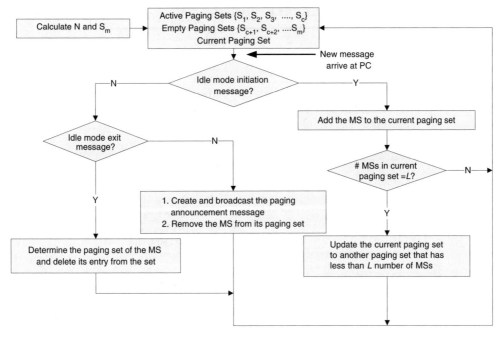

Figure 7.4   Flow chart showing the operations of the proposed paging algorithm [6]

PAGING_OFFSET. The idle mode users that are addressed in this MOB-PAG-ADV message initiate their network re-entry procedures and perform initial ranging in the nDISP available initial ranging opportunity.

7.3.3 Operation of the proposed paging algorithm

The operational steps of the proposed algorithm are illustrated in the flow chart shown in. A PC calculates the maximum number of paging sets (S_m) and the optimum number of members for each paging set (L). When a new idle mode initiation message arrives at the PC, the PC adds the new idle-mode MS to the current paging set and updates the current paging set if the size of the current paging set becomes equal to its optimum capacity L after adding the new MS. When an MS terminates its idle mode, the PC removes that MS from the corresponding paging set. Moreover, the PC carries out the paging operations to locate the desired MSs.

7.4 Performance Evaluation

We considered the parameters shown in Table 7.2 to investigate the performance of our proposed paging algorithm for a specific deployment scenario of mobile WiMAX networks. It should be noted that these parameters depend on a particular network deployment scenario and may vary from one network deployment to another. However, the proposed paging algorithm can be used for all such scenarios. The results may vary depending on the deployment scenario.

Table 7.2   Parameters used for simulation

Parameter name	Value
D_{pb} (defined in Eq (7.6))	20 ms
γ (defined in Eq (7.6))	10 ms
D_r (defined in Eq (7.6))	40 ms
D_a (defined in Eq (7.6))	50 ms
δ (defined in Eq (7.6))	500 ms
PAGING_CYCLE	1 s
K	10
r	2 km
Ω	0.6
η	0.9
P	500–5000 per sq km
T	1.5 s
λ	1/3600, 1/1800, and 1/900

For the parameters specified in Table 7.2, using $\tau = T$ in Eq (7.6), the upper limit for initial ranging delay D_{ir} is 880 ms. Therefore, the number of MSs that are trying to perform initial ranging at the same time, α, should be such that D_{ir} is less than or equal to 880 ms. As we are interested in finding the maximum possible value (optimal value) for $\alpha = R$, we determine R such that $D_{ir} = 880$ ms. R has so far not been discussed explicitly. NDISP, we investigate the correct value for R.

In our simulations to investigate the relationship between the D_{ir} and α, we consider 64 CDMA codes for the initial ranging sub-channel of the OFDMA-PHY frame [1]. However, the number of resolvable codes, C, in one initial ranging sub-channel depends on the cross-correlation properties of these codes. We consider three different values for the number of resolvable codes; 8, 10, and 12. We consider that binomial exponential back off algorithm with maximum contention window exponent of 10 is used for contention resolution. Moreover, we consider that the average number of initial ranging requests from MSs other than those transitioning from idle mode to active mode is S = 4. This is realistic based on our experience with mobile WiMAX system level simulations. Figure 7.5 shows the relationship between D_{ir}

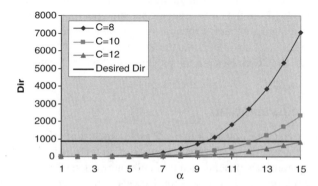

Figure 7.5   Average paging latency for different cardinality of a paging set [6]

Table 7.3 L for different scenarios

C	λ	L
8	1/3600	180000
	1/1800	90000
	1/900	45000
10	1/3600	324000
	1/1800	162000
	1/900	81000
12	1/3600	396000
	1/1800	198000
	1/900	99000

and α for different number of resolvable codes, C. As discussed earlier for the considered scenario, desired $D_{ir} = 880$ which is shown as a solid line in Figure 7.5. Therefore, the point of intersection of the solid line with the D_{ir} curves indicates the maximum number of MSs that can perform initial ranging at the same time without increasing D_{ir} beyond the desired value of 880 ms. It may be noted that this number includes MSs that are performing initial ranging during their idle mode to active mode transition and other MSs that may perform initial ranging for other purposes such as after lost connectivity or after powering up for the first time. As pointed out earlier, we considered that the average number of latter category of MSs is S = 4. Thus, the optimum number of idle-mode MSs that can perform initial ranging at the same time can be calculated by subtracting S from the value of α at the intersection points. Thus, R = 9 − 4 = 5 for C = 8, R = 13 − 4 = 9 for C = 10 and R = 15 − 4 = 11 for C = 12.

After we determine R, we calculate the maximum cardinality of paging sets, L, using (7.9). We present L for different C and λ in Table 7.3. Results show that for a particular value of C, L decreases as λ increases. Moreover, for a particular value of λ, L increases as C increases. Table 7.4 shows the number of paging sets (m) required for different deployment scenarios. We want to point out that a small number of paging sets are enough to accommodate most of the deployment scenarios.

Table 7.4 Number of paging sets required for different scenarios

ρ per sq km	N using (7.11)	m for C = 10, $\lambda = 1/3600$	m for C = 10, $\lambda = 1/1800$	m for C = 10, $\lambda = 1/900$
500	33928	1	1	1
1000	67856	1	1	1
1500	101784	1	1	2
2000	135712	1	1	2
2500	169641	1	2	3
3000	203569	1	2	3
3500	237497	1	2	3
4000	271425	1	2	4
4500	305353	1	2	4
5000	339282	2	3	4

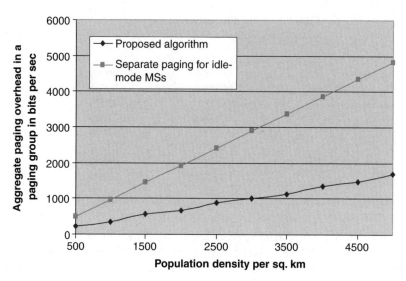

Figure 7.6   Aggregate paging overhead in a paging group for different population density [6]

Once PC determines the maximum cardinality of paging sets for a given C and λ, it creates paging sets and allocates idle MSs into these sets as discussed in Section 7.3. Then during paging it aggregates the paging information of all the MSs that belong to one paging set into one MOB-PAG-ADV message. Such paging information aggregation achieves paging overhead reduction as shown in Figure 7.6. Figure 7.6 shows that using the proposed paging algorithm aggregated paging overhead (summation of the paging overhead of all the idle-mode MSs that are paged) in a PG can be reduced up to 66% compared when the idle-mode MSs are paged individually.

7.5 Conclusion

In this chapter, a novel paging algorithm was proposed to carry out efficient paging in mobile WiMAX networks. The proposed paging algorithm strikes a very good balance between two important parameters of paging procedure: paging signaling overhead and paging latency. Using the proposed algorithm a PC groups the idle-mode MSs into different paging sets. PC aggregates the paging information of the MSs that belong to a particular paging set in one MOB-PAG-ADV message. The proposed algorithm also provides a method to decide the maximum cardinality of each paging set for different network deployment scenarios. Performance investigation shows that such aggregation of paging information of idle-mode MSs is efficient in reducing paging signaling overhead without increasing paging latency beyond a specified upper limit.

References

[1] IEEE Standard 802.16e-2005 Amendment to IEEE Standard for Local and Metropolitan Area Networks – Part 16: Air Interface for Fixed Broadband Wireless Access Systems – Physical and Medium Access Control Layers for Combined Fixed and Mobile Operation in Licensed Bands.

[2] H.S. Roh, S. Lee, and S. Lee, 'Paging Scheme for High-Speed Portable Internet (Hpi) System', The 8th International Conference on Advanced Communication Technology, ICACT 2006, pp. 1729–1732, vol. 3, Feb. 2006.

[3] 'WiMAX End-To-End Network System Architecture', Stage 3: detailed protocols and procedures, WiMAX Form, Aug. 2006.

[4] 'WiMAX Forum^{TM} Mobile System Profile v1.0.0,' WiMAX Forum^{TM}, Technical Working Group, April 2006.

[5] http://www.census.gov/population/www/documentation/twps0027.html.

[6] S. Mohanty, M. Venkatachalam, and X. Yang, 'A Novel Algorithm for Efficient Paging in Mobile WiMAX'. *Proceeding IEEE Mobile WiMAX Symposium*, Orlando, 2007.

8

All-IP Network Architecture for Mobile WiMAX

Nat Natarajan, Prakash Iyer, Muthaiah Venkatachalam,
Anand Bedekar, and Eren Gonen

8.1 Introduction

The IEEE 802.16e-2005 standard [1], approved as an amendment to the 802.16-2004 [2] standard is the physical layer (PHY) and Medium Access Control (MAC) radio specification for combined Fixed and Mobile WiMAX operation. The specifications are limited to the air interface between a client device and a base station (BS). The notion of Convergence Sub-layer (CS) in the standard permits multiplexing various types of network traffic into the MAC layer. The standard defines PHY and MAC air link primitives (between a client device and a BS) for functions required for a mobile broadband wireless access system such as network discovery/selection, network entry and exit, Quality of Service (QoS) signaling and management, security, mobility (handover management), power management modes (Active/Sleep/Idle states) and more. However, the specification of the network architecture is beyond the scope of the IEEE 802.16 standard. For instance, while MAC signaling primitives for handover procedures between a Mobile Station (MS) and Base Station (BS) are specified, network procedures between base stations to facilitate a handover are left unspecified. The WiMAX Forum (http://www.wimaxforum.org) is a non-profit industry group devoted to the global adoption and interoperability of WiMAX systems. In January 2005, the WiMAX Forum constituted the Network Working Group (NWG) to specify the complementary end-to-end interoperable network architecture as shown in Figure 8.1.

Note: This chapter was not prepared by the WiMAX Forum and does not represent an approved position of the WiMAX Forum. Certain figures and text are reproduced with permission of the WiMAX Forum, which owns the copyright in the draft specification for Mobile WiMAX. 'WiMAX', 'Mobile WiMAX', 'WiMAX Forum' and 'WiMAX Forum Certified' are trademarks of the WiMAX Forum. A condensed version of this chapter was presented at the IEEE Mobile WiMAX Symposium, Orlando, Florida, March 2007.

Mobile WiMAX   Edited by Kwang-Cheng Chen and J. Roberto B. de Marca.
© 2008 John Wiley & Sons, Ltd

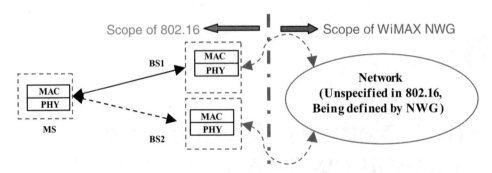

Figure 8.1   Relationship between the scopes of WiMAX NWG and IEEE 802.16 [4]

8.2 WiMAX Network Architecture Principles

The WIMAX Network specification targets an end-to-end (ETE) all-IP architecture optimized for a broad range of IP services. In this section, we examine the network design principles underlying the NWG architecture in some detail.

8.2.1 4G System Characteristics

4G+ systems pose new design challenges and requirements. Three major ETE requirements are summarized below.

1. *Limited spectrum* – Spectrum is a constrained resource which in turn requires a number of innovative techniques to boost system capacity and coverage. A combination of pico/micro/macro-cellular access nodes will imply heterogeneity in access node capabilities which must be accommodated and developed independent of specific operator types and core networks these access nodes integrate with.
2. *Support for different topologies* – With system cost (CapEx and OpEx) and scale-to-demand becoming significant considerations, it is likely that access networks will increasingly employ heterogeneity of configurations (fully flat, partially centralized and fully centralized) and interconnectivity technologies. Any system design must therefore accommodate known and future innovations in this space without requiring significant redesign.
3. *Service heterogeneity* – 4G radio access technologies should be generically capable of delivering different types of IP services – ranging from narrow-band to broadband, real-time and non-real-time, unicast, multicast and broadcast applications. When combined with the need to accommodate different levels of user mobility (fixed, nomadic, pedestrian, low and/or high speed vehicular mobility), access systems are required with advanced capabilities for radio resource management and QoS.

8.2.2 Design Principles for the WiMAX Network

The WiMAX end-to-end network architecture adopted the following design principles to fulfill the major requirements outlined above.

- *IP services optimized Radio Access Service Network (ASN)* – The wireless industry is converging toward OFDM multiple-access combined with advanced, adaptive modulation and coding schemes, Forward Error Correction schemes, flexible UL/DL channel allocation and advanced antenna systems such as MIMO and beamforming as the broadband air-interface foundation to simultaneously deliver a broad range of IP services – real-time voice, store-and-forward unicast and multicast/broadcast of multimedia and TCP-based services.

- *IP-interconnected ASNs* – The WiMAX network should enable flexibility in network design and realization. Flatter ASN architectures with functional autonomy pushed to the radio access edge are to be enabled. It should also allow IP-based interconnectivity enabling inherent redundancy and scalability while accommodating the use of a variety of low-cost and high bandwidth wireline or wireless backhaul interconnectivity options. Functional architectures permitting different physical implementations and topologies based on internet design principles should be described in the specifications.

- *Logical separation* of Access Service Networks (ASN), Connectivity Service Networks (CSN) and Application Service Networks – Support for Access Network sharing among two or more connectivity service operators should be enabled. Support for a connectivity service operator to offer broadband IP services over access service networks deployed by two or more operators should also be possible.

- *Network of ASNs* – Future mobile broadband systems must accommodate a heterogeneous mix of access technologies. Consistent interfaces based on open, interoperable industry-standard IP protocols are necessary to ensure seamless access and mobility within and across access technologies. While the access and connectivity service network lines blur, open IP-based interfaces between radio access components and core IP service functions will permit the independent evolution and migration to future mobile broadband access technologies while leveraging investments in core IP infrastructure.

8.2.3 Adopting a Functional Architecture Model

Traditional 2.5G and 3G systems have adopted physical system design principles basing access and core system designs and interoperability on extremely detailed specifications of network entities. While this approach enables precise characterization of system performance, it stifles innovation in access systems. The Internet, on the other hand, developed based on functional design principles where interoperability is based on protocols and procedures with virtually no considerations for physical instantiations thereof. The WiMAX Network WG has defined a functional architecture consisting of functional entities.

A prominent feature of the NWG specification is the extensive use of IP and IETF-standard protocols. The focus is on enabling IP access for mobile devices. Networking functionality requirements for client devices consist of just standard IP protocols like DHCP, Mobile IP and EAP protocols. IP connectivity is assumed between all interacting entities in the network. Mobile IP is used as the mechanism for redirection of the data as a mobile device moves from one ASN to another ASN, crossing IP subnet/prefix boundaries. Mobility support for mobiles that are not Mobile-IP capable is provided by the use of Proxy Mobile IP. On the network side, IP address pool management is provided through IETF standard mechanisms like DHCP. Decomposition of protocols across reference points enables interoperability while accommodating flexible implementation choices for vendors and operators.

The rest of the chapter is organized as follows: Section 8.3 provides the description of the network architecture for mobile WiMAX as well as the associated reference interfaces and access network configurations. Section 8.4 illustrates the MS session control procedures and describes state transitions from registered state to deregistered state and the impact on the network due to these transitions. Section 8.5 illustrates salient aspects of a two-tiered approach to mobility management. Section 8.6 describes principles of QoS and policy management. Section 8.7 describes an overview of network discovery and selection approaches. Section 8.8 describes how network interoperability is approached within Mobile WiMAX network deployments. Section 8.9 identifies topics not addressed here and concludes the chapter.

8.3 Network Architecture

The WiMAX network architecture can logically be represented by a Network Reference Model (NRM), which identifies key functional entities and reference points over which a network interoperability framework is defined. The WiMAX NRM (Figure 8.2) generally consists of few logical entities, namely Mobile Station, Access Service Network and Connectivity Service Network and their interactions through reference points R1–R5.

At a high level the WiMAX NRM differentiates between Network Access Providers (NAPs) and Network Service Provider (NSPs). A Network Access Provider (NAP) is a business entity that provides WiMAX radio access infrastructure that is implemented using one or more ASNs. A Network Service Provider (NSP) is a business entity that provides IP connectivity and WiMAX services to WiMAX subscribers according to some negotiated service level agreements (SLAs) through contractual agreements with one or more NAPs. The NSP may have control over the CSN.

8.3.1 Network Functional Entities

Each of the MS, ASN and CSN represents a logical grouping of functions as described in the following:

- *Mobile Station (MS)*: Generalized mobile equipment set providing wireless connectivity between a subscriber station and the WiMAX network. The subscriber station may be a host or a customer premise equipment of fixed device that supports multiple hosts.

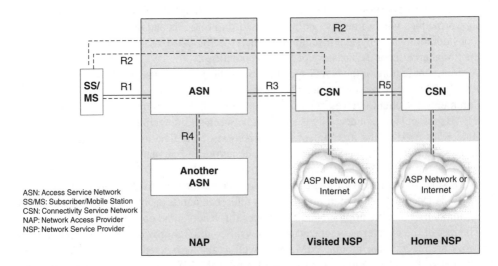

Figure 8.2   Network reference model [4]

- *Access Service Network (ASN)*: The ASN represents the point of entry for WiMAX MS equipment into a WiMAX network, and as such, must support a complete set of network functions required to provide radio access to the MS. The following functions are mandatory for all ASNs regardless of mobility support:
  - 802.16 Layer-2 (L2) connectivity with a WiMAX MS;
  - Transfer of AAA messages to WiMAX subscriber's home network for Authentication, Authorization and Accounting (AAA) for subscriber sessions;
  - Network discovery and selection of the WiMAX subscriber's preferred NSP
  - Relay functionality for establishing Layer-3 (L3) connectivity with a WiMAX MS (i.e. IP address allocation);
  - Radio Resource Management (RRM);
  - QoS and policy management;
  - ASN-CSN and ASN-ASN tunneling.

Additionally, to support mobility the ASN also supports the following functions:

  - ASN-anchored mobility
    CSN-anchored mobility and
    Paging and Location Management.
- *Connectivity Service Network (CSN)*: The CSN is defined as a set of network functions that provide IP connectivity services to the WiMAX subscriber(s). A CSN typically comprises of several network elements such as routers, AAA proxy/servers, user databases and Interworking gateway devices. A CSN provides functions such as:
  - MS IP address and endpoint parameter allocation for user sessions
  - Internet access
  - AAA services
  - Policy and Admission Control based on user subscription profiles
  - ASN-CSN tunneling support
  - WiMAX subscriber billing and inter-operator settlement
  - Inter-CSN tunneling for roaming
  - Inter-ASN mobility
  - Connectivity to WiMAX services such as IP multimedia services (IMS), location-based services, peer-to-peer services and provisioning.

Each function may require interaction among two or more functional entities. Also, each of the functions identified within a logical entity may be realized in a single physical device or distributed over multiple physical devices. All such realizations are compliant WiMAX networks as long as they meet the functional and interoperability requirements across exposed reference points.

8.3.2 Inter-ASN Reference Points (RPs)

In the WiMAX NRM, each reference point is a logical interface aggregating the functional protocols between different functional entities on either side of it. Different protocols associated with a RP may originate at and/or terminate in different functional entities across that RP. The

WiMAX NRM defines the following RPs:

- R1: Protocols and procedures between the MS and ASN. This includes the IEEE 802.16 standard specified PHY and MAC layers [1], [2] as well as L3 protocols and procedures related to control and management plane interactions and any bearer plane traffic terminating at the ASN.
- R2: Protocols and procedures between the MS and the CSN mainly associated with Authentication, Services Authorization and IP Host Configuration management. The authentication part of R2 runs between the MS and CSN in the home NSP, while the ASN and CSN in the visited NSP may partially support in the process. The IP Host Configuration management runs between the MS and the CSN (in either the home or visited NSP).
- R3: Control plane protocols as well as the IP Bearer plane between the ASN and the CSN. This RP supports AAA, policy enforcement and mobility management capabilities as well as necessary tunneling to transfer user data between the ASN and the CSN.
- R4: Control and Bearer plane procedures between ASNs such as RRM, MS mobility across ASNs and idle mode/paging. The various protocols used over R4 may originate at and/or terminate in different functional entities of the communicating ASNs. R4 serves as *the* interoperability RP across any pair of ASNs regardless of their internal configuration profiles.
- R5: Control and Bearer plane protocols needed to support roaming between the CSN operated by a home NSP and that operated by a visited NSP.

The combination of RPs R1, R2, R3 and R4 support interoperability across functions rendered in the MS, one or more ASNs and the CSN anchoring the ASN(s).

8.3.3 ASN Logical Entities

The WiMAX NRM defines the ASN as a logical aggregation of functional entities and protocols associated with the access services, creating a flexible yet interoperable framework for implementing WiMAX radio access network architecture. Using the same functional protocols, one can arrive at different mappings of different functional entities to different logical entities within an ASN to optimize the ASN design to different usage and deployment models. Different ways of mapping the set of ASN functions into physical network elements give rise to a set of ASN profiles. For instance, in one particular realization, the ASN is decomposed with functions selectively mapped into two specific physical entities – called a Base Station (BS) and an ASN Gateway (ASN-GW) as depicted in Figure 8.3. With this ASN profile, a decomposed ASN may consist of one or more Base Stations (BS) and at least one instance of an ASN Gateway (ASN-GW). The BS and ASN-GW functions can be described as follows:

Base Station (BS) is a logical entity primarily consisting of radio related functions of ASN – it represents a full instantiation of the 802.16 MAC and PHY features amended by applicable interpretations and parameters defined in WiMAX Forum System profiles. In this definition each BS represents one sector with one frequency assignment. A BS may also incorporate additional implementation specific functions such as downlink and uplink scheduler. Typically multiple BSs may be logically associated with an ASN. Also a BS may be logically connected to more than one ASN Gateway to enable load balancing and redundancy.

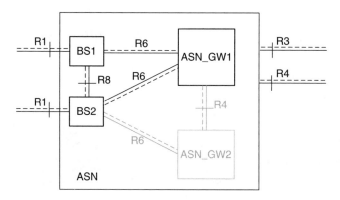

Figure 8.3 Reference model for decomposed ASN into BS and ASN GW entities [4]

ASN Gateway (ASN-GW) is a logical entity that represents an aggregation of control plane functional entities that are either paired with a corresponding function in the ASN (e.g. BS instance), a resident function in the CSN or a function in another ASN. The ASN-GW also performs bearer plane routing or bridging functions. There are other kinds of decomposition of ASN-GW functions, as discussed next.

8.3.4 Intra-ASN Reference Points

The following reference points are defined within an ASN and they are only applicable to, and normative with respect to, a decomposed ASN configuration profile.

- R6: Includes all control and bearer plane protocols between the BS and the associated ASN-GW. The control plane consists of QoS, security and mobility-related protocols such as paging and data path establishment / release and it may include radio resource management. The bearer plane represents the intra-ASN data path between the BS and ASN-GW.
- R7: Optional reference point separating decision and execution functions within a decomposed ASN-GW. If supported, R7 consists of an optional set of control plane protocols within an ASN GW for AAA and Policy coordination as well as the coordination between the two groups of functions involved over R6.
- R8: Optional reference point between Base Stations to ensure fast and seamless handover through direct and fast transfer of MAC context and data between Base Stations involved in handover of a certain MS. If supported, the handover context and related control plane messages on R8 should be consistent with protocols defined in IEEE 802.16-2005 and 802.16g specifications.

Note that R6, in combination with R4, may serve as a conduit for exchange of MAC states information between BSs that cannot interoperate over R8. Also note that if the ASN consists of multiple ASN-GWs the interactions between those ASN-GWs follow the R4 protocols, fully compatible with the Inter-ASN equivalent specifications.

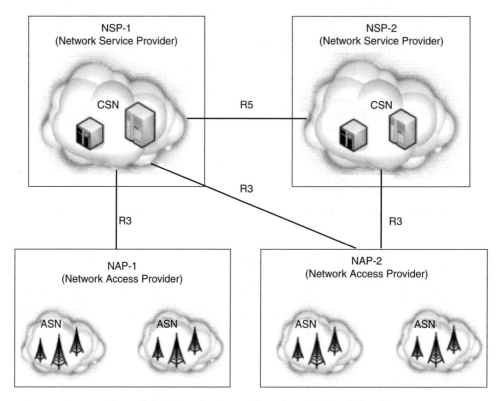

Figure 8.4   Network access and service provider relationship

8.3.5 Network Access and Service Provider Relationships

As mentioned earlier, the WiMAX network architecture differentiates the domains and business entities associated with Network Service Providers (NSPs) from those of Network Access Providers (NAPs). This is to address the fact that in 4G networks the access network consisting of one or more ASN and the core/connectivity service network CSN may not be owned by the same operator.

As shown in Figure 8.4, an NSP can provide service through different NAPs in the same or different geographic areas of coverage. Likewise, a NAP may have business relationship with, and thus is shared by, multiple network service providers. The case where multiple operators share the same ASN constitutes an example of unbundled access networking.

8.3.6 Comparison with 3G System Architectures

By adopting a functional approach to network specification, the WiMAX NWG has enabled flexibility in physical realizations with different mappings of functions into physical entities. The NWG has developed an approach to inter-operability through a model of functional entities and protocols and a consistent framework for inter-operability between implementations of the functional architecture. The interactions amongst the functional entities of a group mapped to

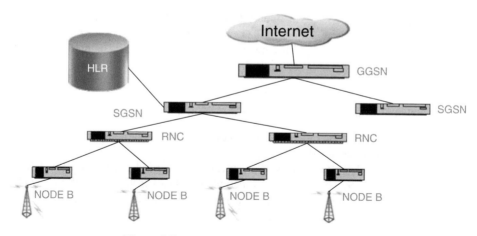

Figure 8.5   3GPP's packet services network [4]

the same physical entity need not be subject to inter-operability. Inter-operability is enabled at the granularity of groups of functional entities rather than at the level of individual functional entities.

There are some key differences in the basic approach to WiMAX network specification compared to 3G systems. Figure 8.5 shows a simplified representation of 3G system architecture. We now point out some of the key characteristics and limitations of traditional 3G architectures.

8.3.6.1 Multiple Levels of Mobility

Cellular networks have multiple levels of mobility. GSM and 3GPP-based systems handle mobility at Intra-RNC, Inter-RNC, Intra-SGSN, and Inter-SGSN levels. CDMA (and 3GPP2-based) systems handle mobility at Intra-BSC, Inter-BSC, Inter-PCF and Inter-PDSN levels. Support of mobility at several levels incurs network complexity without necessarily adding user perceived benefits. In contrast to the cellular system approach, a simplified two-tier integrated seamless mobility management framework has been adopted for mobile WiMAX networks with the following key characteristics. In Mobile WiMAX networks,

- Intra subnet/prefix mobility management is optimized to accommodate multiple physical ASN configurations, and
- Inter-subnet/prefix mobility management is complementary to the above and seamlessly accommodates mobile client and/or network-based mobility management protocols.

8.3.6.2 Latency

A network with multiple levels of mobility has associated with it multiple control points. The end result of having a packet processed with state information at each of these elements is additional delay. The latency problem becomes more pronounced when considering roaming scenarios when a larger number of control points are involved. In addition, the latency issue makes convergence of the voice, video, and data applications onto the same network more difficult. Mobile WiMAX network design has enabled flat network architectures with fewer control points and number of hops (or network elements) in the bearer path. The simplified design has significant benefit to reducing the latency.

8.3.6.3 Overlap of Functionality across Various Elements

There is a considerable degree of overlap in functionality across various elements in cellular networks, two of which are noted here.

The first example relates to security. Authentication is carried out by the BSC/RNC as well as the PDSN/GGSN. The overlap of functionality makes it very difficult to introduce new functionality in the network. A comprehensive end-to-end security model requires good control of the access and the core networks and this is not easy to do in current cellular network architectures.

The second example relates to end-to-end Quality of Service which requires correct identification of admission control points and policy enforcement points in the network. The current split of functionality across multiple elements in cellular networks makes it increasingly complex to support end-to-end Quality of Service. Quality of Service requires admission control both from the radio access link as well as at the network elements. This requires effective coordination and management of the quality of service admission framework.

The WiMAX network architecture has sought simplification in support of many of critical network functions. In the remainder of this chapter, we discuss some of the functions in more detail.

8.4 MS Session Control Procedures

The key function of the mobile WiMAX network is to comprehend the state transitions of the MS and react correspondingly. Minimizing MS power consumption and maximizing the battery life are important aspects of a mobile technology and this presents certain challenges on the network side. In mobile WiMAX, there are two power-save modes that are possible, the 'idle mode' and the 'sleep mode' as shown in Figure 8.6. The WiMAX network design has identified specific transitions and associated procedures for effective functional state transitions. These are described next.

There are macro and micro states for a MS. The key macro states are: Power OFF, 'Registered' and 'Deregistered'. The MS powers on from the Power OFF state and performs network entry to switch to the 'Registered' state. The registered state has two micro states, 'awake' and 'sleep'. In the 'awake' state, the MS has active connections and is transmitting and/or receiving data on these connections. Handovers occur during the 'awake' state. During temporary periods of inactivity, the MS may elect to switch to the 'sleep' micro-state or alternatively, the BS may elect to put the MS into this state. The sleep functions are specific to the airlink only and there is no impact on the network elements. Upon extended periods of inactivity, the MS may elect to switch to the 'deregistered' state (also known as idle mode) from the 'registered' state or the BS may elect to put the MS into this state. When entering the idle mode, the context of this MS may be stored in the network, so as to enable fast switchback from the deregistered state to the registered state.

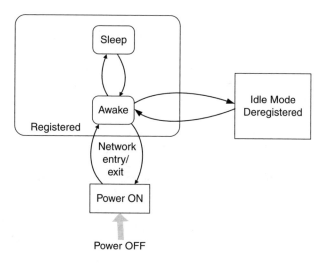

Figure 8.6 MS state transitions [4]

8.4.1 Powering ON and Network Entry

When the MS powers on, it performs a number of Layer-1 and Layer-2 procedures, including scanning for available service, etc. Once the MS is registered, it obtains a basic connection (commonly called basic CID), which can be used for further communication at L2 on the airlink. The next steps would be for the MS to establish the data CIDs at L2. Once the MS has obtained a data CID, it performs DHCP operations to obtain a home IP address. Given that the MS may move away from the subnet once it has obtained the home IP address, NWG uses the mobile IP protocol to address this situation. If the MS supports mobile IP, it can perform a Mobile IP registration with the foreign agent in the new ASN, using the home IP address it obtained via DHCP. If the MS does not support mobile IP, there could be a proxy mobile IP client in the ASN, which could perform the mobile IP registration with the foreign agent on behalf of the MS. The MS now switches to the 'Registered state' shown in Figure 8.6.

8.4.2 Registered State and Deregistered (Idle State)

In the registered state, when the MS is actively receiving traffic on the downlink and sending traffic on the uplink, it will reside in the 'awake' micro-state. When it moves from one BS to another in the registered state, it performs handovers. L2 handovers occur when the MS moves from one BS to another within the same ASN. When the MS crosses ASN boundaries, L2 handovers may occur in conjunction with L3 mobile IP handovers. Minimizing power consumption at the MS is an important aspect of wide area wireless networks. When the MS has brief periods of inactivity in the 'awake' micro state, it may negotiate sleep cycles with the BS to save power.

When the MS is inactive for longer periods of time (i.e. when the MS resides in the 'sleep' micro state for a considerable duration), the MS may negotiate to enter the deregistered state, also commonly called idle mode. The advantage of switching to this state is that the MS will be able to save more power, as it can turn off its processing for significantly longer amounts of time as compared to the sleep micro state; the tradeoff being it will require more radio resources and time to switch back to the 'awake' micro state, when there is data awaiting the MS.

In order to enter this state, the MS relinquishes all its connections, including those of the basic CID and data CIDs. This connection information could, however, be stored in an entity in the network called the 'Paging controller', due to the simple reason that storing this information can help cut down a number of steps and the associated latency, when the MS wants to re-enter the network. The address of the paging controller is passed on to the MS before the MS enters the idle state, so that the MS may report any changes in its location back to its Paging Controller.

It is to be noted that although the L2 connections are relinquished by the MS, the MS does not relinquish its home IP address. This is because the fact that the MS has gone idle should be transparent to an external entity that is trying to reach this MS using its home IP address (for example, a VoIP client or an instant messaging server).

The data path for the MS entering the idle mode, is removed till the FA. However the R3 tunnel between the FA and HA is maintained. The FA is informed of the Paging Controller and vice versa when the MS enters idle mode. The FA is required to know the address of the Paging Controller of the MS, so that it may request the MS to be paged, when data packets arrive for the MS. The Paging controller is required to know the IP address of the FA to enable state transition between idle and active modes.

8.4.3 Idle Mode Mobility

In the idle mode, the MS is expected to move. Instead of tracking the exact location of an idle mode MS at the granularity of a BS, at all times, the IEEE 802.16 specification describes procedures to only keep track of its approximate location – designated by a Paging Group (PG). Typically, a PG comprises a cluster of one or more Base Stations (BSs) and administered by a paging controller (PC). The Paging Controller to which the MS is assigned, is expected to maintain the latest PG of the idle mode MS. When an idle mode MS moves away from its current PG and enters a new PG, it detects the change in the PG by comparing the PG stored in the MS with the PG broadcast on the airlink. Upon detecting the change, it updates the Paging controller of its latest PG by means of the 'location update procedure'. A more complete description of active and idle mode transitions, including paging and location update procedures, may be found in the NWG specifications.

8.5 Mobility Management

Mobility mechanisms in the WiMAX network architecture are designed to achieve minimal handover latency and packet loss. The IEEE 802.16 standard specifies the over-the-air messaging and procedures for handovers. Taking these procedures as a foundation, the WiMAX NWG has designed network messaging and procedures on top of this base to provide the network support of mobility. For example, the IEEE 802.16 standard specifies over-the-air link information a mobile client can acquire about the neighboring base stations and report back to the serving base station during a handover. These are complemented by intra/inter ASN backbone and ASN–CSN procedures to complete a handover.

WiMAX network architecture utilizes a combination of the IETF standard Mobile IP protocol and special protocols defined by WiMAX NWG to handle mobility. The use of standard Mobile IP makes it possible to use off-the-shelf components such as Home Agent in addition to simplifying the interface to the rest of the IP world. Alongside IETF Mobile IP, WiMAX specific protocols are used to provide optimization and flexibility in handling mobility. For a given MS, the Mobile IP Home Agent (HA) resides in a CSN and one or more Foreign Agents (FA) reside in each ASN. Data for this MS are transported through the Mobile IP tunnel, which is terminated at an FA in an ASN. Once the Mobile IP tunnel is terminated at the FA, WiMAX specific protocols (i.e. Data Path Functions) take over and transport the data from the FA to the serving base station, to which the MS is attached. This design offers multiple levels of anchoring for the user data plane path during handovers. Mobile IP is used to provide a 'top' level of anchoring of the data flow for a mobile. WiMAX specific Data Path Function (DPF) provides further levels of anchoring 'below' Mobile IP.

Thus, mobility management is specified as a two-tiered solution: ASN-anchored mobility and CSN-anchored mobility. ASN-anchored mobility refers to handover event without changing the Mobile IP tunnel termination point (i.e. top-level anchoring at the FA). ASN-anchored mobility management procedures are executed on every handover and include several localized optimizations such as extending the data path from a previous serving ASN entity to the new serving ASN entity, thereby postponing or avoiding the impact of layer-3 handover delays when applicable, as shown in Figure 8.7.

Extension of the data path with the DPF minimizes the handover latency by keeping the handover communication local in the ASN operator's domain. A data path may continue to be extended upon subsequent handovers without changing the anchor point of Mobile IP. This is desirable to ensure seamless mobility in fast mobility scenarios.

On the other hand, it is desirable to keep the Mobile IP anchor point as close as possible to the MS in order to reduce the length of the data path, which otherwise may become less efficient in terms of topology. This is when CSN-anchored mobility comes into play, in order to bring back the system to a more efficient state. In most cases, a CSN-anchored handover will be executed subsequent to ASN anchored handovers when the time is right to initiate a new

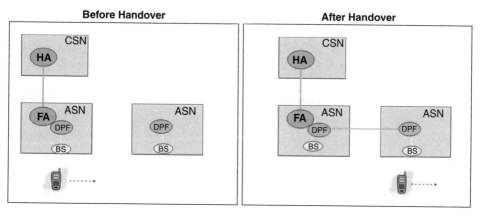

Figure 8.7 ASN Anchored Mobility: Extension of data path via data path function (DPF) during handover [4]

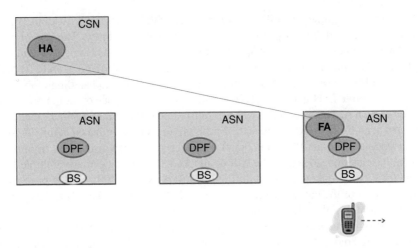

Figure 8.8   CSN anchored mobility: relocation of foreign agent to a closer location to the MS [4]

Mobile IP registration with a different FA. The choice of optimum time depends on several factors such as whether the MS is actively doing data transfer or is idle, how stationary the MS currently is, how the system load is distributed and related factors. If a CSN-anchored mobility procedure was initiated following the scenario described in Figure 8.7, the result would look like Figure 8.8.

Note that many mobile clients today such as PDAs and laptops are not Mobile-IP-capable; instead they work with simple IP. WiMAX network architecture makes sure that those clients, together with Mobile-IP capable clients, are fully able to use services offered by the network. This is achieved by employing Proxy Mobile IP Client (PMIP + Client), which acts as a proxy for the mobile client, in the network and handles the Mobile IP procedures in lieu of the mobile client in a transparent fashion.

8.6   QoS and Policy Architecture

Mobile WiMAX architecture supports Quality of Service on an end-to-end basis, including both the 802.16 air interface and the WiMAX network. We next describe briefly the basic QoS concepts and the support provided in the air interface. This is followed with a discussion on the network support for QoS.

The IEEE 802.16 specification defines a QoS framework for the air interface consisting of four key elements:

1. connection-oriented service;
2. five data delivery services at the air interface (shown in Table 8.1);
3. provisioned QoS parameters for each subscriber;
4. a policy requirement for admitting new service flow requests.

The above elements are supported respectively by UGS, rtPS, ertPS, nrtPS and BE scheduling services.

Table 8.1 Example applications and Quality of Service categories

QoS category	Example applications
UGS – Unsolicited grant service	VoIP
rtPS – Real-time packet service	Streaming audio/video
ertPS – Extended real-time packet service	Voice with activity detection (VoIP)
nrtPS – Non-real-time packet service	File transfer protocol
BE – Best-effort service	Data transfer, web browsing, etc.

In the 802.16 MAC layer, QoS is provided via service flows as illustrated in Figure 8.9. Each *service flow* is a unidirectional flow of packets (MAC Service Data Units) on a connection that is provided with a particular set of QoS parameters. Associated with each service flow is a unique identifier (32 bits long) that identifies it to both the subscriber station and base station (BS). Before providing a certain type of data service, the base station and user-terminal first establish a unidirectional logical link between the peer MACs called a *connection*. Connections are identified by a *connection identifier* (CID). Connections are of two kinds: (a) *transport connections* are used to transport user data; and (b) *management connection* are used for the purpose of transporting MAC management messages, including primary management connection, broadcast connection, initial ranging connection, etc.

The service flow parameters can be dynamically managed through MAC messages to accommodate dynamic variations in service demand. The service flow-based mechanisms apply to both DL and UL to provide QoS in both directions. Mobile WiMAX supports a wide range of data services and applications with varied QoS requirements. A QoS parameter set is associated with each service flow identifier (SFID) and it defines scheduling behavior (i.e. the transmission ordering and scheduling over the air interface) of uplink and downlink service

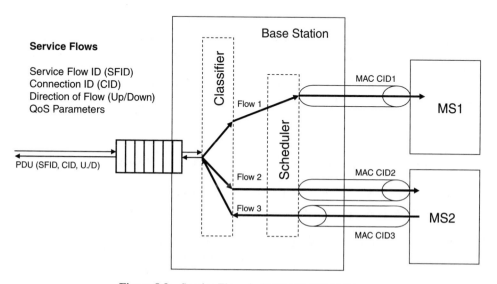

Figure 8.9 Service Flows in IEEE 802.16 MAC layer

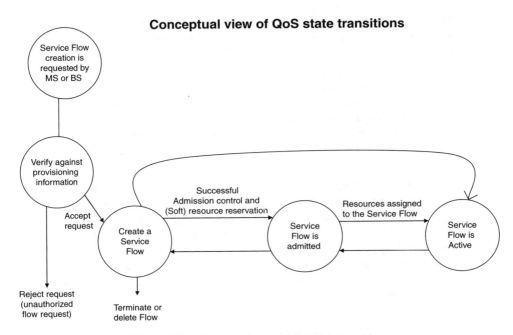

Figure 8.10   Conceptual view of QoS state transitions

flows associated with the transport connections. Priority settings are associated with packets belonging to specific flows.

Under the 802.16 specification, a subscription could be associated with a number of service flows characterized by QoS parameters. This information is presumed to be provisioned in a subscriber management system (e.g., AAA database), or a policy server. Under a static service model, the subscriber station is not allowed to change the parameters of provisioned service flows or create new service flows dynamically. Under a dynamic service model, an MS or BS may create, modify or delete service flows dynamically. A dynamic service flow request is evaluated by the BS against the provisioned information to decide whether the request could be authorized. More precisely, the following steps are envisioned in the process for dynamic service flow creation. Figure 8.10 summarizes the main QoS state transitions for service flows.

1. Permitted service flows and associated QoS parameters are provisioned for each subscriber via the management plane.
2. The provisioned service flow information is made available to an Authorization Module (AM) in the BS (or an external entity that the AM can communicate with, such as a policy server).
3. A service flow request initiated by the SS or a BS is evaluated against the provisioned information, and the service flow is created if permissible.
4. A service flow transitions from a provisioned to an admitted, and finally to an active state either due to SS or BS action. This is possible under both static and dynamic service models.

5. Transition to the admitted state requires the invocation of admission control and (soft) resource reservation, and transition to active requires actual resource assignment for the service flow.
6. A service flow can also transition from an active to an admitted to a provisioned state.
7. A dynamically created service flow may also be modified or deleted.

Differentiated services are realized in the network using IP layer QoS mechanism. IP packets are marked with *diffserv* code points at network entry, and the network elements enforce relative priority of packets based on their code points. Network resources are reserved for classes of traffic, rather than individual flows. This chapter does not address the provisioning aspects of QoS in the ASN and CSN networks. There are many possibilities for enforcing QoS in Layer 2 and Layer 3 networks, and operators may require and utilize specific Layer 2 and/or Layer 3 interfaces in ASN network elements to use known methods for mapping IP traffic onto these networks. The QoS framework established in the IEEE 802.16 specification is currently being extended to the WiMAX network architecture with QoS support on an end-to-end basis. The end-to-end path includes both the 802.16 air interface and the network and may consist of subsystems provided by more than one service provider or third-party carrier.

A rich array of procedures and functions facilitates QoS support, and includes:

- differentiated levels of QoS – coarse-grained (per user/terminal) and/or fine-grained (per service flow per user/terminal);
- QoS profile authorization;
- QoS admission control;
- policy definition and management as defined by various operators for guaranteeing QoS based on Service Level Agreements, with appropriate use standard IETF mechanisms for managing Policy Decision Functions (PDF) and Policy Enforcement Points (PEPs) on a per user or user group basis;
- policing and monitoring;
- QoS parameter mapping across different QoS domains, etc.

Based on the IEEE 802.16 specification and the WiMAX Network architecture, the following points are worth noting about the QoS functional model under development.

- The WiMAX network supports ASN-initiated creation, modification and deletion of service flows. An MS may have this capability as an option. However, the MS must respond appropriately to ASN-initiated service flow actions.
- There is a Home Policy Function (HPF) and an associated policy database. It maintains information such as general policy rules and application dependent policy rules at the H-NSP.
- A Service Flow Management (SFM) logical entity resides in the BS (ASN). The SFM entity is responsible for the creation, admission, activation, modification and deletion of 802.16 service flows. It consists of an admission control function, and associated local resource information. The admission control function is used to decide whether a new service flow can be admitted based on existing radio and other local resource usage.
- An AAA server holds the user's QoS profile and associated policy rules.

- Service Flow Authorization (SFA) logical entities reside in the ASN. In case the user QoS profile and policy rules are downloaded from an AAA server into the SFA at network entry (authentication and authorization procedures) phase, the SFA is responsible for evaluating the service request against user QoS profile.

8.7 Network Discovery and Selection

The WiMAX network allows various network access and service provider relationships (see Figure 8.4).

NSP discovery and selection procedures are typically executed when a subscriber station is used for the first time, on initial network entry, network reentry, or when the subscriber station transitions across NAP coverage areas. The network supports the following two forms of NSP selection:

- Manual selection – the MS receives the information about all available NSPs, and indicates its NSP preference by manual means.
- Automatic selection – the MS will automatically select its NSP based on the detected wireless environment and its configuration settings without explicit user intervention.

The WiMAX network accommodates a variety of subscriber usage models, including, nomadic, portable and full mobility at vehicular speeds. In a WiMAX network deployment, an MS may encounter one or more of the following three scenarios:

- an ASN that is managed by a single NSP administrative domain (aka 'NAP+NSP' deployment);
- an ASN that is managed by a NAP but shared by two or more NSPs (aka 'NAP sharing' deployment);
- a physical geographic region that is covered by two or more ASNs.

In the NAP-sharing ASN deployment scenario, connectivity for the data, control and management planes is shared directly with multiple NSPs. The network enables the MS to discover all accessible NSPs, and indicate the NSP selection during connectivity to the ASN.

The network architecture supports a set of procedures for the discovery and selection of network access and service providers. In particular, the following are supported:

- NAP Discovery – a process by which a MS discovers the available NAPs in the current wireless environment of the user.
- NSP Discovery – a process by which a MS discovers the available NSPs in the coverage area.
- NSP Enumeration and Selection – a process by which the MS chooses its most preferred NSP at the current location and a candidate set of ASNs to attach. The selection process is based on dynamic information obtained in the discovery phase as well as stored configuration information, such as user preferences, etc. at the MS.
- ASN Attachment – a process by which a MS submits its Network Access Identifier and indicates its selection by registration at its chosen ASN associated with the selected NSP.

The selection, enumeration and discovery processes for NSP may occur in parallel with the NAP discovery process. Furthermore, the discovery and selection processes need not be exhaustive, i.e., a MS may stop the discovery process and proceed with ASN attachment as soon as it discovers a NAP and NSP meeting its NSP Enumeration and Selection criteria.

8.8 Network Interoperability

The various functions performed by the ASN and CSN of a WiMAX network were outlined in Section 8.3. The functional approach to WiMAX network reference model (Figure 8.2) permits considerable freedom in the physical placement and packaging of WiMAX network functions into physical entities. Individual functions can be either co-located in a physical device or distributed across multiple physical devices. Thus different physical ASN configurations are possible. Implementation flexibility is enabled without sacrificing inter-operability of different realizations. We discuss key aspects of WiMAX network interoperability in this section.

An ASN Profile defines a particular mapping of ASN functions into specific physical entities. The WiMAX NWG has identified three ASN profiles corresponding to three possible implementations of the ASN. There is no mandate a vendor should support all three profiles. If a vendor chooses to implement any given profile, then the vendor's implementation should conform to the chosen profile. The intent of an ASN profile is to describe intra-ASN reference points (R6, R8) for intra-ASN interoperability within the context of that profile. An ASN of any profile should interoperate with an ASN of any other ASN profile through the inter-ASN reference point R4. Inter-ASN interoperability through reference point R4 is independent of the specifics of the ASN profile.

Profiles A and C define two physical entities, called Base Station (BS) and ASN Gateway (ASN-GW) and represent centralized architectures. Protocols and messages over the exposed reference points are identified. These profiles define an essentially centralized ASN architecture, and expose reference point R6. The salient attributes of Profile A are that the radio resource management control for one or more BS is in the ASN-GW. Inter-BS handover is managed in both control and data planes via the ASN-GW. The ASN anchored mobility among BSs is achieved by utilizing R6 and R4 physical connections.

Profile C has the following salient characteristics. Radio resource management control is delegated to the BS. In addition, the ASN-GW relays radio resource management messages sent from BS to BS via the R6. Thus the inter-BS handover control plane is managed by the BS while the data plane can be routed directly across BSs or via the ASN-GW.

An ASN conforming to Profile B is sometimes referred to as *flat* or *very* flat architecture. Profile B does not mandate two physical entities (such as BS and ASN-WG) to house all the network functions. The architecture tends to be fully or partially distributed among multiple instances of multiple types of physical entities. Profile B ASNs are characterized by unexposed intra-ASN interfaces. No intra-ASN interoperability is specified. However, Profile B ASNs interoperate with other ASNs of any profile type via R3 and R4 reference points. Inter-ASN anchored mobility is achieved via reference point R4. Mapping of ASN functions is not specified for Profile B ASNs. Hence there can be several different realizations based on Profile B. In a very flat architecture, all ASN functions are located within a single physical device type, namely, the Base Station. In a typical flat architecture, the ASN functionality is distributed over multiple physical nodes. The WiMAX network architecture lets the vendor choose their own

Profile B specification of physical entities, interfaces and protocols as long as they conform to R3 and R4 for inter-ASN interoperation.

NWIOT (Network Interoperability Testing) is a program under the auspices of the WiMAX Forum. Two aspects of the NWIOT program are:

- Network Conformance Testing (NCT) – In addition to the air-interface oriented tests (such as Radio Conformance Tests and Protocol Conformance Tests), tests will be conducted between an MS and WiMAX network (ASN + CSN). These tests go beyond the customary air-interface oriented tests. The goal of the tests is to ensure functional interoperability between any MS and any two or more operator networks.
- Infrastructure IOT – This ensures interoperability between network functional entities. The primary purpose is to test interoperability within ASNs (for profiles A and C compliant ASNs) and across ASNs.

By adopting and vigorously promoting testing and certification at both the air interface and the network level, mobile WiMAX systems bring the benefits of multi-vendor interoperability and its associated benefits to operators as well as end user community.

8.9 Conclusion

This chapter presented a brief description of some of the major functions of a WiMAX Network architecture currently being designed and specified in the WiMAX Forum NWG. A few critical functions integral to the initial network specification are not addressed in this chapter. These functions include radio resource management, security and inter-working with other networks (such as DSL, cable, 3GPP, 3GPP2, WLAN etc.). They are described in some detail in the NWG Release 1 specifications [3]. We plan to address the above functions in a future paper as well as support for more advanced IP services (such as, Location Based Services, Broadcast and Multicast services for multimedia traffic, IMS and commercial grade VoIP services, etc.) being developed for the next revision of the network specification.

References

[1] IEEE Standard 802.16e-2005 Amendment to IEEE Standard for Local and Metropolitan Area Networks – Part 16: Air Interface for Fixed Broadband Wireless Access Systems – Physical and Medium Access Control Layers for Combined Fixed and Mobile Operation in Licensed Bands.
[2] IEEE Standard 802.16-2004 – IEEE Standard for Local and Metropolitan Area Networks – Part 16: Air Interface for Fixed Broadband Wireless Access Systems.
[3] WiMAX Forum End-to-End Network Systems Architecture Stage 2–3, Release 1.0.0, March 2007 (www.wimaxforum.org/technology/documents).
[4] P. Iyer, N. Natarajan, M. Venkatachalam, A. Bedekar, E. Gonen, K. Etemad, and P. Taaghol, 'All-IP Network Architecture for Mobile WiMAX'.

Part Three

Multi-hop Relay Networks

9

Aggregation and Tunneling in IEEE 802.16j Multi-hop Relay Networks

Zhifeng Tao, Koon Hoo Teo, and Jinyun Zhang

9.1 Introduction

IEEE 802.16 [1] and 802.16e [2] recently have gained tremendous momentum in the industry as the primary technology for broadband wireless access (BWA). In 2006 alone, the commitment to WiMax [5] deployment pledged by service providers within United States has already totaled US$ 4 billion [6], [7].

However, due to significant loss of signal strength along the propagation path and the transmit power constraint of IEEE 802.16/16e mobile stations (MSs), the sustainable coverage area for a specific high data rate is often of limited geographical size. In addition, blocking and random fading frequently result in areas of poor reception or even dead spots within the coverage region. Conventionally, this problem has been addressed by deploying base stations (BSs) in a denser manner. However, the high manufacturing and maintenance cost associated with BSs and potential aggravation of interference, among others, render this approach less desirable. As an alternative, a relay-based approach can be pursued, wherein low cost relay stations (RSs) are introduced into the network to help extend the range, improve quality of service (QoS), boost network capacity, and eliminate dead spots, all in a cost-effective fashion [8].

In March 2006, a new task group *802.16j* was officially established [9], which attempts to amend the current IEEE 802.16e standard [2] in order to support mobile multi-hop relay (*MMR*) operation in the wireless broadband network.

The new mobile multi-hop relay-based network architecture imposes a demanding performance requirement on relay stations. These relays will functionally serve as an aggregating point on behalf of the BS for traffic collection from and distribution to the multiple MSs

Mobile WiMAX   Edited by Kwang-Cheng Chen and J. Roberto B. de Marca.
© 2008 John Wiley & Sons, Ltd

associated with them, and thus naturally incorporates a notion of 'traffic aggregation'. How-ever, the legacy concept of *connection* and the associated packet construction mechanism defined in IEEE 802.16/16e standard [1], [2], if applied on the relay link directly, may render a potential bottleneck and significantly limit the overall network capacity.

In this chapter, we describe a technique called tunneling, which has been recently proposed for the IEEE 802.16j MMR network [3], [4]. The tunneling essentially leverages concatena-tion and involves the intrinsic notion of 'aggregation', thereby combating potential efficiency deterioration on relay links, and significantly simplifying the traffic handling at intermediate relay stations.

The rest of the chapter is organized as follows. A brief description of current IEEE 802.16e OFDMA protocol [2] and its deficiency is provided in Section 9.2, aiming to supply necessary background and motivate the ensuing discussion. The new MAC tunneling mechanism is elaborated in Section 9.3, while the associated performance evaluation results are presented in Section 9.4. The chapter is completed with a brief conclusion in Section 9.5.

9.2 Background and Motivation

9.2.1 The IEEE 802.16/16e Protocol

The IEEE 802.16 protocol, initially standardized to provide fixed broadband wireless service (i.e., IEEE 802.16-2004 [1]) and subsequently amended to support mobility (i.e., IEEE 802.16e [2]), comprises a physical (PHY) and a medium access control (MAC) layer, as shown in Figure 9.1. The MAC layer of 802.16 can be further partitioned into a convergence sublayer (CS), a MAC common part sublayer (MAC CPS) and a security sublayer.

Upon the arrival of service data units (SDU) from the network layer, the convergence sublayer within MAC first classifies these SDUs and associates them with an appropriate *connection*. Given the fact that MAC CS is service-specific, multiple CS specifications are provided to interface with corresponding higher layer protocols. The MAC common part sublayer resides beneath the MAC CS, and is responsible for the core MAC functions, including medium access, connection management, QoS scheduling, and ARQ operations. The security sublayer is the lowest sublayer within MAC, and provides encryption/decryption, authentication and secure key exchange functions.

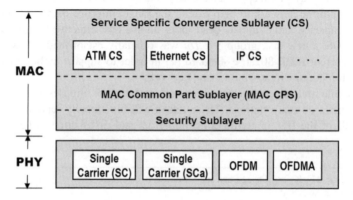

Figure 9.1   The IEEE 802.16/802.16e protocol stack

According to the IEEE 802.16 standard [1], a connection is defined as a unidirectional mapping between the base station and mobile station MAC peers for the purpose of transporting a service flow's traffic. A *service flow* here refers to as a unidirectional flow of MAC SDUs on a connection that is provided a particular QoS. A 16-bit-long connection identifier (CID) is used to uniquely identify a connection in a network, while the service flow identifier (SFID) that this CID is mapped to helps specify QoS parameters of the service flow associated with this connection.

The notion of *connection* is of prime importance to the operation of IEEE 802.16 protocol, as *all* traffic is carried on connection, even for service flows that implement connectionless protocols (e.g., IP). Data are sent over associated transport connections, while management messages are transmitted via three special connections, namely basic connection, primary management connection and secondary management connection. These three management connections are established between an MS and BS during the network entry and ranging process. The basic connection is used to transport delay-intolerant MAC management messages. Meanwhile primary and second management connections are created to carry delay-tolerant MAC management messages and standards-based (e.g., SNMP, DHCP, etc.) messages, respectively.

As shown in Figure 9.1, IEEE 802.16/16e currently supports four physical layer technologies, namely single-carrier (WirelessMAN-SC), single-carrier *type-a* (WirelessMAN-SCa), OFDM (WirelessMAN-OFDM) and OFDMA (WirelessMAN-OFDMA). Of them, WirelessMAN-SC operates on the spectrum of 10–66 GHz for line-of-sight (LOS) application, while the other three are designed for non-line-of-sight (NLOS) environment in the frequency range below 11 GHz.

As compared to other wireless standard (e.g., IEEE 802.11 [10]), one salient feature of IEEE 802.16 protocol is the close coupling between MAC and PHY modules. More specifically, the QoS scheduling and bandwidth allocation performed at MAC CPS and network topology are all heavily dependent on the specific PHY type and parameters. For instance, WirelessMAN-OFDMA, the prevailing physical layer technology defined in the current IEEE 802.16 standard for NLOS environment, can only support single hop point-to-multipoint (PMP) topology, wherein a BS has the sole authority to manage and coordinate the communications initiated by or terminated at MS that are in the direct transmission range of the BS. Regardless of whether the communication is between two MSs that are directly associated with the BS, or is between an MS and an external network entity, all the traffic has to pass through the BS.

Given the complexity of the 802.16/16e system, vast amounts of literature [11], [12], [13], [14] have been dedicated to expounding the detailed protocol mechanisms, to which the interested readers are strongly recommended to refer.

9.2.2 An Overview of the IEEE 802.16j

In order to improve capacity and extend coverage range without compromising the backward compatibility with the legacy of MSs, the IEEE 802.16j task group has been concentrating on designing a minimal set of functional enhancement and extension to support mobile multi-hop relay capability.

An envisioned topology of future IEEE 802.16j MMR network is depicted in Figure 9.2, wherein RSs help BS communicate with those MSs that are either too far away from the BS (e.g., MS3) or placed in an area where direct communication with BS experiences unsatisfactory level of services (e.g., MS1).

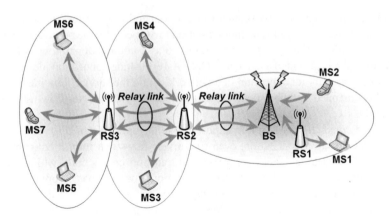

Figure 9.2   Envisioned topology of a mobile multi-hop relay network

Note that the relay operation is not confined to two-hop only. Indeed, IEEE 802.16j intends to support multi-hop relaying function, wherein a BS and multiple RSs can form a multi-level tree topology and the footprint of such an 802.16 network thus can be significantly expanded in a highly economical manner.

Thanks to its superior performance in the NLOS environment, orthogonal frequency division multiple access (OFDMA) has been adopted by IEEE 802.16j as the primary physical layer and channel access mechanism for communications in the frequency bands below 11 GHz. In OFDMA, separate sets of orthogonal tones are allocated to multiple users so that these users can engage in communication in parallel. The basic unit of resource for allocation is a *slot*, which is comprised of a number of OFDMA symbols in the time domain, and one or multiple subchannels in the frequency domain. The BS divides the timeline into contiguous *frames*, each of which further consists of a downlink (*DL*) and an uplink (*UL*) *subframe*. As illustrated in Figure 9.3, a DL subframe starts with a preamble, which helps MSs perform synchronization and channel estimation. In the OFDMA symbol that immediately follows the preamble, BS transmits a downlink MAP (DL-MAP) and an uplink MAP (UL-MAP) message to notify MSs of the corresponding resources allocated to them within the current frame in the DL and UL direction, respectively. Based upon the schedule received from the BS, each MS can determine when (i.e., OFDMA symbols) and where (i.e., subchannels) should it receive from and transmit to the BS. Receive to transmit gap (RTG) and transmit to receive gap (TTG) are inserted between two consecutive *subframes*, in order to give mobile stations sufficient time to switch from the reception mode to transmission mode or vice versa, and to avoid DL and UL subframe interferences in a multi-cell deployment.

9.2.3 Challenges in IEEE 802.16j

The ramification of introducing RSs into an 802.16e network is far and wide. Certainly, numerous new research challenges related to relay system, such as frame structure for multi-hop network, path selection and management, network entry, to just name a few, have to be properly addressed, before a 16j MMR system can actually be deployed in the field.

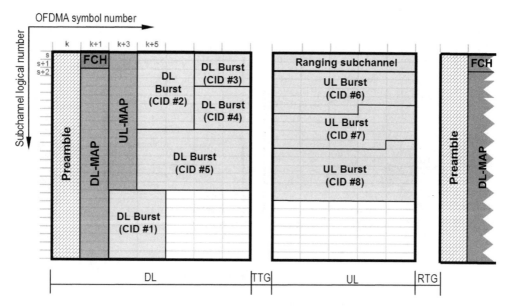

Figure 9.3   Legacy 802.16e OFDMA frame structure

One natural consequence of relay is traffic 'aggregation' on the relay links between a BS and an RS, or between a pair of RSs, as vividly illustrated in Figure 9.2. As RSs have to carry traffic originating from and destined for a multitude of MSs, the notion of *aggregation* is simply inherent. Generally speaking, the closer an RS is to the BS in the network topology, the higher the degree of aggregation it has to endure.

Given the critical role RSs play in an MMR network, it is imperative to strengthen the performance of RSs so that they can deliver high capacity, low delay, good reliability and meet the stringent requirement posed by traffic aggregation.

Unfortunately, in the current point-to-multipoint (PMP) network topology, many key functions (e.g., resource allocation, QoS management, etc.) are performed by BS on a per connection basis, and all the MSs are treated more or less equally. This is a sensible design for a single-hop PMP network, but by no means the most efficient one for a multi-hop tree topology.

For instance, it has already been shown in [16], [17] that as the number of connections increases, the overhead entailed thereby can cost as much as over 50% MAC efficiency degradation. The primary culprits of the performance deterioration are twofold:

1. *Data plane*: The resource allocated to each individual connection sometimes cannot be fully consumed, because the actual data bits do not map exactly to the assigned OFDMA symbols and subchannels. Due to this mapping inefficiency, variable number of padding bits will be appended at the end of the last data bit, leading to resources waste as depicted in Figure 9.4.
2. *Management plane*: In the current standard, each connection is handled individually in the management plane. For example, one downlink MAP information element (DL MAP IE) in IEEE 802.16 normally contains the schedule of one connection only. This design can become highly cumbersome, complicated and inefficient, when a very large number of connections with diverse traffic demands have to be dealt with.

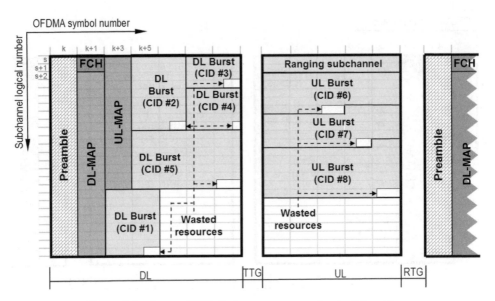

Figure 9.4   Current OFDMA frame structure and potential problem

The aforementioned problem is exacerbated when the current IEEE 802.16e OFDMA protocol is applied on the relay link between BS and an RS, or between a pair of RSs, as significant number of connections will be aggregated therein.

To curb the waste and improve the performance of the current IEEE 802.16e protocol on relay links, a tunnel mechanism was proposed in [3], which directly addresses the problem in the data and management planes.

To facilitate the discussion that follows, several terminologies are defined here. A station is called an *access station*, if it is at the point of direct access into the network for a given MS or RS [15]. Note that an access station can be a BS or an RS. An RS is a *subordinate RS* of another station, if that station serves as the access station for that RS. The wireless link that directly connects an access station with its subordinate RS is defined as a *relay link*, while the one that connects an MS with its access station is called an *access link*.

9.3 Tunneling and Aggregation

9.3.1 Definition of a Tunnel

To support aggregation, a new type of connection called a tunnel connection has been introduced. A tunnel connection is a unidirectional connection between the BS and an RS established to aggregate management or transport traffic. However, note that a specific tunnel connection can carry MPDUs from either management or transport connections, but not both. Separate tunnels may be created for different service types in order to provide a QoS treatment with finer granularity.

Note that it is not required that all connections must pass through a tunnel connection. MPDUs from connections that do not pass through a tunnel are forwarded based on the CID of the connection. Tunnel connections may pass through one or more intermediate RSs.

To uniquely identify the tunnel and its QoS parameters, a service flow identifier (SFID) is assigned to a tunnel. Moreover, a tunnel connection identifier (T-CID) and a management tunnel CID (MT-CID) are also assigned to a transport tunnel and management tunnel, respectively. The T-CID and MT-CID are drawn from the same space as CIDs associated with individual service flow. MT-CID is assigned during the network entry process using REG-RSP or SBC-RSP messages, while transport tunnel connections are created using the dynamic service addition (DSA) messages.

Figures 9.5 and 9.6 provide an illustration of management and transport tunneling, respectively. More specifically, as depicted in Figure 9.5, three management tunnels, namely basic management tunnel, primary management tunnel and secondary management tunnel, have been established between an access RS (e.g., RS1 and RS2) and the BS. Not only the corresponding management messages collected from the associated MSs (e.g., MS1 and MS2), but also those generated by RS (e.g., RS2) are transmitted via each such tunnel. Note that although the basic and primary management connection initiated at access RS are included in the corresponding management tunnel in Figure 9.5, it is not necessarily so in general. Moreover, due to the traffic nature, no secondary management connection is expected to be set up between an RS and the BS. In Figure 9.6, a tunnel for each QoS level is created between the access RS and BS to aggregate the data of those individual transport connections with same QoS requirement.

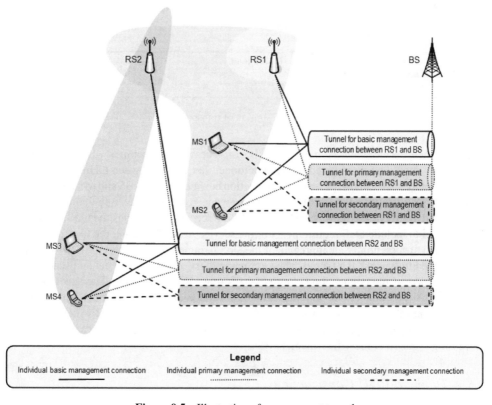

Figure 9.5 Illustration of management tunnel

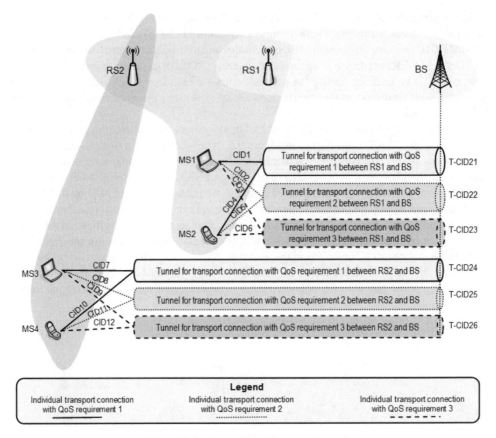

Figure 9.6   Illustration of transport tunnel

When a new service flow is created, the BS or access RS determines whether or not the service flow should be transported by using a tunnel that exists between them. If the service flow does not traverse a tunnel, a separate connection between the MR-BS and MS is established to carry MPDUs from the flow. If the service flow wants to traverse the tunnel, the BS assigns an SFID and T-CID to the flow. Moreover, the BS or access RS modifies the QoS parameters of the tunnel to include QoS requirements of the service flow. Therefore, the QoS parameters of the tunnel are essentially an aggregate of the QoS parameters of the service flows that have been assigned to traverse the tunnel. The QoS parameters of both the tunnel and service flow are sent as part of the connection setup messages (e.g., DSA message). The access RS and BS use the QoS parameters of both the individual service flow and the tunnel in performing admission control and resource reservation. The intermediate RSs traversed by the tunnel may ignore the QoS parameters of the individual service flows. Apparently, this can simplify the processing required at these intermediate RSs.

9.3.2 Tunnel MPDU Construction

Two different MPDU construction methods, namely *encapsulation mode* and *burst mode*, have been designed in order to provide both efficiency and flexibility in relay system.

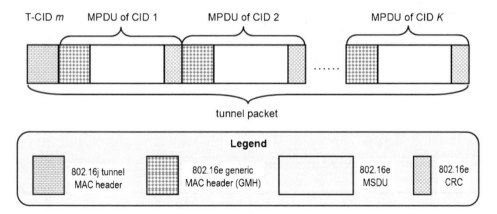

Figure 9.7   Illustration of the encapsulation mode

Encapsulation Mode

In the encapsulation mode, MPDUs of those individual connections that traverse the same tunnel can be concatenated and attached with an additional 802.16j tunnel MAC header, which eventually form a tunnel packet. As indicated in Figure 9.7, the CID of each individual connection is included in the 802.16e generic MAC header (GMH) of each constituent MPDU, while the 802.16j tunnel MAC header contains the tunnel CID (T-CID).

The station at the ingress of the tunnel is responsible for encapsulating the MPDUs into tunnel packets, and the station at the egress of the tunnel is responsible for removing the tunnel header and forwarding the encapsulated MPDUs based on their individual CIDs. Stations through which a tunnel traverses simply forward the tunnel packets based on the T-CID in the tunnel header. When tunnel packets are transmitted with tunnel headers, the T-CID may appear in the map IE that describes the allocation in which the burst is transmitted. Alternately, the T-CID can be omitted from map IEs and the RSs can determine the T-CID of a packet by parsing the tunnel header. In this mode, multiple tunnel packets, potentially from different tunnels traversing an RS can be concatenated into a single PHY burst.

The 802.16j tunnel MAC header can assume the format of the generic MAC header defined in IEEE 802.16e. As an alternative, a relay MAC header recently proposed in IEEE 802.16j [18] can be used as tunnel MAC header. Figure 9.8 portrays the format of this new relay MAC header, and further highlights those fields that should be interpreted differently from the legacy 802.16e generic MAC header. More specifically, a bit in the header called the *Relay MAC Indicator* is dedicated to distinguish this relay MAC header from the legacy 802.16e header. Other reserved bits in this new header can be further defined to support a wide variety of new functions specifically introduced in 802.16j MMR network. For example, the current *length field* (LEN) can be extended leftward to be 12 bits or longer, which then becomes able to represent an MPDU size of even wider range. Given the fact that multiple legacy MPDUs will be aggregated to produce a tunnel packet on relay link, the extension of the length field could be imperative. Moreover, one bit can be taken from the 11 reserved bits in the relay MAC header to indicate whether or not this MPDU is constructed using the encapsulation mode.

One key design constraint for IEEE 802.16j MMR network is backward compatibility with the legacy MS, which essentially prohibits any change on access link. Since the tunnel header

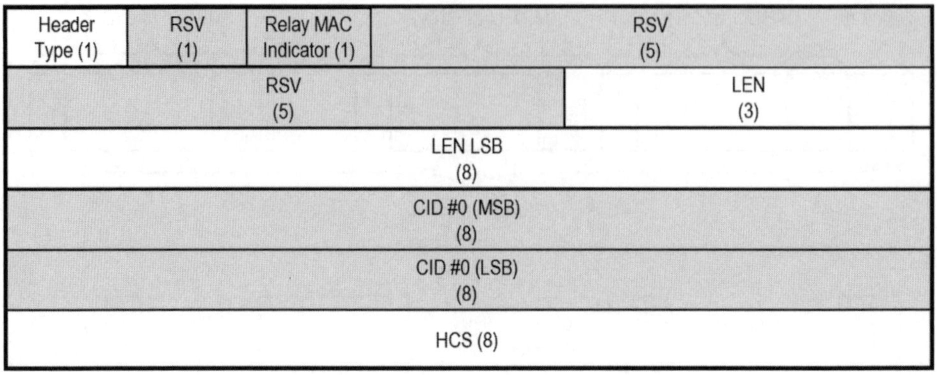

Header Type (1)	RSV (1)	Relay MAC Indicator (1)	RSV (5)		
RSV (5)				LEN (3)	
LEN LSB (8)					
CID #0 (MSB) (8)					
CID #0 (LSB) (8)					
HCS (8)					

Figure 9.8 Newly proposed relay MAC header

is attached and removed by the BS and access RS, it is transparent to the MS, the backward compatibility can still be maintained.

Burst Mode
Alternatively, MPDUs belonging to the connections that travel through the same tunnel can be directly concatenated into a PHY burst, without attaching a tunnel header. This burst mode can be implemented based upon the concatenation operation that was initially defined in 802.16 [1] and further refined in [19] for 802.16j. In essence, IEEE 802.16 concatenation is equivalent to an aggregation at MPDU level.

As compared to the encapsulation mode, the burst mode can save bandwidth and reduce MPDU processing time. Nonetheless, there is no field where T-CID can be placed in the resultant burst, since no additional tunnel header is attached. As a solution, the T-CID of the tunnel can be specified in the MAP information element (IE) within the associated MAP message to identify the tunnel on which the PHY burst should be transmitted.

Regarding the backward compatibility requirement, the burst mode does not make any compromise either, since the MAP IE that contains T-CID is included, consumed and removed by BS and access RS only.

9.3.3 Tunnel-in-Tunnel

It is worthwhile noting that tunnel connections should be used to transport MPDUs from one or more connections that terminate in the *BS* and pass through the RS. This definition in fact can accommodate a wide variety of possible tunneling scenarios, as further discussed in [20], [21]. One case that is especially interesting is called tunnel-in-tunnel, which essentially is a recursive tunnel encapsulation. More specifically, since a tunnel is composed of connections, while a tunnel itself is yet another connection, it is perfectly legitimate to construct a tunnel using a mix of individual connections and tunnels.

Figure 9.9 provides an illustration of tunnel-in-tunnel operation, wherein individual connection with CID1 and CID2 are first aggregated at RS3 into a tunnel connection with T-CID11,

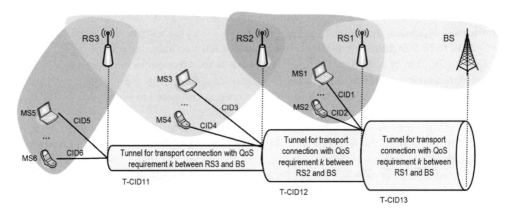

Figure 9.9 Illustration of the tunnel-in-tunnel operation

which is further aggregated with individual connection CID3 and CID4 to form a bigger tunnel, T-CID12.

Apparently, the tunnel encapsulation operation has a direct impact on MPDU construction, as explained in [21] and illustrated in Figure 9.10. For an MMR network topology depicted in Figure 9.9, Figure 9.10 shows that a relay MAC PDU for RS1 is created by BS and transmitted on relay link to RS1. After RS1 receives this relay MAC PDU, it extracts the MPDUs for MS1 and MS2 and blasts them to MS1 and MS2 on the access link. Meanwhile, RS1 retrieves relay MAC PDU for RS2 and forwards it on the relay link to RS2. This process repeats until the very end of the relay path.

Note that although the tunnel-in-tunnel operation is supported by the current definition of tunneling, its actual usage should be justified on a case-by-case basis.

9.3.4 Traffic Prioritization with Tunneling

In order to support scheduling service for various uplink service flows, eight levels of *traffic priority* have been introduced in IEEE 802.16e [2] for each type of service, namely real-time

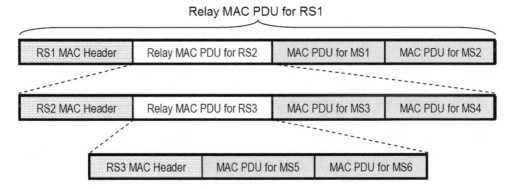

Figure 9.10 Illustration of MPDU construction for the tunnel-in-tunnel operation

polling service (rtPS), non-real-time polling service (nrtPS), best effort (BE) service. This traffic priority, ranging from 0 (i.e., lowest priority) to 7 (i.e., highest priority), can be used by scheduler to further prioritize different data flows that fall into the same service type.

One application of such finer traffic prioritization is the voice service. For instance, when network resource is under constraint, the E911 service mandated by the federal government in recent years has to be treated with more preference as compared to the regular residential phone calls, although both of them are of the same service type.

In an MMR network where tunnels are established for traffic of the same service type, 3 bits can be taken from the reserved bit field in the relay MAC header defined in Figure 9.8 to implement the *traffic prioritization* function [22], especially if scheduling is performed in a distributed manner at each RS. More specifically, each access RS can follow the procedure described in Section 0 to create multiple tunnels, one of which is for a particular service type. MPDUs from individual MS data flows of identical traffic priority and same service type are aggregated into one relay MAC PDU by the access RS. The access RS assigns a traffic priority value to the resultant relay MAC PDU, which is transported in the corresponding tunnel. When this relay MPDU traverses the tunnel, each and every intermediate RS can derive the service type based upon the tunnel CID, learn the priority information instantaneously from the 3-bit traffic priority field in the relay MAC PDU header, and thus provide buffering and scheduling to each relay MAC PDU accordingly.

9.4 Performance Evaluation

As clearly revealed in previous sections, the introduction of tunneling into relay network is primarily motivated by its efficiency improvement effect. For traffic forwarding and QoS management, if individual CIDs are used in the intermediate RSs, the corresponding routing table and QoS mapping table would potentially contain a tremendous number of entries. However, if all individual connections can be aggregated into tunnel connection, and processed accordingly, the size of these tables at intermediate RSs would shrink significantly. To further quantify the benefit of traffic aggregation, the performance of tunneling is evaluated in this section [23].

In order to compare the performance of the legacy 802.16e approach that is based upon individual connection with that of the tunneling mechanism, MAC protocol efficiency *Eff* defined in Equation (9.1) will be used as primary metrics.

$$Eff = \frac{B}{T} \times \frac{1}{R} \times 100\% \qquad (9.1)$$

where B, T, and R denote the total number of MSDU bits, time to transmit these bits, and the actual physical layer transmission rate, respectively.

To concentrate on the proposed schemes, an error-free channel condition is assumed. The network under investigation only includes one BS and one RS, and all the connections are established on the relay link. Moreover, suppose each connection has infinite traffic supply, and thus always has packets to transmit during the slots assigned to it. Other key PHY and MAC parameters used in evaluation are summarized in Table 9.1.

First of all, the size of MAP message is depicted in Figure 9.11 as a function of number of connections for both the legacy IEEE 802.16e and the tunneling scheme proposed in 802.16j.

Table 9.1   Key PHY and MAC parameters

DL permutation *PUSC*	UL permutation *PUSC*		FFT size *1024*	Channel bandwidth *20 MHz*
MCS (data) *64 QAM ³/₄*	MCS (MAP and preamble) *QPSK ¹/₂*		Cyclic prefix (G) *1/32*	Sampling factor (n) *28/25*
Period for UCD/DCD *Every 10 frames*	Frame duration *20 ms*		Number of UL BW/RNG subchannels *6*	
	RTG *10 μs*	TTG *10 μs*		

DL permutation	PUSC
UL permutation	PUSC
FFT size	1024
Channel bandwidth	20 MHz
MCS (data)	64 QAM ³/₄
MCS (MAP and preamble)	QPSK ¹/₂
Cyclic prefix (G)	1/32
Sampling factor (n)	28/25
Period for UCD/DCD Every	10 frames
Frame duration	20 ms
Number of UL BW/RNG subchannels	6
RTG	10 μs
TTG	10 μs

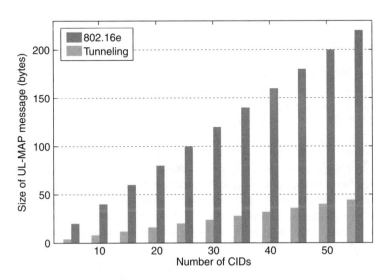

Figure 9.11   Size of MAP message

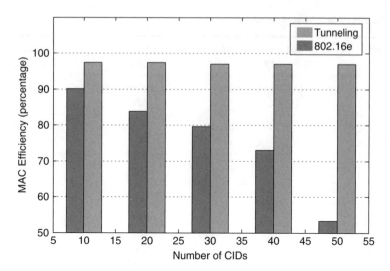

Figure 9.12   MAC efficiency (MPDU = 1000 bytes) [23]

In the simulation, it is assumed that each tunnel aggregates five individual connections. Evidently, the adoption of aggregation always results in smaller management plane overhead, as fewer MAP information elements would be needed for tunneling operation. In addition, the overhead reduction becomes more pronounced, as the number of parallel connections grows. For example, the saving achieved can reach as high as 80%, when the relay station has to simultaneously support 55 connections or more. Although the actual MAP message size is highly dependent on the number of individual connections to be aggregated in a tunnel, the results reported in Figure 9.11 provide a good indication of how much efficiency improvement the tunneling scheme can achieve.

Figure 9.12 further illustrates the relation between MAC efficiency and number of connections. It can be observed in Figure 9.12 that a tunnel in burst mode can sustain a stable MAC efficiency, while the legacy protocol yields a serious degradation in efficiency as the number of connections grows. This highly desirable feature of insensibility is particularly indispensable for the 802.16j MMR network, as the relay links will experience many increases in the number of connections.

Figures 9.13 and 9.14 portray the same relation as, but focus on MPDUs of smaller size (i.e., 500 and 100 bytes). A simple comparison between these three figures suggests that both the MAC efficiency and the corresponding improvement enabled by the proposed tunneling heavily rely on the packet size. A closer examination of the performance results reveals that as the MPDU size decreases, it becomes more likely to occupy most of the allocated slots by fitting in small packets, thereby lowering the waste caused by *mapping inefficiency* to a lesser but still appreciable level.

All the results discussed so far are obtained assuming tunnel burst mode is adopted on the relay link. As tunneling can also be implemented by using packet mode, the efficiency of these two modes is further compared in Figure 9.15. More specifically, Figure 9.15 depicts the ratio of total MSDU size to total tunnel MPDU size, as the length of the each individual MSDU grows

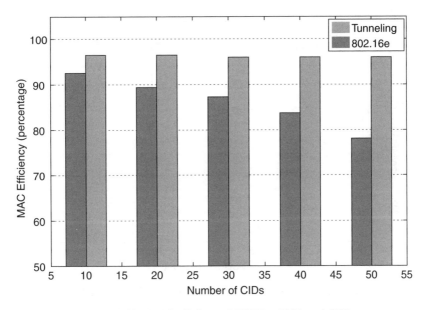

Figure 9.13   MAC efficiency (MPDU = 500 bytes) [23]

from 100 bytes to 1500 bytes, which covers the range of packet size for typical application traffic. Each tunnel contains five individual connections in the simulation. As expected, packet mode would entail higher protocol overhead than the burst mode, as an additional tunnel header will be appended in front of a number of concatenated MPDUs. Nevertheless, given the small size of the tunnel header, the difference in efficiency for these two modes is not pronounced.

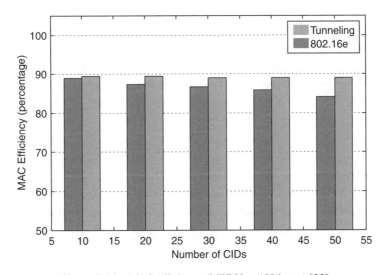

Figure 9.14   MAC efficiency (MPDU = 100 bytes) [23]

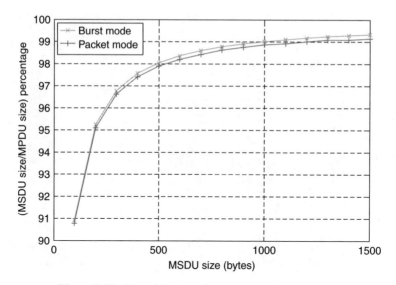

Figure 9.15   Tunnel burst mode versus tunnel packet mode

9.5 Conclusion

In this chapter, we have explained the dire need for protocol efficiency improvement in the next generation mobile multi-hop relay networks. As a solution, we have described a new scheme called *tunneling*, which is designed specifically to leverage the inherent notion of 'aggregation' in relay links and has recently been introduced into the IEEE 802.16j standard for mobile multi-hop relay network. The tunneling mechanism can significantly simplify the routing, QoS management and RS handover at the intermediate RSs along the relay path. Moreover, the tunnel scheme, regardless of being implemented in encapsulation mode or burst mode, can still maintain backward compatibility with the legacy IEEE 802.16e mobile stations, which is a requirement for any scheme to be accepted by IEEE 802.16j.

The performance evaluation results further confirm that the tunneling mechanism can help the system sustain a stable protocol efficiency, and avoid the dismal degradation that plagues the legacy of IEEE 802.16e, thereby providing a comprehensive solution to improve the efficiency in application on relay links.

References

[1] 'IEEE Standard for Local and Metropolitan Area Networks – Part 16: Air Interface for Fixed Broadband Wireless Access Systems', *IEEE Computer Society and the IEEE Microwave Theory and Techniques Society*, October 2004.

[2] 'IEEE Standard for Local and Metropolitan Area Networks – Part 16: Air Interface for Fixed Broadband Wireless Access Systems, Amendment 2: Physical and Medium Access Control Layers for Combined Fixed and Mobile Operation in Licensed Bands', *IEEE Computer Society and the IEEE Microwave Theory and Techniques Society*, February 2006.

[3] J. Sydir et al., 'Proposal on Addresses, Identifiers and Types of Connections for 802.16j', *IEEE 802.16j-06/274r6*, http://www.ieee802.org/16/relay/index.html, January 2007.

[4]   'Baseline Document for Draft Standard for Local and Metropolitan Area Networks Part 16: Air Interface for Fixed and Mobile Broadband Wireless Access Systems – Multi-hop Relay Specification', IEEE 802.16j-06/026r4, http://www.ieee802.org/16/relay/index.html, May 2007.

[5]   'Worldwide Interoperability for Microwave Access', http://www.wimaxforum.org/home/.

[6]   'Sprint Nextel Announces 4G Wireless Broadband Initiative with Intel, Motorola and Samsung', http://www2.sprint.com/mr/news_dtl.do?id=12960.

[7]   'Clearwire Secures $900M in Financing Round Led by Intel Capital and Announces the Sale of NextNet Wireless to Motorola', http://www.clearwire.com/company/news/07_05_06.php.

[8]   R. Pabst, B.H. Walke, D.C. Schultz, P. Herhold, H. Yanikomeroglu, S. Mukherjee et al. 'Relay-Based Deployment Concepts for Wireless and Mobile Broadband Radio', *IEEE Communications Magazine*, **42**, September 2004, 80–89.

[9]   'IEEE 802.16j Mobile Multi-hop Relay Project Authorization Request (PAR)', *Official IEEE 802.16j Website:* http://standards.ieee.org/board/nes/projects/802-16j.pdf, March 2006.

[10] 'Part 11: Wireless LAN Medium Access Control (MAC) and Physical Layer (PHY) Specifications', *ANSI/IEEE Std 802.11,1999 Edition*, 1999.

[11] C. Eklund, R.B. Marks, E.L. Stanwood, and S. Wang, 'IEEE standard 802.16: A Technical Overview of the WirelessMan Air Interface for Broadband Wireless Access', *IEEE Communications Magazine*, **40**, June 2002, 98–107.

[12] I. Koffman and V. Roman, 'Broadband Wireless Access Solutions Based on OFDM Access in IEEE 802.16', *IEEE Communications Magazine*, **40**, April 2002, 96–103.

[13] A. Ghosh, D.R. Wolter, J.G. Andrews, and R. Chen, 'Broadband Wireless Access with WiMax/802.16: Current Performance Benchmarks and Future Potential', *IEEE Communications Magazine*, **43**, February 2005, 129–136.

[14] C. Eklund, R.B. Marks, S. Ponnuswamy, K.L. Stanwood, and N.J. V. Waes, 'WirelessMAN: Inside the IEEE 802.16 Standard for Wireless Metropolitan Area Networks', May 2006.

[15] 'Harmonized Definitions and Terminology for 802.16j Mobile Multi-hop Relay', *IEEE 802.16j 06/014r1*, October 2006, http://www.ieee802.org/16/relay/docs/80216j-06_014r1.pdf.

[16] S. Redana, M. Lott, and A. Capone, 'Performance Evaluation of Point-to-multipoint (PMP) and Mesh Air-Interface in IEEE Standard 802.16a', in *Proceedings of IEEE 60th Vehicular Technology Conference (VTC 2004-Fall)*, Los Angeles, CA, September 2005.

[17] A.E. Xhafa, S. Kangude, and X. Lu, 'MAC Performance of IEEE 802.16e', in *Proceedings of IEEE 62nd Vehicular Technology Conference (VTC 2005-Fall)*, Dallas, Texas, September 2005.

[18] J. Tao, et al., 'Proposal for Relay MAC PDU Format', *IEEE 802.16j-07/198r8* http://www.ieee802.org/16/relay/, May 2007.

[19] Y. Zhou, et al., 'MAC PDU Concatenation in RS', *IEEE 802.16j-07/009r2* http://www.ieee802.org/16/relay/, January 2007.

[20] A. Chindapol, et al., 'Connection Management and Relay Path Configuration', *IEEE 802.16j-07/241r5* http://www.ieee802.org/16/relay/, March 2007.

[21] H. Qu, S. Cai, M. Chion, Y. Liu and Y. Chen, 'A Proposal for Construction and Transmission of Relay MAC PDU in 16j Network', *IEEE 802.16j-07/267r4* http://www.ieee802.org/16/relay/, May 2007.

[22] H. Zhang, et al., 'QoS Control Scheme for Data Forwarding in 802.16j', *IEEE 802.16j-07/309r4* http://www.ieee802.org/16/relay/, May 2007.

[23] Z. Tao, K.H. Teo and J. Zhang, 'Aggregation and Concatenation in IEEE 802.16j Mobile Multi-hop Relay (MMR) Networks', IEEE Mobile WiMAX Symposium, March 2007 (Mobile WiMAX 2007), Orlando, FL, March 2007.

10

Resource Scheduling with Directional Antennas for Multi-hop Relay Networks in a Manhattan-like Environment

Shiang-Jiun Lin, Wern-Ho Sheen, I-Kang Fu, and Chia-Chi Huang

10.1 Introduction

Next generation mobile cellular systems are envisioned to provide high-data-rate multi-media services to users any time, anywhere at an affordable cost [1]. The systems are expected to operate at a higher frequency than 2 GHz because of the very crowded usage of the frequency bands below and around 2 GHz by the current 2G, 3G and other systems. With a fixed transmit power, both high data rate and high operating frequency indicate a smaller cell coverage for a given error rate performance. In other words, it requires more base stations (BSs) to cover a given area in the future cellular system than in the traditional cellular network, and that increases the system cost enormously.

Multi-hop relay (MR) cell architecture is a cost-effective cell architecture for the next generation wireless communication systems [2–5]. The basic concept is to set up low-cost relay stations (RSs) in a cell of the traditional cellular network to relay information from a BS to mobile stations (MSs), and vice versa. Therefore, the cell coverage can be kept the same without deploying extra BSs even though the data rate requirement and the operating spectrum are higher in the new mobile systems. In addition, since the transmit antennas of both ends of a relay link (BS ↔ RS) and/or an access link (BS, RS ↔ MS) are closer to each other, per user throughput can be improved as well [6–9]. Furthermore, the RS has no direct backhaul connection to the network, so it is much simpler and easier to deploy than the BS [2–4].

Mobile WiMAX   Edited by Kwang-Cheng Chen and J. Roberto B. de Marca.
© 2008 John Wiley & Sons, Ltd

MR networks can be classified into different categories depending on their characteristics. For example, a wireless MR network can be classified as homogeneous relaying or heterogeneous relaying [3]. Homogeneous relaying uses the same radio access technology for all its connections including relay links (BS ↔ RS) and access links (BS, RS ↔ MS), while heterogeneous relaying uses different access technologies for these links. Another example of classification is that RSs can be categorized as either amplify-and-forward (AF) or decode-and-forward (DF) RSs according to the forwarding strategy [3]. An AF-RS simply scales and retransmits the received analog signal, while a DF-RS decodes the received signal and re-encodes before re-transmission. The AF-RS has a simpler signal processing and a lower transmission delay but suffers from noise enhancement, as compared with the DF-RS. Finally, RSs can also be distinguished as fixed/nomadic RSs and mobile RSs depending on their mobility [4]. In this chapter, we are only concerned with a homogeneous MR network with fixed RSs in the decode-and-forward mode.

In recent years, more and more research has been devoted to the design of MR networks [6–15]. In [16], an MR network was adopted as an amendment to the IEEE 802.16e standard [17, 18] for cell coverage extension, user throughput improvement and/or system capacity enhancement. In [7–10], a scenario of MR network, Scenario 1, was proposed for the Manhattan-like environment, where four RSs are deployed outside of the BS's coverage in a cell in order to extend the cell coverage, which is illustrated in Figure 10.1. In order to achieve frequency reuse factor of 1, the multi-cell setup and the transmission frame structure as illustrated in Figure 10.2 and Figure 10.3 are proposed in [10]. Through proper coordination between adjacent cells, when BSs in cell group A serve RSs and MSs in the phase 1, BSs in cell group B keep

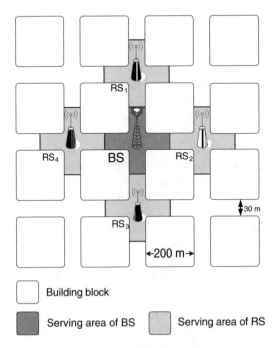

Figure 10.1   MR cell architecture in a Manhattan-like environment; Scenario 1

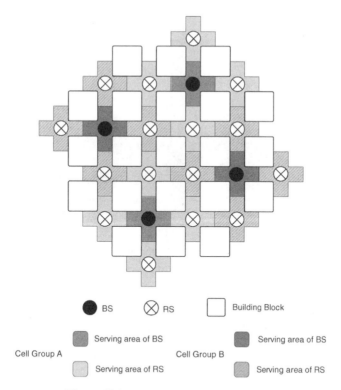

Figure 10.2 Multi-cell setup of Scenario 1

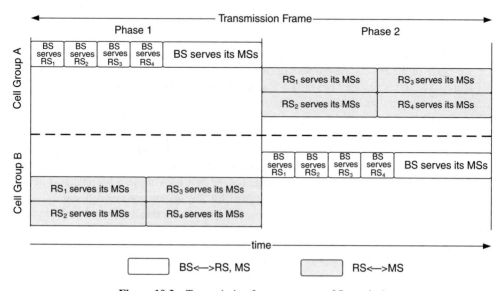

Figure 10.3 Transmission frame structure of Scenario 1

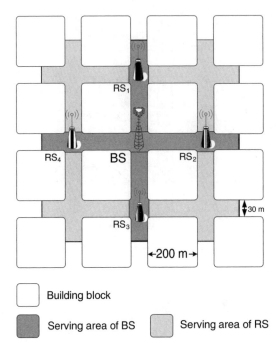

Figure 10.4   MR cell architectures in a Manhattan-like environment; Scenario 2

silent and RSs serve their MSs. Similarly, the BSs in cell group B and the RSs in cell group A become active in phase 2. In addition, by utilizing the spatial isolation which is inherent in the Manhattan-like environment, two relay stations (e.g., RS_1 and RS_2, or RS_3 and RS_4) within the same cell can be scheduled to be active simultaneously. However, as shown in Figure 10.3, since there are always some inactive BSs in every transmission phase, the radio resource is not fully utilized in this design.

Another MR network, Scenario 2, was proposed in [6], where RSs are deployed within the service range of the BS for the purpose of user throughput enhancement, which is illustrated in Figure 10.4. In this design, both the BS and RSs employ omni-directional antennas to serve users and to communicate with each other. As a consequence, the frequency reuse factor has to be at least 2 in order to avoid severe co-channel interference, which is shown in Figure 10.5, and this reduces the system capacity.

Previous research has mostly been focused on the aspects of coverage extension and end-to-end user-throughput enhancement of an MR network [6–8]. In this chapter, we consider the overall system capacity enhancement issue for the MR network in a Manhattan-like environment. New scheduling methods are proposed for the system with directional antennas equipped at both the base station and relay stations. Simulation results show that the system throughput can be dramatically increased by the proposed methods, as compared to the system with omni-directional antennas.

The rest of this chapter is organized as follows. The system setup and the radio propagation models are described in Section 10.2. The proposed scheduling methods are detailed in Section 10.3. Simulation results are presented in Section 10.4, and finally, conclusions are given in Section 10.5.

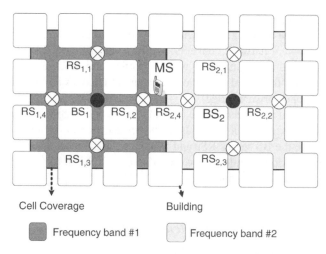

Figure 10.5 Frequency reuse pattern in the multi-cell environment of Scenario 2

10.2 System Setup and Propagation Models

10.2.1 System Setup

An MR network with an MR cell structure consisting of one BS and four RSs in the Manhattan-like environment is considered in this section. As shown in Figure 10.6, the BS is located at the main street-crossing with four RSs deployed at the four street corners. The side-length of a street block and the street width are set as 200 m and 30 m, respectively.

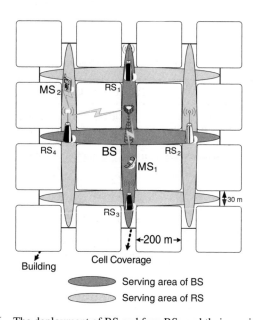

Figure 10.6 The deployment of BS and four RSs and their serving areas [15]

One MR cell covers nine blocks and thus the coverage of each cell is about 690×690 square meters. According to the technical requirements set out by IEEE 802.16j baseline document [16], there is no backhaul network connection to RSs, and BS and RSs are assumed to use the same air interference to communicate with each other and MSs (homogeneous relaying). In addition, RSs decode the signal from the source and re-encodes it before forwarding to the destination (decode-and-forward mode).

To improve the system performance, the BS and RSs are equipped with directional antennas to communicate with users and/or each other. The service area of each station is also shown in Figure 10.6, where the BS serves RSs and MSs (e.g., MS_1) in its line-of-sight (LOS) area with single-hop connections, while RSs serve MSs (e.g., MS_2) in the BS's non line-of-sight (NLOS) area through two-hop connections.

10.2.2 Propagation Models and Antenna Pattern

In this section, the propagation loss model proposed in [19] is considered, where the path-loss and shadow fading models for the urban micro-cell environment are adopted. Table 10.1 summarizes the model parameters for both the LOS and NLOS cases, where f_c is the carrier frequency, d, d_1, and d_2 are the distances as indicated in Figure 10.7. Note that the probability of having LOS depends on the distance value d.

The antenna pattern proposed in [20] is also adopted here, which is

$$A(\theta) = -\min\left[12 \cdot \left(\frac{\theta}{\theta_{3dB}}\right), A_m\right] dBi \tag{10.1}$$

where $-180° < \theta \le 180°$, θ is the angle between the direction of interest and the steering direction of the antenna, $\theta_{3dB} = 70°$ is the 3dB beam width, and $A_m = 20dB$ is the maximum attenuation.

Table 10.1 The propagation loss model for the urban micro-cell environment

Path loss model		
LOS	$PL(d, f_c) = 41 + 22.7\log_{10}(d) + 20\log_{10}(f_c/5.3)$	
NLOS	$PL(d_1, d_2, f_c) = 0.096d_1 + 65 + (28 - 0.024d_1)\log_{10}(d_2) + 20\log_{10}(f_c/5.3)$	

Standard Deviation of log-normal shadow model	
LOS	2.3 dB
NLOS	3.1 dB

Probability of LOS
$P_{LOS}(d) = \begin{cases} 1 & , d \le 15m \\ 1 - (1 - (1.56 - 0.48\log_{10}(d))^3)^{1/3}, & d > 15m \end{cases}$
where $d = \sqrt{d_1^2 + d_2^2}$
Note: f_c is in GHz, d_1, d_2, d_3 are in metres

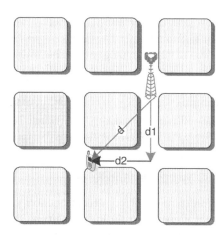

Figure 10.7   The relevant distances from BS to determine the path loss and the probability of having LOS [15]

10.3 Resource Scheduling Methods

In this section, resource scheduling methods for MR network in a Manhattan-like environment are presented. The scheduling with omni-directional antennas will be discussed first in Section 10.3.1 and it is used as a baseline for comparison. Then proposed scheduling methods with directional antennas will be discussed in Section 10.3.2. In these methods, we assume that RSs are multiplexed in the time domain. In other words, time domain relaying is considered here.

10.3.1 Scheduling with Omni-directional Antennas

Since the BS and RSs use the omni-directional antennas, at least six phases of transmissions are needed to complete the two-hop connections between the BS and MSs, as shown in Figure 10.8.

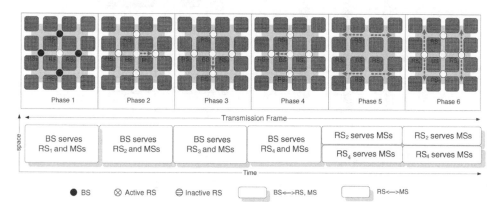

Figure 10.8   Resource scheduling with omni-directional antennas [15]

Each phase includes both downlink and uplink transmissions. Here we assume that all the transmission phases are completed in a frame. The BS takes turns to serve RSs and MSs in its LOS area in the first four phases. To take advantage of the shadowing effect in the environment, RS_1 and RS_3 serve MSs in their service areas simultaneously in Phase 5, and RS_2 and RS_4 do the same in Phase 6.

Now consider the multi-cell structure. Because of the use of omni-directional antennas, the reuse factor of at least 2 is required to avoid the severe inter-cell interference as clearly shown in Figure 10.5, and that decreases the overall system capacity.

10.3.2 Scheduling with Directional Antennas

To improve the system performance as mentioned in Section 10.3.1, resource scheduling methods with directional antennas are proposed. In this scenario, the BS and RSs are equipped with directional antennas to further exploit the advantage of shadowing effect in a Manhattan-like environment. As shown in Figure 10.9, the BS and RSs are equipped with four directional antennas pointed to different directions, respectively. In different antenna directions, the radio resource can be reused completely. Accordingly, RSs are divided into groups according to their mutual interference levels, and those in the same group are scheduled to transmit in the same transmission phase. In Figure 10.9, RS_1 and RS_3 are grouped as Group-A, and RS_2 and RS_4 as Group-B. Since there are two groups in a cell, two transmission phases are needed to complete the two-hop transmissions. Again, each phase comprises both downlink and uplink transmissions.

10.3.2.1 Method 1

In this design, as shown in Figure 10.9(a), the BS serves the Group-A RSs (RS_1 and RS_3) and MSs in its corresponding LOS area in Phase 1 (both up- and down-link). At the same

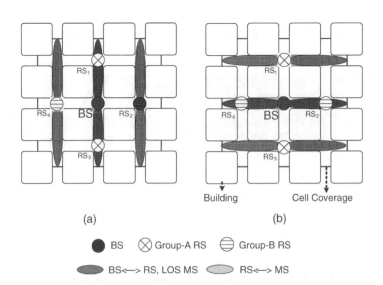

Figure 10.9 Two phases of transmissions in Method 1, (a) Phase 1 and (b) Phase 2 [15]

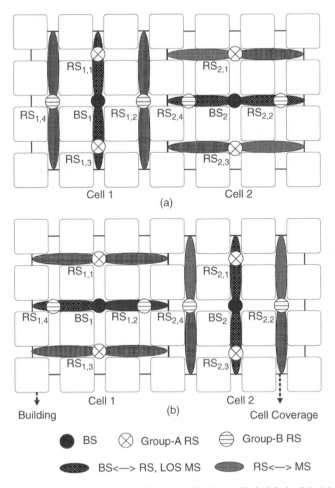

Figure 10.10 Two phases of transmissions of the neighboring cells in Method 1, (a) Phase 1 and (b) Phase 2

time, the Group-B RSs (RS$_2$ and RS$_4$) serve MSs in their service areas. Alternatively, in Phase 2, as illustrated in Figure 10.9(b), the BS turns to serve the Group-B RSs and MSs in its corresponding LOS service areas, while the Group-A RSs serve MSs in their service areas.

When considering the multi-cell setup, the frequency reuse factor of 1 (universal frequency reuse) can easily be achieved as follows. As represented in Figure 10.10(a), in Phase 1, when BS$_1$ serves the Group-A RSs and MSs in the respective directions, BS$_2$ in the neighboring cell serves the Group-B RSs and the MSs in the indicated directions. Of course, at the same time, the Group-B RSs in the cell 1, and the Group-A RSs in the cell 2 provide services to MSs in their service areas. In Phase 2, shown in Figure 10.10(b), the operation is the same by changing the role of Group-A and Group-B in each cell. As clearly seen, with directional antennas indicated in the proposed scheduling scheme, the universal frequency reuse can be achieved very easily. The corresponding frame structure for Method 1 is shown in Figure 10.11.

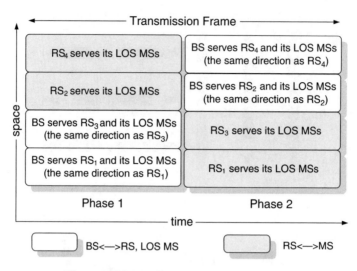

Figure 10.11   The frame structure of Method 1

10.3.2.2 Method 2

In Method 1, only two antennas at the base station are activated in each transmission phase. Obviously, the two idle antennas can also be activated to use the radio resource more efficiently. However, the transmit power of these two antennas needs to be properly controlled so that they will not cause too much interference to ongoing transmissions. Except for this added feature, Method 2 is the same as Method 1. Each phase of transmissions in a cell is shown in Figure 10.12, and the scheduling for the multi-cell setup to achieve frequency reuse factor of 1 is shown in Figure 10.13.

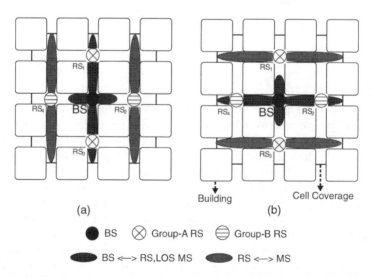

Figure 10.12   Two phases of transmissions in Method 2, (a) Phase 1 and (b) Phase 2

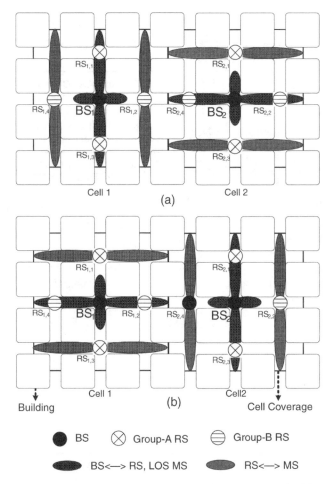

Figure 10.13 Two phases of transmissions of the neighboring cells in Method 2, (a) Phase 1 and (b) Phase 2

The frame structure of Method 2 is shown in Figure 10.14. We can see that since four directional antennas of the base station are active simultaneously, the number of BS's transmissions in each phase is four. The spectrum efficiency of Method 2 is ideally twice that of Method 1. Note that the data from/to the BS constitute the effective throughput of the MR-cell; RSs only relay data between the BS and MSs.

10.4 Numerical Results

The parameters of the simulated system are set according to the OFDMA mode in the IEEE 802.16 standard [17, 18], which are summarized in Table 10.2. The required signal to noise plus interference ratio (SINRs) to achieve 10^{-6} bit-error-rate (BER) for the various modulation-and-coding schemes (MCSs) considered are given in Table 10.3 [17]. The scheduling method for the system with omni-directional antennas is also simulated to serve as the base-line scheme for

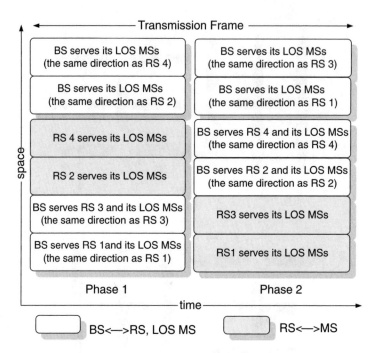

Figure 10.14   The frame structure of Method 2

Table 10.2   OFDMA parameters for system-level simulation

Parameter	Value
Carrier frequency	3.5 GHz
Bandwidth	6 MHz
FFT size	2048
Sub-carrier frequency spacing	3.348 kHz
OFDM symbol duration (including 1/8 cyclic-prefix)	336 μs
Duplexing	TDD
Frame duration	20 ms
Permutation mode of each sector	FUSC (full usage of sub-channels)
Number of sub-channels	32
Number of sub-carriers per sub-channel	48
Max. transmit power of BS/RS	100 mW
BS/RS antenna gain	14 dBi
MS antenna gain	0 dBi
MS/BS noise figure	7 dB/5dB
Noise power	$-174 + 10\log B$ dBm
BS-to-BS distance	690 m
Number of MR-cells (two-tier interfering cells)	25

Table 10.3 The used MCS

Modulation	Coding rate	Receiver SINR (dB) to achieve 10^{-6} BER
BPSK	1/2	6.4
QPSK	1/2	9.4
	3/4	11.2
16-QAM	1/2	16.4
	3/4	18.2
64-QAM	2/3	22.7
	3/4	24.4

performance comparisons. A multi-cell setup with two-tier interfering cells is considered. The locations of BS and RSs are fixed, while users are uniformly distributed within each cell. The data traffic of each user is full buffered. A fixed transmission power is used in the downlink and adaptive rate control is executed every frame to select an appropriate modulation and coding scheme based on the perfect channel state information.

Figure 10.15 shows the simulated CDFs of SINR for different scheduling methods. To emphasize on the capacity gain obtained from the new scheduling methods, the transmission power in our simulation of the onmi-directional antenna case is increased to offset the antenna gain obtained with the directional antennas. Clearly, the scheduling scheme with omni-directional

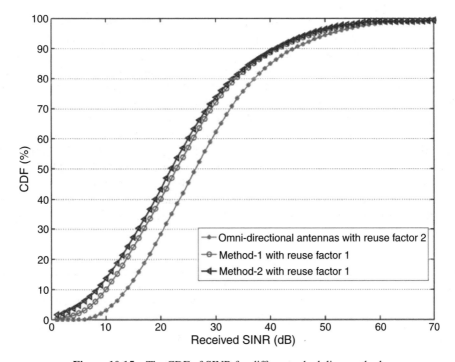

Figure 10.15 The CDF of SINR for different scheduling methods

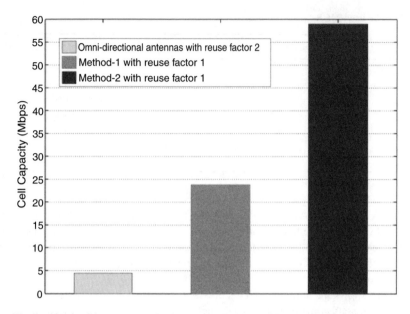

Figure 10.16 Comparisons of cell capacity between different scheduling methods

antennas has the best SINR performance, while Method 2 has the worst because it reuses the frequency spectrum more aggressively.

The capacity simulation result is shown in Figure 10.16. The cell capacity is increased enormously by the proposed methods. The capacity gain is approximately 6 and 12 times for Method 1 and Method 2, respectively, as compared with the system with the omni-directional antennas. The reason for the capacity gain can be simply explained as follows. First, as shown in Figure 10.10 and Figure 10.13, Method 1 and Method 2 can achieve a frequency reuse factor of 1, while in Figure 10.5, the reuse factor of at least 2 (here the case of reuse factor 2 is simulated) is required for the system with omni-directional antennas. Hence, at least twice the capacity gain is obtained as compared to the omni-directional antenna case. Second, from Figure 10.8 only 2/3 of the transmission phases in a frame are used for the BS's transmission. On the other hand, 2 and 4 BSs' transmissions in each phase are possible for Method 1 and Method 2 respectively, as shown in Figure 10.11 and Figure 10.14. That results in another three and six times the capacity gain over the case of the omni-directional antennas. Table 10.4 summarizes where the capacity gains occur for Method 1 and Method 2 as compared with

Table 10.4 The analysis of the capacity gain

	Reuse factor	Effective data frame transmitted in a frame	Analysed capacity gain
Omni-directional antennas	2	2/3	1
Method-1	1	2	6
Method-2	1	4	12

the omni-directional antenna case. However, due to the higher interference level observed in Figure 10.15, the simulated capacity gain of Method 1 over the system with omni-directional antennas is slightly less than six times, as our simplified analysis predicts. On the other hand, the simulated capacity gain of Method 2 over Method 1 is larger than 2 because the former provides more higher-data-rate connections.

10.5 Conclusion

Multi-hop relay (MR) cell architecture is a promising candidate for the next generation of wireless communication systems. It has been adopted as an amendment to the IEEE 802.16e standard for cell coverage extension, user throughput improvement and/or system capacity enhancement. In this chapter, we investigate the important issue of resource scheduling for MR networks in a Manhattan-like environment. New resource scheduling methods are proposed for the MR networks with directional antennas equipped at both the base station and relay stations. By taking advantage of the effect of high degree shadowing in the Manhattan-like environment, the system throughput can be increased by nearly 6 and 12 times, respectively, by the proposed methods, as compared to the system with omni-directional antennas.

References

[1] Recommendation ITU-R, 'Framework and Overall Objectives of the Future Development of IMT-2000 and Systems Beyond IMT-2000', *International Telecommunication Union*, June 2003.

[2] A. Adinoyi et al., 'Description of Identified New Relay Based Radio Network Deployment Concepts and First Assessment by Comparison against Benchmarks of Well Known Deployment Concepts Using Enhanced Radio Interface Technology', *IST-2003-507581 WINNER D3.1*, Nov. 2004.

[3] A. Adinoyi et al., 'Description Of Identified New Relay Based Radio Network Deployment Concepts and First Assessment by Comparison Against Benchmarks of Well Known Deployment Concepts Using Enhanced Radio Interface Technologies', *IST-2003-507581 WINNER D3.2 V.1* , Feb. 2005.

[4] M. Nohara et al., 'IEEE 802.16 Tutorial: 802.16 Mobile Multihop Relay', Official Document of IEEE 802.16 Mobile Multihop Relay Study Group, *IEEE 802.16mmr-06/006*, March 2006.

[5] B. Walke, and R. Pabst, 'Relay-based Deployment Concepts for Wireless and Mobile Broadband Cellular Radio', *WWRF/WG4 Relaying Subgroup White Paper*, June 2003.

[6] N. Esseling, B. Walke, and R. Pabst, 'Performance Evaluation of A Fixed Relay Concept for Next Generation Wireless Systems', *Proc. of IEEE International Symposium on Personal, Indoor and Mobile Radio Communications*, **2**, Sept. 2004, 744–751.

[7] D. Schultz, R. Rabst and T. Irnich, 'Multi-hop Radio Network Deployment for Efficient Broadband Radio Coverage', *Proc. of WPMC*, **2**, Oct. 2003, 377–381.

[8] R. Pabst et al., 'Relay-based Deployment Concepts for Wireless and Mobile Broadband Radio', *IEEE Communications Magazine*, **42**(19), Sept. 2004, 80–89.

[9] T. Irnich, D. Schultz, R. Pabst, and P. Wienert, 'Capacity of a Relaying Infrastructure for Broadband Radio Coverage of Urban Areas', *Proc. of IEEE Vehicular Technology Conference*, **5**, Oct. 2003 2886–2890.

[10] D. Schultz, B. Walke, R. Rabst and T. Irnich, 'Fixed and Planned Based Radio Network Deployment Concepts', *Proc. of WWRF*, Oct. 2003.

[11] B. Walke, H. Wijaya, and D. Schultz, 'Layer-2 Relays in Cellular Mobile Radio Networks', *Proc. of IEEE Vehicular Technology Conference*, **1**, 2006, 81–85.

[12] H. Viswanathan, and S. Mukherjee, 'Performance of Cellular Networks with Relays and Centralized Scheduling', *IEEE Trans. on Wireless Communications*, **4**(5), Sept. 2005, 2318–2328.

[13] J. Cho, and Z.J. Hass, 'On the Throughput Enhancement of the Downstream Channel in Cellular Radio Networks through Multihop Relaying', *IEEE Journal on Selected Areas in Communications*, **22**(7), Sept. 2004, 1206–1219.

[14] I.K. Fu, W.H. Sheen and F.C. Ren, 'Shadow-assisted Resource Reuse for Relay-augmented Cellular Systems in Manhattan-like Environment', *International Journal of Electrical Engineering*, **14**(1), Feb. 2007, 11–19.

[15] S.J. Lin, W.H. Sheen, I.K. Fu and C.C. Huang, 'Resource Scheduling with Directional Antennas for Multi-hop Relay Networks in Manhattan-like Environment', *IEEE Mobile WiMAX Symposium*, March 2007, pp. 108–113.

[16] 'Baseline Document for Draft Standard for Local and Metropolitan Area Networks, Part 16: Air Interface for Fixed and Mobile Broadband Wireless Access Systems: Multihop Relay Specification', *IEEE 802.16j-06_026r3*, Apr. 2007.

[17] IEEE Standard for Local and Metropolitan Area Networks, Part 16: Air interface for fixed broadband wireless access systems, IEEE 802.16-2004, Oct. 2004.

[18] 'IEEE Standard for Local and Metropolitan Area Networks, Part 16: Air Interface for Fixed and Mobile Broadband Wireless Access Systems, Amendment for Physical and Medium Access Control Layers for Combined Fixed and Mobile Operation in Licensed Bands', *IEEE 802.16e-2005*, Feb. 2006.

[19] D. S. Baum et al., 'Final Report on Link Level and System Level Channel Models' *IST-2003-507581 WINNER D5.4*, V. 1.00, Sept. 2005.

[20] 'Spatial Channel Model for Multiple Input Multiple Output (MIMO) Simulations', *3GPP TR 25.996 V6.1.0*, Sept. 2003.

11

Efficient Radio Resource Deployment for Mobile WiMAX with Multi-hop Relays

Yong Sun, Yan Q. Bian, Andrew R. Nix, and Joseph P. McGeehan

11.1 Introduction

WiMAX (Worldwide Interoperability for Microwave Access) represents an IEEE metropolitan access standard that provides broadband wireless IP (internet protocol) access to fixed and mobile terminals. The standard can be used to expand the coverage of existing wired IP networks to facilitate network access in more remote or suburban areas. In 2004, the IEEE 802.16d standard [1] was published for fixed wireless access (FWA) applications. In December 2005, the IEEE ratified the 802.16e [2] amendment, which supports mobile wireless access (MWA) with seamless network coverage. Consequently, as 802.16e is commercialized, WiMAX will enable the migration from FWA to MWA. Considerable interest currently exists in the exploitation of mobile WiMAX; this is mainly the result of lower infrastructure costs and higher data transfer rates compared to current 3G. The mobile WiMAX air interface adopts a technology known as scalable orthogonal frequency division multiple access (SOFDMA) to achieve enhanced multi-user performance in non-line-of-sight (NLoS) multipath environments.

The fixed WiMAX standard is often quoted as combining long transmission ranges with high data capacities. While this may be true for an isolated FWA base station (BS), for cellular MWA applications, power and spectral efficiency is key to a successful deployment [3]. For cellular deployments the wireless medium are a precious infrastructure commodity. MWA standards must critically conserve the use of scarce radio resource. It is well known that radio relays may be used to enhance coverage (and capacity in some cases) [4]. Radio relays can address many of the challenges faced in a mobile WiMAX deployment. They are currently under consideration for early implementation in the broadband wireless market. Relay deployment

Mobile WiMAX   Edited by Kwang-Cheng Chen and J. Roberto B. de Marca.
© 2008 John Wiley & Sons, Ltd

targets one of the biggest challenges in next generation MWA, namely the provision of high data rate coverage in a cost-effective and ubiquitous manner.

The concept of multi-hop relaying is already well-developed in the fixed telecommunication world. Modern relay technology dates back to the 1940s. In 1945, the first experimental microwave relay system was introduced by Western Union between New York and Philadelphia. This distribution system transmitted communication signals via radio along a series of towers. Since then, microwave radio relays have been widely used to transmit digital and analog signals over long distances, with examples including telephony and broadcast television. In these markets, radio relays achieve long distance communications at a fraction of the cost of coaxial cable. With the evolution of mobile networks, wireless relays have been further developed for cellular transmission. Analog repeaters are sometimes used in cellular systems to extend coverage into regions that are not covered by the standard network [5]. However, the use of analog processing boosts not only the wanted transmission, but also the background noise. An integrated cellular and ad-hoc relay system was presented in [6]. Research on intelligent relays for GSM is also reported in [7]. Digital relaying for cellular applications was initially investigated in [8], to enhance coverage for delay insensitive traffic. Relay technology is being promoted in the IEEE 802.16j multi-hop project [9] to enhance aspects of broadband wireless transmission. If adopted by the 802.16 standard, this will represent the first truly global standardization of relay technology.

Basically, if no constraints are imposed on the radio resource, it is relatively simple to implement relay transmissions to enhance the coverage of a cellular network. Relays can be used to fill 'holes' in the coverage map and thus enhance coverage and data rate at (or even beyond) the BS cell-edge. As shown in Figure 11.1, a multi-hop link can be formed between the BS and a distant mobile station (MS) using a number of intermediate relay stations (RS). The RSs can operate in either an amplify-and-forward mode (similar to a repeater), or a more advanced decode-and-forward mode. In order to avoid interference between the links, the simplest approach is to assign unique radio resources to each link (as shown in Figure 11.1). However, using this approach, the multi-hop users will rapidly drain the system of valuable

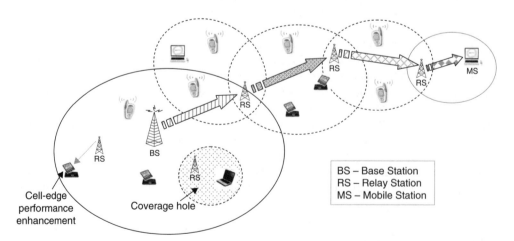

Figure 11.1   A general relay deployment for cellular coverage extension and performance enhancement

radio resource. In turn, this results in poor radio resource efficiency, and hence low system capacity. This approach can only be used for a very small number of very important MSs, or in applications where spectral efficiency is not vital, such as military or disaster relief communication networks. A more comprehensive definition of usage scenarios can be found in [10].

Given its commercial applications, relaying in the context of WiMAX must conserve radio spectrum and emphasize the need for high spectral efficiency. Consequently, enhancing radio resource efficiency is a key challenge in the competitive business development of a mobile WiMAX relay system. Radio resource reuse is a critical requirement of the 802.16j. To quantify this reuse efficiency it is necessary to monitor the radio interference introduced by relays [11]. Hence, interference measurement and management techniques must be considered at the design stage. To meet these goals, smart antennas are expected to play a key role. Mobile WiMAX has been designed to exploit a smart antenna-based form of SOFDMA that can work together with relaying technology. The combination of flexible channelization with adaptive modulation and coding (AMC) can lead to improved system coverage and capacity while maintaining high power and spectrum efficiency.

This chapter provides a thorough analysis of relay efficiency in the context of mobile WiMAX. A directional distributed relaying architecture is introduced for highly efficient radio resource sharing. This architecture is based on both interference cancellation and interference avoidance. The method leads to a high data throughput with reduced demands on radio resource. In order to provide a comprehensive insight into the application and efficiency of relays, the system performance of a mobile WiMAX network is presented. This is followed by the definition and analysis of effective system capacity gain. Using a theoretical analysis of interference, a directional distributed relaying architecture is introduced and case studies are presented based on the use of a site-specific ray-tracing model [12] in a practical urban environment. These scenarios demonstrate the potential of relays to enhance capacity, coverage and efficiency in a mobile WiMAX system.

11.2 System Performance and Enhancement

The performance of a Mobile WiMAX system based on the currently published IEEE 802.16 standard has been widely evaluated [3], [13]. This section focuses on the system performance following a number of enhancements. Using a statistical approach, it is possible to demonstrate the potential benefits of a relay-enhanced mobile WiMAX deployment.

It is well understood that spectrum efficiency is the key to good system design. Without loss of generality, the normalized effective system spectral efficiency can be defined as

$$Effective\ system\ efficiency = \frac{System\ data\ throughput}{Total\ radio\ resources\ allocated} \tag{11.1}$$

Two clear approaches emerge to improve the effective efficiency. First, the system data throughput can be improved by using methods such as MIMO with high level AMC. Second, improvements can be made by reducing the amount of radio resource required in the system. The IEEE 802.16 standard fully supports AMC together with a range of smart antenna technologies, which include space time block coding (STBC), spatial multiplexing (SM) and array processing (e.g. beamforming/pre-coding). The inclusion of MIMO link adaptation alongside flexible

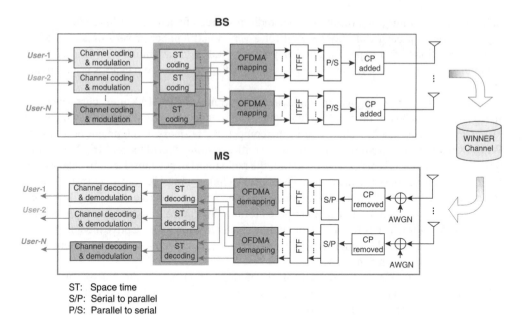

ST: Space time
S/P: Serial to parallel
P/S: Parallel to serial

Figure 11.2   MIMO OFDMA PHY simulator

sub-channelization enables mobile WiMAX technology to improve system coverage and capacity. If systems are correctly configured, these benefits will result in power and spectrum-efficient terminals.

Based on the published standard and system profile, a MIMO mobile WiMAX physical layer (PHY) simulator has been developed, as depicted in. Following the WirelessMAN-OFDMA PHY air-interface, three sectors are employed at the BS and a frequency reuse factor of one is used between the sectors (to satisfy reliability, coverage and capacity requirements). Each sector accesses one-third of the total number of subcarriers. A number of initial operational profiles have been defined within the WiMAX Forum Technical Working Group [14]. This section focuses on the exploitation of the 512-FFT PUSC (partial usage of sub channels) OFDMA time division duplex (TDD) profile operating with a 5 MHz channel bandwidth. For the specified 512-FFT downlink (DL) OFDMA profile, a total of 30 physical clusters (or 15 subchannels) are mapped (after renumbering and permuting the logical clusters) in one OFDMA symbol. There are three groups (one per sector), with 5 subchannels in each group, and each sub-channel comprises 24 data subcarriers and 4 pilot subcarriers. Since the subchannel is the smallest allocation unit for a user on the DL, this profile can support up to 15 users per slot. The use of a cyclic prefix (CP) eliminates inter-symbol interference (ISI) for delay spreads less than or equal to the CP length. A CP length of 1/8th of the OFDMA symbol period is defined to ensure that up to 11.2µs of delay spread can be tolerated. This introduces an overhead of approximately 10%. A detailed list of the PHY parameters can be found in [15], [16].

The above WiMAX PHY simulator is used within a Monte Carlo simulator to evaluate system performance in terms of expected throughput and outage capacity for various scenarios

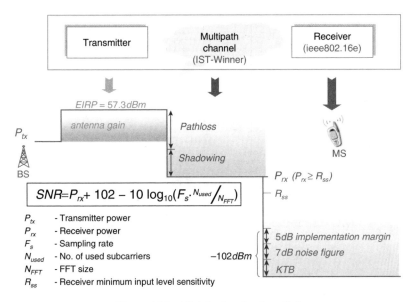

Figure 11.3 Link-budget for downlink

(e.g., urban microcells and macrocells). We assume that the system has: (1) fair scheduling such that all users have an equal opportunity to access the BS (users are uniformly selected without any other constraints, e.g., SNR); (2) ideal channel estimation; and (3) optimal MIMO link adaptation (see the following paragraph). Figure 11.3 illustrates a link-budget for the DL. This assumes that a maximum transmit power of 42.3 dBm is applied to the BS antenna port. We further assume a sector antenna gain of 15 dBi to produce an effective isotropic radiated power (EIRP) of 57.3 dBm [13].

The ideal MIMO link adaptation scheme selects the best operating mode (in terms of modulation and coding) and space time coding method (in terms of STBC or SM) to maximize the link throughput at each location (while constraining the PER to be less than 10%). In this chapter the selection is based on an exhaustive throughput simulation of all possible combinations. A suitable MIMO-orientated link adaptation strategy is critical to mobile WiMAX to fully exploit the wide range of MIMO systems and channel conditions that can occur. In general, SM combined with higher order modulation schemes can be used to increase the peak throughput of a link. However, such schemes require extremely high SNR levels, and are thus difficult to exploit near the cell edge. In contrast, STBC exploits the MIMO structure to offer high levels of spatial diversity. This results in a scheme that offers strong performance at lower received signal levels. STBC is used together with AMC to generate capacity improvements [17], [18].

In a mobile transmission environment, the received signal strength is not only dependent on the available transmit power, but also on the structure of the mobile channel. In a practical urban environment, the radio channel linking the BS to the MS is unpredictable; however, it plays a key role in the evaluation of transceiver configurations, such as link adaptation and multi-user scheduling. For FWA scenarios the SUI (Stanford University interim) channel model [19] is commonly applied, where it is assumed that the customer premises equipment (CPE) is

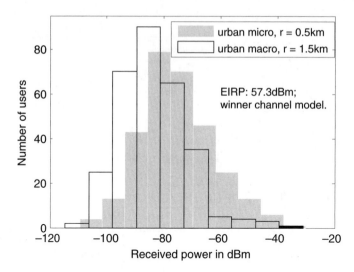

Figure 11.4 Histogram of received power

located at a fixed roof top location. Unlike the FWA case, in the MWA scenario mobile terminals are allowed to move freely at street level. Local surrounding buildings result in deep shadowing and severe multipath. To represent this situation, the WINNER channel model was developed as part of the European Union (EU) IST-WINNER project [20]. This model builds on the well-known 3GPP2 spatial channel model (SCM) [21].

The WINNER model defines three basic environment scenarios, namely urban micro (cell radius, r, is not greater than 0.5 km, as $r \leq 0.5$), urban macro, and sub-urban macro ($r \approx 1.5$ km). For mobile transmissions operating in licensed spectrum, the EIRP is specified by the radio regulator. From [13] the maximum permitted EIRP for mobile WiMAX operation is 57.3 dBm on the DL and 22 dBm on the MS uplink (UL). It is important to characterize the expected received power levels at a mobile receiver by a given maximum transmit EIRP at the BS and a specific application environment. Figure 11.4 presents a histogram of the number of users receiving signals in a range of power bands. Results are compared for urban microcells and macrocells (note: for the macrocell the radius is three times that of the microcell). We assume a fixed BS transmission power and 300 uniformly distributed MSs over a circular region. As confirmed by Figure 11.4, the average received power at the MS in the urban-macro cell is much reduced compared to that of the microcell. This occurs as a result of the larger cell radius and the more severe shadowing assumed in the macrocell.

In order to study the achievable receiver performance, three zones are defined according to the BS-MS separation distance. The first zone covers BS-MS separation distances up to 0.5 km. The second zone covers BS-MS distances between 0.5 km and 1.0 km. The third zone covers BS-MS distances between 1.0 km and 1.5 km. The shadowing and pathloss models are configured differently in these zones. An urban micro channel model is employed in the first zone, while an urban macro channel model is used in the outer two zones. Figure 11.5 illustrates the AMC behavior for the DL case. The 512-FFT DL OFDMA PHY described earlier was used based on three users, where each user is supported from a unique sector of the BS. As shown in Figure 11.5, an average link throughput of 11.7 Mbps per sector is achieved within

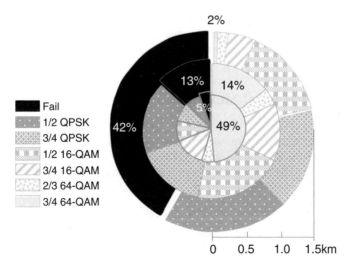

	Radius range	Channel environment	Average throughput
1st zone	0~0.5km	Micro-cell	11.7 Mbps
2nd zone	0.5~1km	Macro-cell	7.29 Mbps
3rd zone	1~1.5km	Macro-cell	3.46 Mbps

Figure 11.5   Performance of different zones for full coverage (2×1 STBC with antenna separation of 10 wavelengths)

the microcell. The outage probability falls to just 5% in the microcell, with 49% of users able to operate at the highest throughput level (3/4 rate 64-QAM). However, for the larger BS-MS separation distances in the macrocell, performance degrades in terms of coverage and throughput. For users 0.5–1.0 km from the BS, it is no longer possible to meet a 90% coverage target using this profile. In the outer zone (MS located 1.0–1.5 km from the BS), the quality of service (QoS) drops such that 42% of users fail to connect. The average link throughput over all users for the outer zone is 3.46 Mbps per sector. Interestingly, for a BS equipped with transmit diversity, satisfactory coverage can be achieved up to a radius of 1 km [15].

The IEEE 802.16 standard states the minimum received power levels for each of the different modulation and coding modes (as a function of channel bandwidth) [1], [2]. Results shown in Figures 11.4 and 11.5 are based on the DL case. For the UL case, the subchannelization gain (achieved by using only a sub-set of the available subchannels) can be used to balance the link. Based on the results shown in Figures 11.4 and 11.5, we conclude that high levels of QoS may be difficult to achieve in a mobile WiMAX network. It is known that radio relays can be used in areas with low throughput or high outage probability. To realize coverage up to a cell radius of 1.5 km in a macrocell, we now consider a relay-based MIMO WiMAX configuration. However, the use of RSs requires the allocation of additional radio resources to support the relaying signals. For commercial applications, radio resource efficiency is critical for relay deployment. A typical operator is unlikely to identify additional spectrum for relays, and hence the radio resource must come from that already allocated to the BS. It is now clear that there are tradeoffs between performance, capacity, coverage, radio resource allocation and interference. The following section explores these tradeoffs for a multi-hop cellular network.

11.3 Effective Efficiency of Multi-hop Relaying

Generally, relays can be applied in either a single or multi-hop architecture. Ideally, for an n-hop relay it is necessary to provide n unique radio resource units to avoid interference between each hop. This need for additional radio resources reduces the overall spectral efficiency; however, since the relays enable links to operate at lower powers and higher throughputs, there is also the potential for higher spectral efficiency. It is clear that the success of relaying is intimately related to the radio resource reuse efficiency. In order to measure the efficiency of a relay deployment, we now introduce an effective system capacity metric (C_{eff}). This is intended to leverage link level capacity gain and system capacity gain. Assuming there are a total of s MSs to be allocated, and p of these users are served via multi-hop relays, C_{eff} can be expressed as

$$C_{eff} = \sum_{i=1}^{s-p} C_i^{BS} + \frac{\sum_{j=1}^{p} C_j^{RS}}{N_{rc}} \qquad (11.2)$$

where, N_{rc}, C^{BS} and C^{RS} denote the number of radio resources employed in the relays, the capacity without relaying (BS access capacity), and the capacity with relaying (relayed capacity) respectively. For simplicity, we assume that the BS-RS and/or RS-RS links are ideal and that no 'bottle neck' problems exist. Under these assumptions, the effective relay efficiency can be derived as

$$\xi_c = \frac{\sum_{i=1}^{s-p} C_i^{BS} + \left(\sum_{j=1}^{p} C_j^{RS}\right) \Big/ N_{rc}}{\sum_{k=1}^{s} C_k^{BS}} \qquad (11.3)$$

It can be seen that several kinds of tradeoff exist in the relay efficiency expression. From a system-level point of view, the effective relay efficiency must be greater than one ($\xi_c \geq 1$). In the case where $\xi_c < 1$, radio resource sharing should be considered. However, resource sharing may introduce interference. If no prior channel knowledge is available at the transmitter, then the theoretic channel capacity for a MIMO system with M transmit and N receiver antennas [22] is given by

$$C = \log_2 \det\left[\mathbf{I}_N + \frac{1}{M} \cdot \frac{P_s}{P_I + P_n} \cdot \mathbf{H}\mathbf{H}^H\right] \quad (bps/Hz) \qquad (11.4)$$

where, P_s, P_I and P_n represent the power of the received signal, interference and noise, respectively; $\mathbf{I}_N$ is an $N \times N$ identity matrix and $\mathbf{H}$ is the normalized channel matrix which is considered to be frequency independent over the signal bandwidth, and $(\bullet)^H$ denotes the transpose conjugate. Using Equation (11.4), Equation (11.3) can now be re-written as shown below for the DL case.

$$\xi_c = \left[1 \middle/ \sum_{k=1}^{s} \log_2 \det \left(\mathbf{I}_N(k) + \frac{1}{M_k} \cdot \frac{P_{s,k}^{BS}}{P_{I,k} + P_{n,k}} \cdot \mathbf{H}_k \mathbf{H}_k^H \right) \right]$$

$$\times \left[\sum_{i=1}^{s-p} \log_2 \det \left(\mathbf{I}_N(i) + \frac{1}{M_i} \cdot \frac{P_{s,i}^{BS}}{P_{I,i} + \sum_{ii=1}^{l_i^{RS}} P_{I,i,ii}^{RS} + P_{n,i}} \cdot \mathbf{H}_i \mathbf{H}_i^H \right) + \right.$$

$$\left. \frac{1}{N_{rc}} \sum_{j=1}^{p} \log_2 \det \left(\mathbf{I}_N(j) + \frac{1}{M_j} \cdot \frac{P_{s,j}^{RS}}{P_{I,j} + P_{I,j}^{BS} + \sum_{\substack{jj=1 \\ jj \neq j}}^{q_j^{RS}} P_{I,j,jj}^{RS} + P_{n,j}} \cdot \mathbf{H}_j \mathbf{H}_j^H \right) \right]$$

(11.5)

where, P_s^{BS} and P_s^{RS} represent the signal power received from the BS and RS respectively; and P_I^{BS}, P_I^{RS} and P_I denote the co-channel interference (CCI) power received from the BS, RS and any other resources (e.g., from other cells) respectively. l_i^{RS} denotes the number of RSs that use the same radio resource as that of the i-th user (BS-MS), and q_j^{RS} represents the number of RSs that use the same radio resource as that of the j-th user (RS-MS). From Equation (11.5) we can deduce the impact of interference and radio resource usage. Increasing any of the interference terms (P_I^{BS}, P_I^{RS} and P_I) or the value of N_{rc} could reduce the relay gain. Furthermore, reducing N_{rc} could also increase the value of P_I, P_I^{BS} and P_I^{RS}. Finally, for MIMO transmissions the channel correlation is an important factor, and this is discussed in the following section.

11.4 Relay Efficiency without Radio Resource Sharing

Without radio resource sharing, the system relies on the use of unique radio resources for each link, including the BS-RS, BS-MS, RS-RS and RS-MS. To clarify, we start with a simple relay deployment for a SISO system with a single user. We define the relay SNR-gain as $G_{SNR} = SNR_{relay} - SNR_{access}$, where the SNR_{relay} and SNR_{access} represent the signal to noise ratio of the last hop between the RS and MS (shortened to relay-SNR) and the BS-access link (directly between the BS and MS, shortened to access-SNR) respectively.

Figure 11.6 shows that the relay efficiency does not linearly increase with G_{SNR}, and for a certain required efficiency the value of G_{SNR} varies according to SNR_{access}. Figure 11.6 also indicates that the requirement of the relay SNR-gain is much lower if the value of SNR_{access} is low, e.g., when SNR_{access} is -10 dB the relay SNR-gain only needs to be around 3.1 dB for 100% relay efficiency. However, in order to maintain a high level of relay efficiency, the relay SNR-gain must exceed 10 dB if the value of SNR_{access} is 10 dB.

For multi-hop systems (i.e. hop numbers greater than 2), the required relay SNR-gain is even higher. A number of examples are shown in Table 11.1. The required relay SNR-gain shows a near linear increase with increasing hop number.

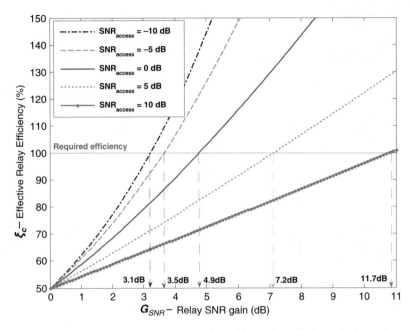

Figure 11.6 Analysis of relay SNR-gain required for high relay efficiency (2-hop with two radio resource, no resource sharing) [11]

Using Equation (11.5) it is possible to produce a comparison of the required relay SNR-gain for different MIMO configurations. This analysis is shown in Figure 11.7. It is interesting to note that MIMO links are more efficient for relay deployments. Compared to a SISO system, MIMO requires a lower relay SNR-gain to achieve a given level of relay efficiency, which is due to the MIMO efficiency. Increasing the number of antennas at each node can therefore increase the relay efficiency. This implies that the combination of MIMO and relaying forms an effective solution for enhanced system efficiency and capacity. Note that in this study the MIMO channels are uncorrelated.

It is well known that MIMO channel correlation is detrimental to the performance of a MIMO system. The capacity of a MIMO channel in the presence of spatially correlation fading (without channel knowledge at the transmitter) is given by

$$C = \log_2 \det \left(\mathbf{I}_N + \frac{1}{M} \cdot \frac{P_s}{P_I + P_n} \cdot \mathbf{R}_r^{1/2} \mathbf{H} \mathbf{R}_t \mathbf{H}^H \mathbf{R}_r^{H/2} \right) \quad (bps/Hz) \qquad (11.6)$$

Table 11.1 Comparison of multi-hop required relay SNR-gain at efficiency of 100%

	Access-SNR = 0 dB	Access-SNR = 5 dB
2-hop	4.86 dB	7.17 dB
3-hop	8.45 dB	13.53 dB
4-hop	11.76 dB	19.76 dB

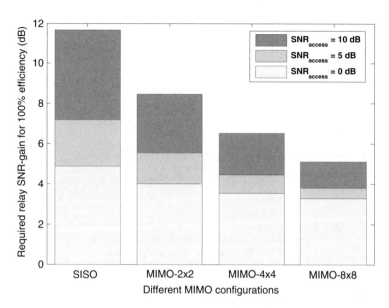

Figure 11.7 Comparison of required relay SNR-gain for different MIMO configurations (2-hop with two radio resource, no resource sharing) [11]

where $\mathbf{R}_r$ and $\mathbf{R}_t$ are positive definite Hermitian matrices that specify the receive and transmit correlation values respectively. Assuming that $M = N$ and that $\mathbf{R}_r$ and $\mathbf{R}_t$ are full rank, at high SNR the capacity of the MIMO channel can be written as [22]

$$C \approx \log_2 \det \left(\frac{1}{M} \cdot \frac{P_s}{P_I + P_n} \mathbf{HH}^H \right) + \log_2 \det (\mathbf{R}_r) + \log_2 \det (\mathbf{R}_t)$$

$$= \log_2 \det \left(\frac{1}{M} \cdot \frac{P_s}{P_I + P_n} \mathbf{HH}^H \right) + \log_2 \det (\mathbf{R}) \quad (bps/Hz) \qquad (11.7)$$

where $\mathbf{R}$ represents the combination of $\mathbf{R}_r$ and $\mathbf{R}_t$. In order to study this degradation, three extreme scenarios are now considered: (1) the BS-MS link is correlated but the RS-MS link is uncorrelated; (2) both the BS-MS and the RS-MS links are correlated with identical correlation values; and (3) the RS-MS link is correlated but the BS-MS link is uncorrelated.

The effective relay efficiency versus the channel correlation value is shown graphically in Figure 11.8. For a 2 x 2 MIMO system we deliberately set both the SNR_{access} and the relay SNR-gain to 6 dB, since this produces a 100% effective relay efficiency value when the BS-MS-link and the RS-MS-link are uncorrelated. For the first scenario, the relay improves the link channel properties and consequently improves the relay efficiency. In contrast, for the third scenario the correlated RS-MS-link severely degrades the relay efficiency. However, it is interesting to see that the second scenario shows that a high channel correlation value (greater than 0.6) can destroy the relay gain and dramatically decrease the relay efficiency.

Fundamentally there are two basic factors that affect MIMO relay applications: (1) the value of the relay SNR-gain; and (2) the condition of the MIMO matrix channel. Spatial correlation has a harmful impact, especially on high order MIMO transceivers with small

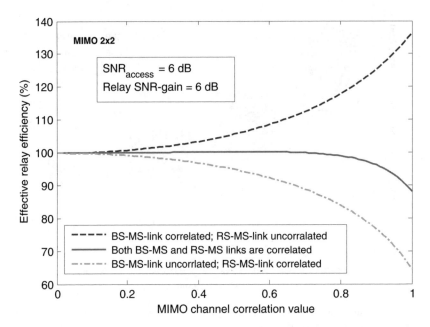

Figure 11.8 Channel correlation impacts on relayed MIMO system [11]

antenna separations. Therefore, both the SNR and the degree of MIMO correlation must be taken into account when quantifying a MIMO relaying application.

11.5 Relay Efficiency with Radio Resource Sharing

Without resource sharing, relaying requires plentiful radio resources and a high relay SNR-gain. Given the scarcity of radio resources, to achieve a highly efficient relay deployment it is necessary to implement radio resource sharing. With resource sharing, the critical issue is the interference introduced into the system. The interference between links must be kept to a minimum. Figure 11.9 illustrates the impact of interference on the relay efficiency based on a SISO system with 2-hops. It is clear that a high signal to interference ratio (SIR) is required if the access-SNR is high.

Figure 11.10 presents the impact of interference on a MIMO system. We have defined two thresholds: one is at 100% efficiency (which is required to compute the minimum SIR value), and the other is at 200% efficiency, which implies the relay deployment has doubled the capacity of the direct BS-MS link. It is interesting to note that the MIMO configuration is more tolerant to interference (compared to SISO). For example, the SISO system requires 4.5 dB more SIR than an 8 x 8 MIMO approach.

So far, we have studied relay deployment with and without resource sharing. Table 11.2 provides a clear comparison for a 2-hop SISO relay link (values refer to Figures 11.6 and 11.9). We assume that two radio resources are available for the relay. In this example, a target of 120% for the effective relay efficiency was assumed. The SNR_{access} is 0 dB and the achievable capacity between the BS and MS is C_{access}.

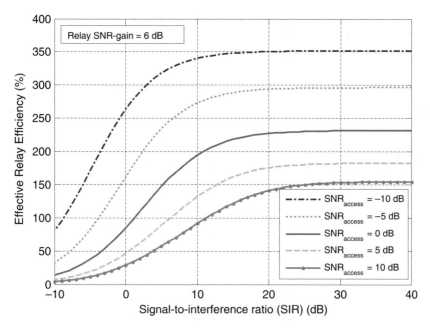

Figure 11.9   Analysis of relay SIR required for high relay efficiency [11]

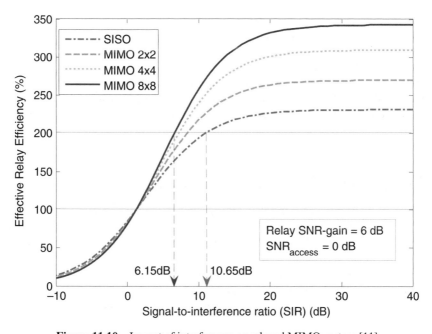

Figure 11.10   Impact of interference on relayed MIMO system [11]

Table 11.2 Comparison of SISO relay with and without sharing

Relay SNR-gain = 6 dB; Access-SNR = 0 dB; SISO

	Achieved efficiency	Minimum SIR requirements	Link relayed capacity	System capacity	Possibly maximum system capacity
Without sharing	120%	N/A	$2.4C_{access}$	$2.4C_{access}$	$2.4C_{access}$
Full sharing*	120%	SIR = 2 dB	$1.2C_{access}$	$2.4C_{access}$	$4.8C_{access}$

*'Full sharing' means no additional resources are used when RS nodes are inserted. C_{access} represents the capacity of the access link from BS to MS without relay

With these assumptions, the relay SNR-gain must be 6 dB (based on Figure 11.6). For the relay link without radio resource sharing, two radio resources are required to support transmission from the BS to the MS via the relay. Based on (11.3), the link relayed capacity must be 2.4 times C_{access} to achieve the target effective relay efficiency (ξ_c). In this case, the system capacity is also $2.4C_{access}$, since the two radio resources are already fully occupied by the relay. Consequently, the maximum system capacity cannot exceed $2.4C_{access}$. In contrast, if one resource is fully shared in the relay (i.e. one resource in both the BS-RS and RS-MS links), interference will arise from the BS to MS. To achieve the same value of ξ_c with sharing, a minimum SIR of 2 dB is required (from Figure 11.9) and the link relay capacity is now $1.2C_{access}$. However, this approach frees one radio resource, which can now be used to support another transmission (either with or without relaying). A system capacity of $2.4C_{access}$ can also be obtained under the same conditions. Furthermore, in an ideal case, if all the interference can be eliminated as $SIR \rightarrow \infty$, the maximum ξ_c of 240% can be achieved by one relay link as shown in Figure 11.9. Hence, the maximum $4.8C_{access}$ system capacity is possible for relays with full sharing of their radio resources. The maximum system capacity with resource sharing is twice that without sharing.

We summarize that there are high potential benefits for relay deployments that consider radio resource sharing. With careful design, it is possible to achieve one radio resource to one user (or one group of users) in average even with multi-hop relay. This will be detailed in next section.

11.6 Directional Distributed Relay Architecture

In order to achieve highly efficient relay deployments for practical applications, appropriate frequency reuse and multi-user access strategies are required to share the valuable and limited radio resource. It is essential to develop criteria for resource sharing and to balance relay efficiency and system performance. As discussed in the previous sections, the main concern with resource sharing is interference. In order to support a highly efficient WiMAX network, any relay system must be based on a topology that fully exploits effective resource assignment based on the spatial separation of nodes. We now propose a directional distributed relaying architecture based on a paired radio-resource transmission scheme [11]. This approach is depicted in Figure 11.11.

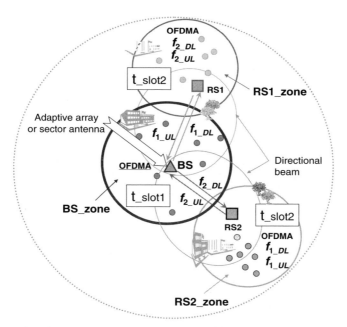

f – *radio resource; DL – Down Link; UL – Up Link; zone – coverage area*
f_{1_DL} – downlink radio resource of pair-1, and so on

Figure 11.11   Directional distributed relaying with paired radio resource

The radio resource (f) can be defined in frequency (e.g., subchannels in an OFDMA symbol) or time (e.g. OFDMA time slots). Transmissions in the BS_zone are the same as standard 802.16e. For relay links, paired transmissions are applied, where the BS forms two directional beams, or uses two sector antennas to communicate with RS1 and RS2 simultaneously. Two paired radio resources are required: $f_{1\_DL} - f_{1\_UL}$ and $f_{2\_DL} - f_{2\_UL}$. The first pair ($f_{1\_DL} - f_{1\_UL}$) is applied to the BS-RS1 link and also to the RS2-MS links (in the RS2_zone); while the second pair ($f_{2\_DL} - f_{2\_UL}$) is applied to the BS-RS2 link and also to the RS1-MS links (in the RS1_zone). Radio resources are shared between the RSs and MSs, and each end-user employs a single pair of radio resources, on average, for the UL and DL.

Using the sharing scheme outlined above, the interference can be controlled at the BS and RS nodes. In this relay configuration there are only two sets of interference, as illustrated in Figure 11.12. The interference between the BS and MS-groups can be detected and controlled by the BS. First, the BS could employ an adaptive array to exploit the spatial separation of the groups. Second, since the received power from each MS in each MS-group is known to the BS, the BS can apply interference avoidance [23], [24] between the two groups. Furthermore, in this scenario the expected level of interference is small since the BS connects to the MSs through a relay, which means the relay SNR-gain will be much higher than the SNR_{access} level. Interference between RSs can be reduced by array processing (including the use of sector antennas) at the RSs.

The topology shown in Figure 11.1 is compatible with the existing 802.16e standard and no modifications are required at the MS. Alternative deployment topologies are also

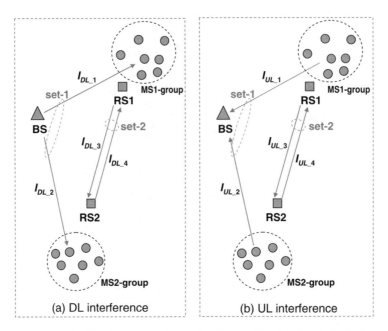

Figure 11.12   Interference illustration for the directional distributed relaying

possible based on the same concept, such as a single RS to cover a 'coverage hole' (see Figure 11.13 in the next section). However, these will result in different interference modes as the radio resource sharing is performed between the RS and its BS. The interference mode and its impact on system performance are actually highly dependent on the application environment. From our experience, given an acceptable SIR level (e.g., 10 dB), interference management can be achieved by spatial separation in the relaying system. To achieve high levels of SIR (e.g., 10–25 dB), array processing (including the use of sector antennas at the RS) is desirable. If required, interference avoidance can be used to further enhance system performance. To demonstrate the potential of the proposed resource sharing framework, a site-specific

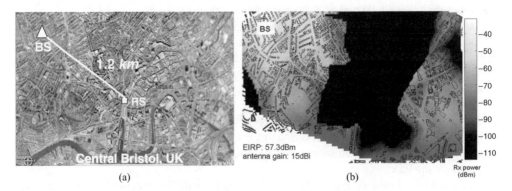

Figure 11.13   Simulated a macro-cell in Bristol, UK

ray-tracing propagation model is introduced in the next section to provide realistic environment specific propagation data for the BS-MS, BS-RS and RS-MS links.

11.7 Case Study of Radio Resource Sharing

A realistic MWA scenario is now analyzed based on a region of central Bristol, UK (Figure 11.13(a)). A single BS location was chosen on the roof-top of a tall central building (30 m above ground level). 100 MS units were distributed over this geographic area at street level with heights of 1.5 m. The BS was assigned an EIRP of 57.3 dBm (based on a 15 dBi 120^0 sector BS antenna).

The raw multipath data is created using the ray tracing tool developed in 0. This model takes individual buildings, trees and terrain contours into account and determines specific multipaths based on scattering and diffraction. The ray tracing tool was verified with measurement data in [25], and has been used in many previous WLAN and WiMAX system evaluations[17], [26]. Based on isotropic ray-traced channel data, the ETSI specific antenna beam patterns (including their side lobes) [27], [28] are incorporated via spatial convolution [29]. Figure 11.13(b) presents the distribution of received power from the BS to the surrounding region at street level. A severe coverage hole is clearly visible, and this is caused by variations in the terrain height (i.e. the hole lies in a valley).

Two methods could be used to achieve acceptable WiMAX coverage in this macrocell. Firstly, a second BS could be deployed in the coverage hole. However, this would add to the infrastructure costs and hence it may be more effective to deploy an RS node. To explore this second option, Figure 11.14 shows the locations of the BS, RS and MS. For each MS a line is drawn to indicate whether communication occurs via the BS or RS. The choice of connection type depends on the received signal to interference plus noise ratio (SINR). It should be noted that in NLoS conditions the MS may connect to an adjacent sector since this is determined by the direction of the strongest multipath component (MPC). Transmit powers of 42.3 dBm and 23 dBm are applied respectively at the BS and RS [13]. An identical three-sector antenna is used at the RS. The MS units are assumed to employ omni-directional antennas.

When radio resource sharing is applied, a number of RS-MS and BS-MS links are supported simultaneously using the same resources. For simplicity, in this study we assume there are a total of three groups for three users, and each group comprises five subchannels, as shown in Figure 11.14. Each group is made up of 140 physical subcarriers, resulting in a total of 420 subcarriers in each OFDMA symbol (360 data bearing carriers and 60 pilot carriers). When an RS is used to connect to an MS, a certain number of timeslots (or alternatively subcarriers) must be assigned in the covering sector to support the BS-RS link. Here we assume that BS-RS-link and BS-MS-link (taken from the sector covering the RS) in group #3. For all the RS-MS links, and also the BS-MS links where the antenna beam is steered away from the RS, it is possible to share group #1. Group #2 is used for those MSs that connect to one of the BS sectors (i.e. but not the sector coving the RS); that can be located near to the RS if required.

Figure 11.15 presents the SINR contribution within the BS coverage hole for the assumptions made in the previous paragraph. As expected, the RS improves the coverage, although users now suffer CCI from the BS because of radio resource sharing. Results shows that without the relay, around 60% of users have an SNR value less than 15 dB. In contrast, only 10% of users have an SINR less than 15 dB when the relay is applied.

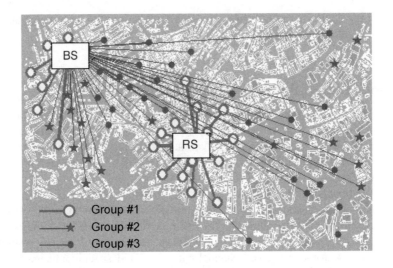

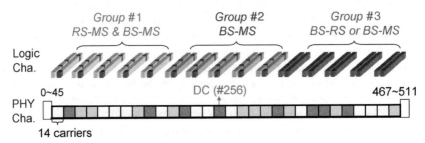

Figure 11.14   BS and RS radio resource location

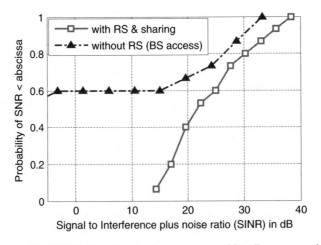

Figure 11.15   SINR in directional relaying system with radio resource sharing

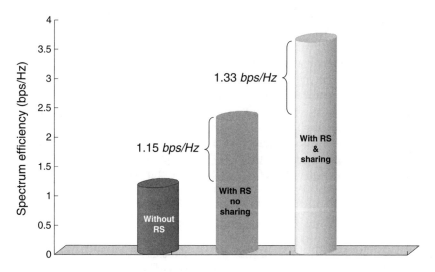

Figure 11.16   Spectrum efficiency comparison

Based on optimal AMC, Figure 11.16 compares the spectral efficiency (in terms of bps/Hz) for three types of SISO deployment. The deployments are called 'without RS', 'with RS, but no sharing' and 'with RS and sharing'. Results indicate that without the relay the capacity is just 1.14 bps/Hz in this area. By using directional relaying to reduce the interference and by exploiting spatial separation, the system is able to share radio resource and hence increase the spectral efficiency (to more than 3.5 bps/Hz in the case of sharing).

11.8 Conclusion

We have seen that relaying is straightforward in certain scenarios, such as military or disaster environments, where there is no strong need for high spectral efficiency. The use of RSs requires additional resources to support the relaying mechanism. Unless new bandwidth is made available, the relays must use (or reuse) spectrum allocated to the BS. In this situation the interference generated by the RS nodes must be carefully managed. Large spectrum allocations for cellular applications are difficult to acquire, particularly on a worldwide basis. For relays to be cost-effective, they must not only extend coverage, but also improve both the capacity and efficiency of the system deployment.

In this chapter we studied and analyzed the effective relay efficiency, focusing on coverage and system capacity for a mobile WiMAX deployment. For relay systems without radio re-source sharing, a higher relay SNR-gain was required. This implies that the system requires a higher transmit power at the RS. In contrast, radio resource sharing offers high potential for a mobile WiMAX network when relays are deployed. First, for the cases considered, results demonstrated that resource sharing has the potential to double system efficiency compared to relay systems without resource sharing. Second, topology control was performed at the BS and RS nodes. Third, relay deployment extends the applicability of adaptive antenna systems

(AAS) to interference control, elimination and avoidance. In addition, results indicate that radio resource sharing is very applicable to MIMO relaying.

To achieve the practical application of radio resource sharing, a directional distributed relay topology was introduced for highly efficient relay deployment. This scheme is fully backward-compatible with the current mobile WiMAX standard. Results were presented for a realistic urban environment. Compared to a relay system without resource sharing, the implementation of sharing improved capacity significantly in this more realistic application scenario.

In general, this chapter has highlighted a technical overview and design strategies in order to understand the potential when integrating the multi-hop relay-based architecture in mobile WiMAX cellular systems. It also focused on the system efficiency which affects the relay deployment to provide design criteria for optimal architecture configuration.

References

[1] IEEE Std 802.16™-2004, 'Part 16: Air Interface for Fixed Broadband Wireless Access Systems', Oct. 2004.
[2] IEEE Std 802.16e-2005, 'Part 16: Air Interface for Fixed and Mobile Broadband Wireless Access Systems', Feb. 2006.
[3] Carl Eklund, Roger B. Marks, Subbu Ponnuswamy, Kenneth L. Stanwood, and Nico J.M. van Waes, *Wireless-MAX: Inside the IEEE 802.16™ Standard for Wireless Metropolitan Networks*, Standards Information Network, New York: IEEE Press, 2006.
[4] Ralf Pabst et al., 'Relay-based Deployment Concepts for Wireless and Mobile Broadband Radio', *IEEE Communications Magazine*, **42**, Sept. 2004, 80–89.
[5] E.H. Drucker, 'Development and Application of a Cellular Repeater', *IEEE Vehicular Technology Conf.*, June 1988, 321–325.
[6] H. Wu, C. Qiao, and O. Tonguz, 'Performance Analysis of iCAR (Integrated Cellular and Ad-hoc Relay System)', *IEEE International Conference on Communications*, **2**, 2001, 450–455.
[7] G.N. Aggelou and R. Tafazolli, 'On the Relaying Capability of Next Generation GSM Cellular Networks', *IEEE Personal Communications*, February 2001, 40–47.
[8] V. Sreng, H. Yanikomeroglu and D. Falconer, 'Coverage Enhancement Through Two-Hop Relaying in Cellular Radio Systems', in *IEEE Wireless Communications and Networking Conference*, 2002, (WCNC2002), **2**, March 2002, 881–885.
[9] http://ieee802.org/16/relay.
[10] IEEE C802.16j-06/043r3, 'Harmonized Contribution on 802.16j (Mobile Multi-Hop Relay) Usage Models', July 2006.
[11] Y. Sun, Y. Bian, A. Nix and P. Strauch, 'Study of Radio Resource Sharing for Future Mobile Wimax with Relay', IEEE Mobile Wimax'07, March 2007.
[12] E. Tameh, A. Nix and M. Beech, 'A 3-D Integrated Macro and Microcellular Propagation Model, Based on the Use of Photogrammetric Terrain and Building Data', IEEE VTC1997, **3**, May 1997, 1957–1961.
[13] WiMAX Forum, 'Mobile WiMAX – Part I: A Technical Overview and Performance Evaluation', White Paper, June, 2006.
[14] WiMAX Forum™ Mobile System Profile, Release 1.0 Approved specification, Nov. 2006.
[15] Y. Bian, Y. Sun, A. Nix and P. Strauch, 'Performance Evaluation of Mobile WiMAX with MIMO and Relay Extensions', in *IEEE Wireless Communications and Networking Conference 2007* (WCNC2007), March 2007.
[16] H.Yaghoobi, 'Scalable OFDMA Physical Layer in IEEE 802.16 Wirelessman', *Intel Technology Journal*, **8**(3), Aug. 2004.
[17] Y. Bian, A. Nix, E. Tameh and J. McGeeham, 'High Throughput MIMO-OFDM WLAN for Urban Hotspots', *IEEE VTC2005* Fall, **1**, Sept. 2005, 296–300.
[18] S. Catreux, V. Erceg, D. Gesbert, and R.W. Heath Jr., 'Adaptive Modulation and MIMO Coding for Broadband Wireless Data Networks', *IEEE Comm. Mag.*, **2**, June 2002, 108–115.
[19] IEEE 802.16a-03/01, 'Channel Models for Fixed Wireless Applications', Jun. 2003.
[20] https://www.ist-winner.org.

[21] 3GPP TR 25.996 v6.1.0, 'Spatial Channel Model for Multiple Input Multiple Output (MIMO) Simulations', Sept. 2003.

[22] A. Paulraj, R. Nabar and D. Gore, *Introduction to Space-time Wireless Communications*. Cambridge: Cambridge University Press, 2003.

[23] C. Rose, S. Ulukus and R.D. Yates, 'Wireless Systems and Interference Avoidance', *IEEE Transactions on Wireless Communications*, **1**(3), July 2002, 415–428.

[24] Shin Horng Wong and Ian J. Wassell, 'Channel Allocation for Broadband Fixed Wireless Access', *IEEE Wirelss Personal Mutimedia Communications*, 2002, **2**, Oct. 2002, 626–630.

[25] K. Ng, E. Tameh, A. Doufexi, M. Hunukumbure and A. Nix, 'Efficient Multielement Ray Tracing with Site Specific Comparisons Using Measured MIMO Channel Data', *IEEE Trans on Vehicular Technology*, **56**(3), May 2007, 1019–1032.

[26] C. Williams, Y. Bian, M.A. Beach, and A.R. Nix, 'An Assessment of Interference Cancellation Applied to BWA', *IEEE Mobile WiMAX* 07, March 2007.

[27] 2004/421/UK, 'UK Interface Requirement 2015; Public Fixed Wireless Access Radio Systems Operating Within the 3 to 11 GHz Frequency Bands Administered by Ofcom', V6.2, Feb. 2005.

[28] E TSI EN302.326-3 V1.1.1, 'Fixed Radio Systems Multipoint Equipment and Antennas; Part 3: Harmonized EN Covering the Essential Requirements of Article 3.2 of the R&TTE Directive for Multipoint Radio Antennas', Oct. 2005.

[29] Y. Bian and A. Nix, 'Throughput and Coverage Analysis of a Multi-Element Broadband Fixed Wireless Access (BFWA) System in the Presence of Co-Channel Interference', *IEEE VTC2006* Fall, Sept. 2006, 1–5.

12

Dimensioning Cellular Multi-hop WiMAX Networks

Christian Hoymann and Stephan Göbbels

12.1 Dimensioning Cellular 802.16 Networks

This section investigates the Carrier to Interference and Noise Ratios (CINRs) in a cellular 802.16 network. For dimensioning purposes, the Uplink (UL) and Downlink (DL) CINR from/at the most distant Subscriber Station (SS) are the interesting parameters. Thus, these worst cases are modeled in the following. In DL, the central Base Station (BS) transmits to the most distant SS, which is located at the cell border. In UL, the SS at the cell border transmits to the central BS. Interference is generated by co-channel cells that utilize the same frequency channel. In the considered network, DL and UL channels are assumed to be perfectly separated either by a Frequency Division Duplex (FDD) or by a fully synchronized Time Division Duplex (TDD) scheme. Thus, in DL, neighboring BSs cause interference, while in UL, interference is generated by SSs of neighboring cells.

12.1.1 Clustering and Sectorization

In order to avoid interference in cellular networks, frequency channels used within one cell can only be reused after a sufficient reuse distance. Hence, cells are combined into clusters, where frequency channels are uniquely assigned to cells. The frequency usage pattern of the entire cluster is regularly repeated throughout the network. Like this, the distance to co-channel cells can be increased. Figure 12.1(a) shows a cellular network with cluster order three. For a cluster of order k, the distance to co-channel cells D is only a function of the cell radius R [1] and is expressed as:

$$D = R\sqrt{3k} \tag{12.1}$$

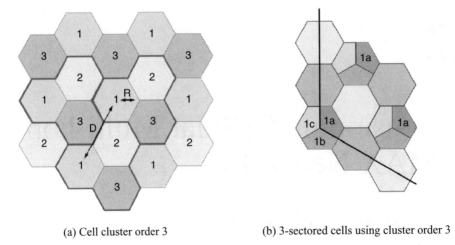

(a) Cell cluster order 3   (b) 3-sectored cells using cluster order 3

Figure 12.1   Cellular network

According to [13], the CIR only depends on the cluster order if noise is ignored and if the neighboring SSs are assumed to be centrally located in their cells. The Carrier to Interference Ratio (CIR) at a central BS receiving a signal from a SS at the cell border increases with cluster order k. With γ as the path-loss component, the CIR can be calculated as:

$$\frac{C}{I} = \frac{1}{6}\left(\frac{D}{R}\right)^{\gamma} = \frac{1}{6}(3k)^{\gamma/2} \tag{12.2}$$

Dividing cells into sectors is an established technique to further reduce the interference level in cellular wireless networks. The cell is subdivided into several sectors. Each sector is covered by a sector antenna. An individual frequency channel is assigned to each sector and an individual MAC protocol controls the access to the wireless channel. The sectorization of cells and the frequency assignment are periodically repeated all over the network. A sophisticated way to allow for sectorization with adaptive antennas is presented in [7].

Because the power that is emitted backwards from the BS sector antenna is minimized, the number of interfering co-channel cells can be reduced. Figure 12.1(b) shows a cellular network with 3-sectored cells. It is illustrated so that only two co-channel cells are visible for the receiving BS sector antenna instead of six in Figure 12.1a. Furthermore, in DL a receiving SS at the border of a cell receives only interference, which is generated by the most distant co-channel BSs. Thus, the number of interferers and the strength of interference can be reduced by sectorization. Analogous to the previous equation, the expected CIR in a sectorized and clustered cell is given by the equation, in which m is the number of sectors [13]:

$$\left(\frac{C}{I}\right)_{\text{sec torization}} = \frac{m}{6}\left(\frac{D}{R}\right)^{\gamma} = m\left(\frac{C}{I}\right)_{\text{non-sec torization}} \tag{12.3}$$

In the example illustrated in Figure 12.1b, sectorization with $m = 3$ sectors per cell increases the CIR by a factor of three. The drawback of sectorization is that the number of channels increases and therewith the number of users per sector decreases. This results in a lower trunking gain. Furthermore, several sector antennas have to be deployed at each BS and the control of each MAC entity per sector causes overhead.

Equations (12.2) and (12.3) do not consider noise and they assume that all co-channel interferers are equally distant. In the following analysis, the effect of noise is considered and positions of interfering BSs and SSs are modeled more accurately. This results in a separate analysis for UL and DL.

12.1.2 Mean Interference Generated by a Distant Cell

During UL transmission, SSs of co-channel cells generate interference. These SSs are randomly distributed within the cell area. From time to time, they are closer to the neighboring cell, and sometimes they are further away. Event-driven simulations are able to consider the current position of a SS precisely. Simulations average their measures over a certain simulation time. During the simulation run, the SS is moving so that in the end, it has appeared nearly everywhere in the cell. Like that, the influence of interference is averaged over time. The following analysis is a snap-shot of an actual CINR situation, so the interference is not averaged over time. Averaging the generated interference by just placing all SSs in the center of the cell is not correct, due to the non-linear influence of the pathloss. SSs, which are close to the neighboring cell, increase the co-channel interference more than distant SSs decrease it. In order to model the influence of co-channel interference accurately, the mean interference generated by a co-channel cell is calculated by assuming a planar transmitter with the shape of the hexagonal cell. The transmit power is equally distributed all over the cell surface area.

A comparable approach to model interference has been developed by Habetha and Wiegert [5]. They also estimated the mean interference as the integral over the cell area. However, Habetha and Wiegert [5] assumed circular cells and they did not consider the effect of noise. Sectorization was also not considered.

Figure 12.2 shows the cell of interest on the right and a co-channel cell that generates interference on the left. The mean interference level that is received by a central BS, which is located at x_0, y_0, can be calculated by assuming that each area element of the hexagonal cell transmits with equal fractions of the transmit power. According to its distance from the BS

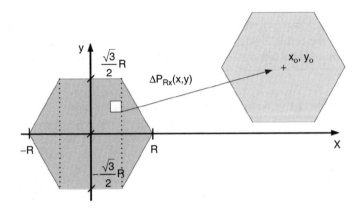

Figure 12.2 Interference received from a co-channel cell

of interest, each fraction of the transmit power is attenuated by the wireless channel. Thus, each area element of the distant cell generates one fraction of the overall receiver power. By integrating over the hexagonal cell area, the mean level of receiver power P_{Rx} can be calculated.

$$\overline{P_{Rx}} = \int_{area} \Delta P_{Rx}(x, y) = \int_x \int_y \frac{P_{Tx} * G_{Tx} G_{Rx}}{area} \text{pathloss}(x, y) \, dxdy \qquad (12.4)$$

Equation (12.4) integrates the receiver power per area element $\Delta P_{Rx}(x, y)$ over the surface area of the cell. In order to gain a location independent mean level of interference, the transmit power of the neighboring SS P_{Tx} is assumed to be uniformly distributed over the surface area. Each fraction of the transmitted power is attenuated by the pathloss and it is amplified by the antenna gain of the transmitter G_{Tx} and the receiver G_{Rx}.

The surface area of a hexagonal cell and the one-slope pathloss model with the pathloss coefficient γ can be calculated as follows:

$$\text{area} = \frac{3}{2}\sqrt{3}R^2 \text{ and pathloss}(x, y) = \beta D^{-\gamma} = \beta \sqrt{(y - y_0)^2 + (x - x_0)^2}^{-\gamma} \qquad (12.5)$$

The coordinates x_0, y_0 depend on the co-channel distance, which depends on the cell radius and the cluster order (refer to Equation (12.1)). Table 12.1 lists the coordinates for the most established cluster orders. The coordinates give the position of one out of six neighboring cells of the first tier of interferers. However, since the geometric arrangement of a hexagonal cell with its six neighbors is identical for each interfering cell, it is sufficient to calculate the interference level of one single cell.

We can sum up the receiver power per area element in Equation (12.6), which only depends on the variables x and y. All other characters are constants. Especially the term in front of the brackets does neither depend on x nor on y. For the integration it could be written outside the integral.

$$\Delta P_{Rx}(x, y) = \frac{P_{Tx} * G_{Tx} G_{Rx}}{\frac{3}{2}\sqrt{3}R^2} \beta \left[(y - y_0)^2 + (x - x_0)^2\right]^{-\frac{\gamma}{2}} \qquad (12.6)$$

Now, the receiver power per area element can be integrated over the cell area. To do so, the surface area of the cell is divided in three parts. The dotted lines in Figure 12.2 show the three parts. The limits of the integral in Equation (12.7) are set accordingly. Note that the inner

Table 12.1 Co-channel distances of various-cluster orders

cluster order	x_0	y_0	$D = \sqrt{x_0^2 + y_0^2} = \sqrt{3k}R$
1	0	$\sqrt{3}R$	$\sqrt{3}R$
3	$3R$	0	$3R$
4	0	$2\sqrt{3}R$	$2\sqrt{3}R$
7	$3R$	$2\sqrt{3}R$	$\sqrt{21}R$
12	$6R$	0	$6R$

y-limits depend on the x-coordinate in the first and the last part of the surface area.

$$\overline{P_{Rx}} = \int_{-R}^{-\frac{R}{2}} \int_{-\sqrt{3}x-\sqrt{3}R}^{\sqrt{3}x+\sqrt{3}R} \Delta P_{Rx}\, dy dx + \int_{-\frac{R}{2}}^{\frac{R}{2}} \int_{-\frac{\sqrt{3}}{2}R}^{\frac{\sqrt{3}}{2}R} \Delta P_{Rx}\, dy dx + \int_{\frac{R}{2}}^{R} \int_{\sqrt{3}x-\sqrt{3}R}^{-\sqrt{3}x+\sqrt{3}R} \Delta P_{Rx}\, dy dx$$

$$(12.7)$$

Unfortunately, the sum of double integrals in Equation (12.7) cannot be resolved into a closed form. Thus, it has been implemented in Matlab in order to be calculated numerically. The chosen granularity of the sum was 1/10000 of the cell radius. This granularity results in a difference of 0.02% between the integral and the sum if simply the size of the surface area is calculated by summation. If a circular instead of a hexagonal cell area is assumed, the integral is solvable for certain γ values [5].

For the pathloss, the suburban C1 Metropol pathloss model from the IST -WINNER project was chosen [12]. The C1 Metropol is composed of two models, a Line-of-Sight (LOS) and a Non Line-of-Sight (NLOS) model. The NLOS model is a one-slope model with the parameters given in Equation (12.8). The LOS case is a two-slope model. However, the antenna height is assumed to be high enough so that the break-point distance between the two slopes is beyond the distances of interest in the following analysis. As a result, a one-slope model with the parameters given in Equation (12.9) remains.

$$\text{NLOS:} \beta = 10^{-\frac{27.7}{10}} \quad \gamma = 4.02 \tag{12.8}$$

$$\text{LOS:} \beta = 10^{-\frac{41.9}{10}} \quad \gamma = 2.38 \tag{12.9}$$

In order to evaluate the interference generated by a distant cell, the mean receiver power P_{Rx} is normalized to the receiver power P_{Rx} of a single SS located at the center of the cell (refer to Equation (12.10)). Thus, a factor results that corrects the wrong assumption of a centered source of interference. This factor is used in the following analysis. There, the UL interference is calculated as if the neighboring SS is located in the center of the cell, and then the interference is corrected by the interference correction factor. However, the interference correction factor could also be used to enhance the calculations in Section 12.1.1. It could be easily integrated into Equations (12.2) and (12.3).

The first line of Table 12.2 (one sector per cell) lists the interference correction in percent for the LOS case. The first line of Table 12.3 shows the NLOS scenario. It has to be noted that the correction factor is independent of the cell radius. This is not directly visible from the calculation. Although the cell radius is part of the limits of the integral, its influence is canceled by the division with the receiver power P_{Rx} in (12.10). On the one hand a large cell radius increases the cell area, so that the difference between a transmitting point source and a

Table 12.2 Interference correction for LOS scenarios in [%]

cluster order → sectors per cell ↓	1	3	4	7	12
1	24.89	7.12	5.22	2.92	1.68
2	−21.48	−15.38	−13.56	−8.52	−8.82
3	−41.04	−27.50	−24.28	−20.90	−15.33
6	−48.18	−33.66	−30.12	−24.04	−19.51

Table 12.3   Interference correction for NLOS scenarios in [%]

cluster order → sectors per cell ↓	1	3	4	7	12
1	89.70	21.83	15.65	8.55	4.86
2	−21.97	−20.75	−18.71	−11.70	−13.37
3	−57.46	−41.00	−36.72	−32.28	−24.16
6	−65.98	−49.37	−44.91	−36.79	−30.45

planar transmitter increases. On the other hand, the large cell radius leads to a large co-channel distance, so that the difference between the point and the planar transmitter decreases.

$$
1 + \text{interference correction} = \frac{\overline{P_{Rx}}}{P_{Rx}} = \frac{\overline{P_{Rx}}}{P_{Tx} * G_{Tx}G_{Rx} * \text{pathloss}(0,0)}
$$

$$
= \frac{1}{\text{pathloss}(0,0)} \int\limits_{x} \int\limits_{y} \frac{\text{pathloss}(x,y)}{\text{area}} dy dx
\tag{12.10}
$$

By comparing the first lines of Tables 12.2 and 12.3, it can be seen that for LOS scenarios the correction is lower than for NLOS scenarios. This is due to the higher pathloss coefficient γ which causes the non-linear behavior of the attenuation. Beside the pathloss coefficient, the correction factor depends on the cluster order. With higher cluster orders, the co-channel distance increases. The further the distant cell, the more it looks like a point source. The influence of the hexagonal surface decreases. Both scenarios, LOS and NLOS show this behavior. In the LOS scenario the values range from 24.89% to 1.68% and in the NLOS case the values are between 89.7% and 4.86 %. That is, the mean interference of a distant LOS cell is between 1.68% and 24.89% larger than the interference generated by a transmitter located in the center of the cell. The mean interference of a NLOS cell is between 4.86% and 89.7% larger than the interference of a single-centered transmitter.

When a cell is covered by sector antennas, the geometry within the cellular network changes. The surface area, which is covered by one frequency channel, is no longer a hexagon. The shape of a sector and the relative position of interfering sectors depend on the number of sectors per cell and on the cluster order. Figures 12.1(b) and 12.3 show examples of different shapes of sectors and different relative positions between the sectors.

Furthermore, the orientation of the surface area of the interfering sector is always oriented away from the BS of interest. SSs of the sector, which are generating UL interference, are always further away from the receiving BS than the center of the co-channel cell. This leads to higher pathloss attenuation and a reduced level of interference.

Analogous to the calculation above, the mean interference generated by one sector can be derived. This results in an interference correction factor that can be applied when modeling UL interference within a sectored cellular network. In order to accurately model such interference, it is assumed to be generated at the center of the co-channel cell and afterwards, the received interference power is corrected by the correction factor.

Tables 12.2 and 12.3 list the interference correction in percent for the LOS and the NLOS cases. Since the sector is always further away than the center of the cell, the correction for two or more sectors per cell is always negative. This means that the mean interference generated by

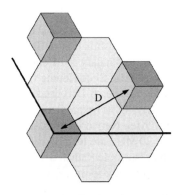

(a) 3-sectored cell using cluster order 4

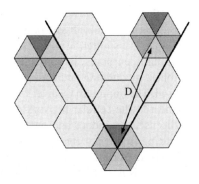

(b) 6-sectored cell using cluster order 7

Figure 12.3   Different shapes and positions of sectors

the planar sector is lower than the interference generated by a single source located at the center of the co-channel cell. Apart from cluster order one, the difference between the mean value and the approximation is larger for sectored cells than for non-sectored cells. For instance, the reduction for sectors is larger than 15.38% for cluster order three and LOS compared to an increase of 7.12% in non-sectored cells.

Analogous to a non-sectored network, the corrections for LOS scenarios are smaller than for NLOS ones. For instance, in Table 12.2 the reduction for two or more sectors per cell lies between 8.52% and 48.18%, whereas Table 12.3 shows reductions between 11.7% and 65.98%. The higher pathloss coefficient causes this effect. Furthermore, the values depend on the number of sectors and on the cluster order. With an increasing cluster order, the reduction approaches one. This is due to higher co-channel distances. With an increased number of sectors per cell, the shape of the sector narrows down so that the correction becomes larger.

For network dimensioning, the modeling of interference is a crucial part. The interference corrections in Tables 12.2 and 12.3 show that the approximation of a centrally located UL interferer is not very precise, especially while using sectorization and with small cluster orders. Now, the interference correction allows modeling of the UL interference more accurately through the mean interference generated by a planar transmitter, which has the shape of the cell or sector.

12.1.3 Cellular Scenario

The considered cellular scenario consists of a hexagonal cell with a central BS. The cell has a certain cell radius, which varies in the analysis. The cell is covered by one to six sectors and the network is clustered in groups ranging from three to twelve frequency channels. The first tier of six interfering co-channel cells is considered. The distance to the co-channel cells depends on the cluster order and the cell radius (refer to Equation (12.1)). The BSs of the co-channel cells are also centrally located. In contrast to Section 12.1.1, the unknown and varying locations of the interfering SSs are modeled by means of the mean interference generated by a planar transmitter (refer to Section 12.1.2).

Unlike in Section 12.1.1, noise is considered in the following analysis. Thermal noise of $-174\,$dBm/MHz and a noise figure of 5 dB is assumed. Antenna gain is ignored at the receiver as well as at the transmitter, it will be considered in the capacity estimation in Section 12.4. For dimensioning purposes, in particular the CINR at the cell boundary is of interest where the most robust modulation and coding scheme, i.e., Binary Phase Shift Keying (BPSK) 1/2 has to be used. The minimum receiver requirement for BPSK 1/2, i.e., 6.4 dB is taken from the 802.16 standard. The cellular network operates in the upper 5 GHz frequency bands, which had been licensed for indoor and outdoor Wireless Local Area Networks (WLAN). Eleven non-overlapping frequency channels with a bandwidth of 20 MHz are located between 5.47 and 5.725 GHz. In Europe, the maximum allowed Equivalent Isotropic Radiated Power (EIRP) within these bands is restricted to 1000 mW or 30 dBm. According to this, BSs and SSs are both transmitting with 1000 mW.

The suburban C1 Metropol pathloss model from the IST-WINNER project was applied during the analysis. The model was developed for the 5 GHz spectrum [12]. The pathloss model consists of two cases, a LOS and a NLOS case. Parameters for both parts are given in Equations (12.9) and (12.8). Equation (12.5) shows the general structure of a one-slope pathloss model. The DL and UL channels of the considered network are perfectly separated. This is accomplished either by synchronized switching points that separate DL and UL transmissions all over the TDD network or by an FDD scheme where channels are separated in the frequency domain. Like this, only the co-channel BSs cause interference in DL, while interference is generated by SSs of co-channel cells in UL. It is assumed that during the transmission of interest, every co-channel cell transmits and thus generates interference. This worst case analysis is valid for the broadcast phases in a synchronized network as well as for DL/UL data transmissions in a fully loaded network.

Figure 12.4 shows the CINR of an example scenario where the cell radius is 1000 m and the cluster order is seven. This results in a co-channel distance of 4583 m. Figure 12.4(a) plots the DL CINR over the surface area of the scenario. The transmitting BS is located in the center while six co-channel cells are located according to the cluster order. It can be seen that the CINR in the middle is quite high but it decreases with increasing distance to the BS. Near the co-channel cells, the CINR decays drastically. Figure 12.4(b) plots the CINR for an SS traversing the cell across the x-axis. One can see the BS position and the cell border. The height

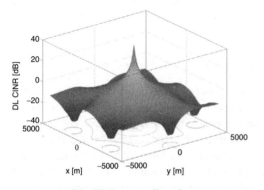

(a) DL CINR over cell surface area

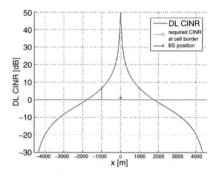

(b) DL CINR while traversing the scenario

Figure 12.4   DL CINR in a cellular network (cell radius 1000m, cluster order 7)

of the two stems at the cell border mark the minimum receiver requirement for BPSK1/2. It can be seen that the actual CINR level at the border, which is 6.46 dB, is shortly above the minimum requirement. Thus, the shown scenario has a sufficient CINR at the cell radius, but there is hardly any CINR margin left at the border.

In the following two sections only the CINR level at the cell border is evaluated. Thereby, the cell radius varies and the corresponding CINR at the cell border is plotted.

12.1.4 Downlink Transmission

In DL, the central BS transmits to the most distant SS, which is located at the cell border. For dimensioning, especially the CINR at the cell border is relevant, so it is plotted versus the cell radius in Figures 12.5 and 12.6.

Figure 12.5 shows the DL CINR at the cell border in a LOS scenario. Figure 12.5(a) illustrates the influence of clustering. With an increasing cluster order, the co-channel distance increases and the interference level decreases. This leads to an increased CINR at the cell border. The size of the cell radius affects the CINR in the same way: the larger the radius, the higher the CINR at the border. However, it can be seen that not all cluster orders are valid in the LOS scenario. Low cluster orders, such as three and four do not provide a sufficient CINR level at the cell border. Even very small radii are not satisfactory, because co-channel cells are very close and thus the level of interference is too high. For high cluster orders, e.g., seven or 12, the cell radius can range up to 1000m respectively 1475m to provide a proper CINR. In scenarios, where interference is the limiting factor, the system is called interference-limited.

In DL, sectorization reduces the number of interferers that are simultaneously receivable by the SS and it reduces the interference power level (refer to Section 12.1.1). Figure 12.5(b) shows the CINR at the cell border in a cellular network with cluster order seven and additional sectorization. The graph illustrates that the coverage area can be extended from 1000 m radius without sectorization to 1625 m with only two sectors per cell. A radius of up to 1775 m can be reached when the cell is covered by six sector antennas.

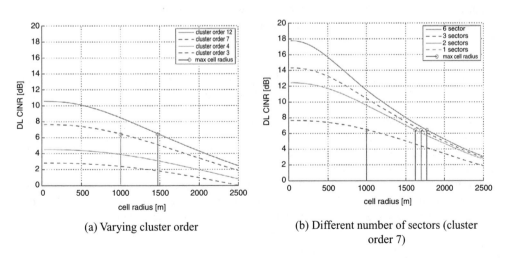

(a) Varying cluster order

(b) Different number of sectors (cluster order 7)

Figure 12.5   DL CINR at the cell border in a cellular LOS scenario

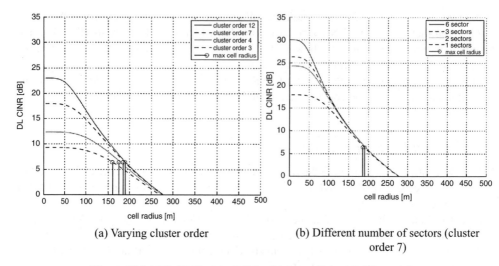

(a) Varying cluster order

(b) Different number of sectors (cluster order 7)

Figure 12.6   DL CINR at the cell border in a cellular NLOS scenario

For cluster order three and four, clustering only could not provide sufficient CINR. Sectorization can increase the CINR at the border to a proper level. This would allow for a valid network deployment with small cluster orders. Furthermore, clustering and sectorization not only increase the CINR level at the cell border, they improve the received signal quality all over the cell area. However, this capacity improvement is not part of this analysis.

In Figure 12.6, the NLOS pathloss model has been applied to the same scenario as before. In the NLOS pathloss model, the pathloss coefficient γ is nearly two times higher than in the LOS case (refer to Equation (12.8)). The high pathloss coefficient attenuates the interference level of distant co-channel cells more than it attenuates the carrier signal of the nearby BS. The CIR is increased. If noise is ignored, Equation (12.2) can be transformed to show the increased (UL) CIR in the NLOS scenario. Inequality 12.11 is always true because, first, the co-channel distance D is always larger than the cell radius R, which leads to D/R > 1 and, second, because the pathloss coefficient of a NLOS scenario is larger than for the LOS case, which leads to $\Delta\gamma > 0$.

$$\left(\frac{C}{I}\right)_{NLOS} = \frac{1}{6}\left(\frac{D}{R}\right)^{(1+\Delta)\gamma} = \left(\frac{C}{I}\right)_{LOS}\left(\frac{D}{R}\right)^{\Delta\gamma} > \left(\frac{C}{I}\right)_{LOS} \qquad (12.11)$$

Figure 12.6(a) shows the DL CINR at the cell border in such a NLOS scenario. The graph shows the increased CIR at small radii, where the carrier signal is much larger than the noise level. All cluster orders are able to cover the entire cell area, at least with cell radii smaller than 160 m. However, the absolute values of the carrier and the interference signals attenuate faster in NLOS scenarios so that the influence of the constant noise level increases. Although the CIR is higher, the CINR decays much faster in NLOS scenarios. In Figure 12.6(a) this effect can be seen at large cell radii. The CINR is lower than Figure 12.5(a). Since the level of interference in the NLOS example is rather low, it does not affect the CINR a lot, the valid cell radii for different cluster orders vary only between 160 and 190 m. In this scenario the system is limited by the noise, so it is called noise-limited.

Figure 12.6(b) plots the DL CINR at the border for a varying number of sectors. For small cell radii, sectorization can increase the CINR at the cell border. For larger radii, the system becomes noise-limited, so that the interference reduction by means of sectorization has nearly no effect on the valid cell radii. Sectorization increases the maximum cell radius of a network with cluster size seven only from 185 to 190m with any number of sectors.

Beside clustering and sectorization, several other features may increase the CINR level and thus extend the DL coverage. They are listed in the following:

- *BS transmit power.* The transmit power of the BS was aligned to the maximum EIRP allowed in the targeted 5 GHz spectrum. If regulations allow increasing the transmit power, all co-channel BSs may increase their transmit power, too. The signal strength of carrier and interference grow the same way and finally, the CIR stays constant. Thus, an increased transmit power will have nearly no effect on the maximum cell radius in scenarios where the system is interference-limited. Nevertheless, the transmit power affects the CINR in noise-limited scenarios. There, it can increase the DL coverage area.
- *DL subchannelization.* The mobility amendment of IEEE 802.16e expands subchannelization to the DL data transmission. If BSs transmit on a subset of subcarriers only, the number of interferers per subcarrier can be reduced [10]. However, the spectral density and thus the transmission range stays constant. This feature is beneficial in interference-limited systems. If BSs transmit in subchannelization mode on all subcarriers, the interference situation does not change at all.
- *BS transmit antenna gain.* During the DL subframe, a BS with adaptive antennas can steer its transmit antenna to the receiving SS so the BS transmit antenna gain improves the signal quality [3]. This reduces the inter-cell interference since less power is emitted in undesired directions. If regulations allow exceeding the EIRP by focusing the transmission power and thus increasing the spectral density, the received signal strength at the SSs is increased. This is additionally useful in noise-limited systems.
- *Network load.* A non-saturated system where not all co-channel BSs are constantly trans-mitting has a reduced level of interference.

The mentioned features to increase the CINR level are only valid during the scheduled DL data transmission. Nevertheless, the synchronized broadcast phase of a cellular network, in which all cells are transmitting omni-directionally on all available subcarriers, is the worst case. A dimensioning approach should focus on this phase.

12.1.5 Uplink Transmission

In UL, SSs transmit to the central BS. For dimensioning, the most distant SS, which is located at the cell border, is most critical. Interference is generated by SSs of neighboring cells, which utilize the same frequency channel. The SSs' position is randomly distributed over the cell area. To model the interference level as accurately as possible, the mean level of interference as was deduced in Section 12.1.2 is used. Figures 12.7 and 12.8 show the CINR perceived at the central BS while the most distant SS is transmitting. The UL CINR is plotted versus the cell radius.

In general, the UL CINR is quite similar to the DL CINR investigated in the previous section. On the one hand, it is a little bit lower because in UL, the receiver is located at the center of

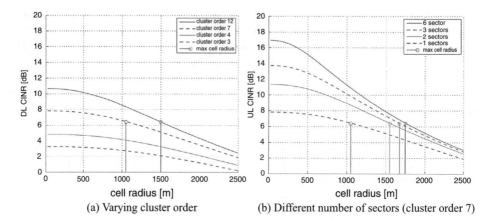

(a) Varying cluster order   (b) Different number of sectors (cluster order 7)

Figure 12.7   UL CINR at the BS in a cellular LOS scenario

the cell and not at the cell border. Hence, the level of interference is slightly reduced. On the other hand, the CINR is reduced because in UL, SSs generate interference and not the central BSs. As outlined in Section 12.1.2, their mean level of interference is slightly higher than that of co-channel BSs.

Figure 12.7(a) illustrates the influence of clustering in a LOS scenario. Cluster orders three and four do not lead to a sufficient CINR. Cluster order seven and 12 allow cell radii of 1050 and 1500 m respectively. Here, the system is interference-limited.

In UL, sectorization reduces the number of SSs that are simultaneously receivable by the BS sector antenna and it reduces the interference power level (refer to Sections 12.1.1 and 12.1.2). Figure 12.7b shows the UL CINR in a cellular network with cluster order seven and additional sectorization. The radius of the coverage area can be extended from 1050 m without sectorization to 1550 m with two sectors per cell. A radius of up to 1750 m is valid with six sectors per cell.

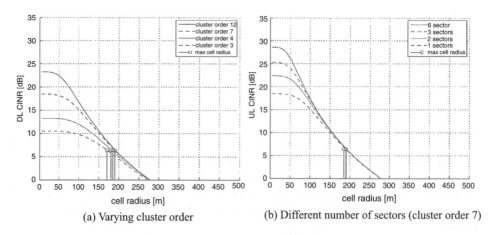

(a) Varying cluster order   (b) Different number of sectors (cluster order 7)

Figure 12.8   UL CINR at the BS in a cellular NLOS scenario

Figure 12.8(a) shows the UL CINR in a NLOS scenario. The CIR increases with small radii and all cluster orders are able to provide proper signal quality with cell radii smaller than 170 m. With larger cell radii, the CINR decays rapidly due to the high pathloss coefficient and the resulting influence of noise. Again, the level of interference in the NLOS case is so low that the valid cell radii for different cluster orders vary only between 170 and 190 m. In the NLOS scenario the system is noise-limited.

The UL CINR for cluster order seven and for a varying number of sectors is plotted in Figure 12.8(b). For small cell radii, sectorization can increase the CINR, but for larger radii, the interference reduction by means of sectorization has nearly no effect on the valid cell radii. Reducing interference does not give any benefit in a noise-limited system. With two sector antennas per cell it can be increased from 185 to 190 m. With more sectors it remains constant.

Beside clustering and sectorization, several other features may increase the UL CINR level and thus extend the UL transmission range:

- *SS transmit power.* The SSs' transmit power was set to the maximum allowed EIRP. Portable and mobile SSs will most probably be battery-powered. Their restricted power consumption may force the devices to reduce the transmit power, which will reduce the carrier strength. If all co-channel SSs transmit with reduced power, too, interference is reduced the same way and the CIR stays constant. In interference-limited systems, the possible link distances are nearly not affected. In noise-limited systems, a reduced transmit power leads to reduced coverage.
- *UL subchannelization.* UL subchannelization is specified for initial ranging, for Bandwidth (BW) requests and for UL data transmission. Subchannelization during ranging and BW request procedures permits focusing the transmit power onto a subset of subcarriers. This increases the spectral density by 12 dB and extends the transmission range significantly [9]. Since this feature increases the carrier signal and reduces interference, it is beneficial in both, interference and noise limited systems. If the transmit power per subcarrier stays constant during UL data transmission, interference limited systems benefit from subchannelization: if all SSs are using a subset of the available subcarriers, the number of interfering stations per subcarrier is reduced.
- *Network load.* In a non-saturated system not all co-channel cells have constantly active transmissions. This reduces the number of interferers.
- *BS receive antenna gain.* During the scheduled part of the UL subframe, the BS can focus its receiver antenna on the transmitting SS so that the BS receiver gain improves the signal quality [Godara, 1997]. Since an adaptive antenna can reduce the received interference and increase the receiver carrier strength, it is useful in all scenarios. During the scheduled part of the UL subframe, the BS can focus its receive antenna to the transmitting SS so that the receiver gain improves the signal quality.

Features to increase the CINR level are applicable during the scheduled UL data transmission (subchannelization and antenna gain) and during the contention-based access (subchannelization). In particular, subchannelization extends the UL range significantly. If this optional feature is implemented by the manufacturer, the UL transmission is most probably not the limiting factor in a cellular 802.16 network. As mentioned in the previous section, a dimensioning approach should focus on the synchronized DL broadcast phase. This is critical.

12.2 Dimensioning Cellular Multi-hop 802.16 Networks

This section investigates the CINRs in cellular 802.16 networks that are enhanced by multi-hop transmissions. Again, the UL and DL CINRs from/at the most distant SS are investigated since they are crucial to dimensioning.

12.2.1 Cellular Multi-hop Scenarios

For the following dimensioning of multi-hop networks, basically, the same parameters as in the single hop case were assumed (for details refer to Section 12.1.3). Hence, hexagonal cells that are clustered according to a given cluster order. Inter-cell interference is generated by the six co-channel cells of the first tier. The network operates in the 5 GHz spectrum using a channel bandwidth of 20 MHz. The transmit power of all stations is restricted to the EIRP, i.e., BSs, SSs, and relays are transmitting with 1 W. LOS and NLOS propagation conditions are taken into account.

DL and UL channels are perfectly separated either by TDD or by FDD schemes. According to the design of the multi-hop-enabled MAC frame transmissions on the first and the second hop are assumed to be perfectly separate in time [8].

The positioning of relay stations may vary according to the intended benefit. Two different scenarios are be distinguished in the following.

Coverage scenario. In order to extend the coverage area of the cell, relays may be placed at the border of the BS's transmission range. The distance between the BS and the relay equals the original BS cell radius. Figure 12.9(a) illustrates such a scenario. It plots the BS's coverage area with dotted lines. Three relays at the corners extend the coverage area by a factor of three.

$$A_{\text{multihop}} = 3\frac{3}{2}\sqrt{3}R^2 = 3 * A_{\text{singlehop}} \tag{12.12}$$

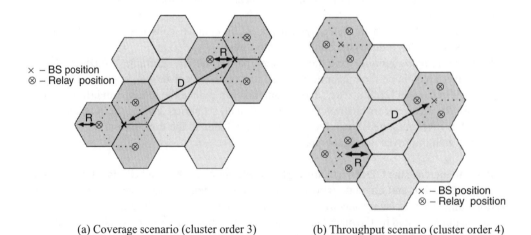

(a) Coverage scenario (cluster order 3)   (b) Throughput scenario (cluster order 4)

Figure 12.9 Multi hop scenarios [15]

Due to the extended coverage of a relay enhanced cell, the co-channel distance is enlarged by a factor of $\sqrt{3}$ in these scenarios. Based on Equation (12.1), the co-channel distance of the coverage scenarios can be calculated as

$$D = 3R\sqrt{k} \tag{12.13}$$

- *Throughput scenario.* Placing relays within the BS's coverage, the CINR and thereby the throughput is increased. The distance between the BS and the relay is half the cell radius. Figure 12.9(b) illustrates this kind of scenario. The co-channel distance equals the one from the single hop scenario given in Equation (12.1). From a dimensioning perspective, the throughput scenario equals the single hop case, since the entire cell area is covered by the BS. Thus, the results given in Section 12.1 can be directly applied to the throughput scenario.

SSs are assumed to utilize omni-directional antennas all the time. Classical sectorization is not regarded in the multi-hop scenarios. However, directive antennas are optionally used at the relay stations to reduce the inter- and intra-cell interference. The antenna angle is set to 240°. The relay antenna look direction always points away from the central BS. The directive antenna's backward signal is suppressed to zero and the forward signal is transmitted undistorted, thus the antenna characteristic is assumed to be perfect. Figure 12.10 illustrates the resulting coverage area of relay stations that use directive antennas. It can be seen that the angle has been chosen so that the other two relays of the same cell are not affected by the transmission. At the same time, the number of co-channel interferers is reduced by a factor of 2/3.

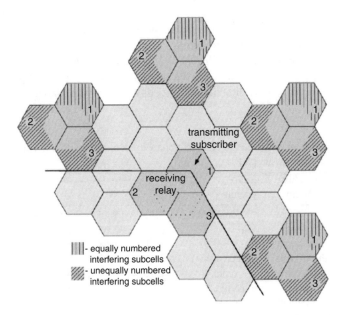

Figure 12.10   Coverage scenario with co-channel interferers (directional antennas, cluster order 4) [15]

12.2.2 Mean Interference Generated by Multi-hop (Sub-)Cells

In DL, co-channel interferers are co-channel BSs, which are centrally located. Relays suffer interference from co-channel relays, whose positions are well known. In UL, co-channel SSs generate interference. Their position is unknown and may vary within the cells coverage area. In order to model the random SS position, in the following UL co-channel interference is assumed to be generated by a planar transmitter instead of a centrally located point source. The planar transmitter has the shape of the interfering BS cell or the relay's subcell.

A comparable model for interference in multi-hop networks has been presented by [14]. However, Wijaya assumed circular cells instead of hexagonal and he did not consider the effect of noise. Directive antennas and simultaneous operation of relays were not covered either.

Analogous to the single hop dimensioning approach, the receiver power of a signal that is emitted by a planar transmitter can be calculated. This receiver power models the interference more accurately than the assumption of centrally located SSs. According to the method derived in Section 12.1.2 the interference correction factors for distant cells and subcells are calculated. The results are listed in Table 12.4. The factors depend on the pathloss model, the scenario type and the antenna characteristic. Apart from cluster order one, the factors for omni-directional antennas are positive and the corrections for directive antennas are negative. Hence, the interference of a hexagonal planar transmitter is higher than the centrally generated interference. Using directive antennas, co-channel SSs are located behind the central position of the relay so that their interference level is lower.

In order to increase cell capacity, relays might operate simultaneously in Space Division Multiplex (SDM). SDM operation means that a distinct relay station (station of interest) transmits and receives in parallel to the other two relays of the same cell and in parallel to all relays of the co-channel cells. Interference is generated by three different sources. In order to explain the sources, the relay stations are numbered in Figure 12.10. Each relay that is on the BS's upper right-hand side is numbered with 1. The relays on the BS's left-hand side are numbered 2, and the lower right relays get the number 3.

First, equally numbered subcells of distant co-channel cells interfere. As in the single hop scenario, each source is equally distant and the subcells' look direction is equal to the subcell of interest, e.g., the antenna look direction of subcells numbered 1 are all pointing to the upper

Table 12.4 Interference correction of equally numbered, distant multi-hop subcells within LOS and NLOS scenarios in [%]

pathloss ↓	scenario ↓	cluster order → antenna ↓	1	3	4	7	12
LOS	coverage	omni	7.12	2.25	1.68	0.96	0.56
LOS	coverage	directive	−0.58	−2.61	−2.02	−1.20	−1.88
LOS	throughput	omni	5.21	1.68	1.26	0.72	0.43
LOS	throughput	directive	−2.16	−2.02	−2.39	−1.50	−1.42
NLOS	coverage	omni	21.82	6.55	4.86	2.74	1.59
NLOS	coverage	directive	7.62	−1.81	−1.51	−0.96	−2.53
NLOS	throughput	omni	15.64	4.86	3.61	2.05	1.19
NLOS	throughput	directive	2.56	−1.51	−2.59	−1.73	−1.93

Table 12.5 Averaged interference correction of unequally numbered, distant multi-hop subcells within LOS and NLOS scenarios in [%]

pathloss ↓	cluster order → scenario ↓	antenna ↓	1	3	4	7	12
LOS	coverage	omni	91.29	22.66	15.42	8.37	4.72
LOS	coverage	directive	205.33	64.75	51.21	33.71	21.82
LOS	throughput	omni	82.74	15.42	11.36	6.13	3.49
LOS	throughput	directive	193.16	51.21	39.11	28.06	19.99
NLOS	coverage	omni	503.49	80.39	50.19	25.79	14.04
NLOS	coverage	directive	1033.30	185.50	126.33	74.60	45.80
NLOS	throughput	omni	500.44	50.19	36.09	18.44	10.25
NLOS	throughput	directive	1086.00	126.00	93.80	59	39.50

right. Correction factors of these sources are listed in Table 12.4. It can be seen that directive antennas reduce the interference compared to a central source, because in general, the SSs are located further away from the station of interest. Using omni-directional antennas, the correction is comparable to the single hop correction listed in the Tables 12.2 and 12.3.

Second, in SDM operation, unequally numbered subcells of distant co-channel cells interfere. Their distance to the station of interest varies and the look directions of these sources are different, since they are located at different positions relative to the central BS. For instance, the coverage areas of subcells numbered 2 are on the left-hand side of the relay. So some interfering SSs are closer, others are further away. Averaged interference corrections that take these effects into account are listed in Table 12.5. Due to the varying distances, these factors are higher than the previous ones.

Third, in SDM operation and with omni-directional antennas, relays of the same cell interfere. This interference is worse since the interfering SSs are close to the relay station of interest. Table 12.6 lists the corrections to model this kind of interference.

12.2.3 Time Division Multiplex of Relay Subcells

This section outlines the dimensioning of the coverage scenario, in which the relays operate in Time Division Multiplex (TDM). Relay stations use omni-directional antennas. Figure 12.11 illustrates the DL CINR for this scenario. The potential subscriber is always connected to the best server, i.e., either the central BS or the relays. Figure 12.11(a) shows four CINR peaks, which results from the central BS and three relays. The perceived DL CINR of a SS that traverses the scenario area is plotted in Figure 12.11(b). It shows the position of the central BS as well as the position of the relay at a distance of 1000 m. The height of the stems at the cell border reflects the minimum required CINR. Since the actually perceived CINR is higher than

Table 12.6 Interference correction of multi-hop subcells of the same cell in [%]

	omni antenna	directive antenna
LOS	24.89	−18.15
NLOS	89.68	−16.47

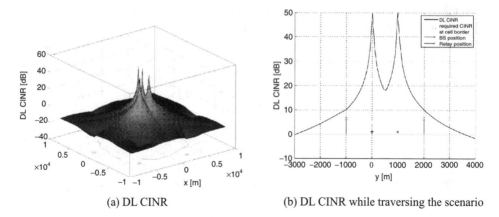

(a) DL CINR   (b) DL CINR while traversing the scenario

Figure 12.11   DL CINR in a cellular multi-hop scenario (TDM, omni antennas, LOS) [15]

the minimum required one, all SSs of that particular scenario are able to receive at least with the most robust modulation and coding scheme.

Figure 12.12 illustrates the best server, i.e., the BS or relay, which provides the best CINR level to the potential subscriber. Additionally, circles are plotted, whose radii are the cell and subcell radii as depicted in Figure 12.9(a). The best server analysis shows that the inner part of the relay-enhanced cell is covered by the BS while the outer areas are covered by the relays. Since the relays use omni-antennas, the points where BS and relay provide equal CINR levels are inside the BS's original cell.

For dimensioning purposes, the signal quality for the most distant station or most interfered with station is crucial. In a relay-enhanced cell, this position is at the outer border of a relay

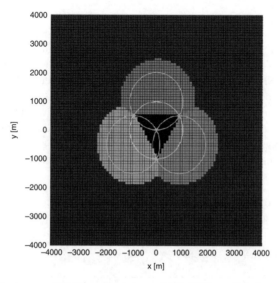

Figure 12.12   DL best server (TDM, omni antennas, LOS, 1000m cell radius, cluster order 7) [15]

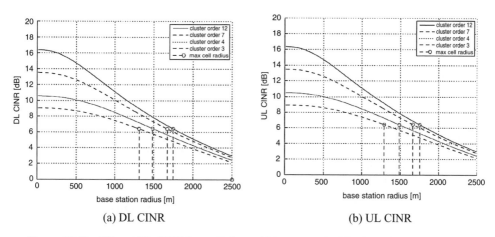

(a) DL CINR　　　　　　　　　　　　　　　　　　　(b) UL CINR

Figure 12.13　DL and UL CINR in a cellular multi-hop scenario (TDM, omni antennas, LOS)

subcell. In Figure 12.12 this is at the coordinates x = 0 m, y = 2*cell radius. In DL, the receiving SS is located at the subcell's border. The transmitter is the central BS or one of the three relay stations, whichever is most beneficial. In UL, the same SS transmits while the BS or one of the relay stations receives.

In Figure 12.13, the DL and UL CINR that is perceived by/from this particular SS is plotted versus the BS radius. Note that the BS radius is the radius of the inner cell. The overall coverage of the relay-enhanced cell is extended by the relay to a maximum radius of twice the original radius (refer to Figure 12.9(a)).

Figure 12.13(a) plots the DL case. For all cluster orders larger than one, a valid cell radius can be given. For the single hop case, only cluster orders larger than four were valid (refer to Figure 12.5(a)). The valid cell radii can be extended in the scenario. For instance, BS radii of up to 1675 m are possible for cluster order of seven. In the single hop case, a radius of 1000 m was the limit. Note that the most distant SS is 3350 m away from the BS. The overall coverage area of the BS is extended from 2.598 km^2 to 21.868 km^2 (refer to (12.12)). Thus, three relays can extend the coverage area by a factor of 8.4.

The UL case is plotted in Figure 12.13(b). The valid BS radii are similar to the one of the DL case. As in the single hop scenario, the LOS scenario is interference-limited. By increasing the cluster order, the inter-cell interference is decreased. This improves the CINR and thus extends the maximum cell radius.

Figure 12.14 plots the coverage scenario with NLOS propagation conditions. The plots for DL and for UL signal quality do not differ much. In general, the valid cell radii are smaller than for LOS conditions. They range only up to 192 m. Compared to the single hop system introduced in Section 12.1 the maximum BS radius cannot be significantly extended (only from 185 m to 192 m for cluster order seven). However, the three relay stations increase the coverage area from 0.0889 km^2 to 0.287 km^2, which is a factor of 3.23.

Using NLOS links, the system is noise limited, because modifying the interference by varying the cluster order does not affect the maximum cell radius. It stays nearly constant. The heavy attenuation of inter-cell interference allows cluster order one to provide valid cell radii. Radii of approximately 150 m are possible.

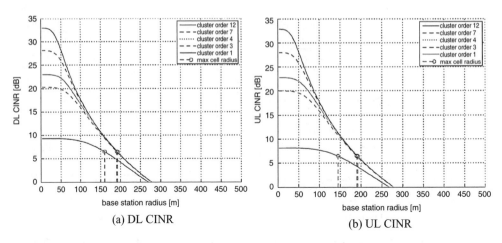

(a) DL CINR   (b) UL CINR

Figure 12.14   DL and UL CINR in a cellular multi-hop scenario (TDM, omni antennas, NLOS)

12.2.4 Space Division Multiplex of Relay Subcells

Operating relays in TDM results in high signal quality and large coverage. However, this approach shortens the portion of the MAC frame that is dedicated to each station. The MAC frame capacity can be increased by operating the relays simultaneously. Like this, the frame needs to be divided only into two parts: one dedicated to the BS, and the other is simultaneously used by the relays.

The investigated scenario is the same as the one in the previous section, except that the three relays and their associated SSs transmit and receive concurrently. Figure 12.15 shows the best server of the relay enhanced cell. Again, the inner part of the cell is covered by the BS. The

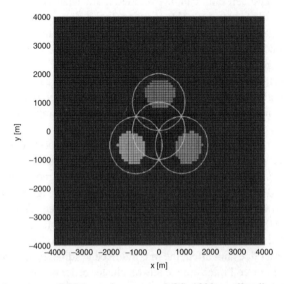

Figure 12.15   DL best server (SDM, omni antennas, LOS, 1000m cell radius, cluster order 7) [15]

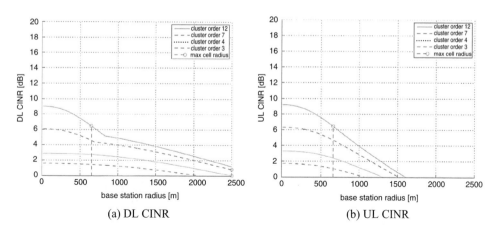

(a) DL CINR   (b) UL CINR

Figure 12.16   DL and UL CINR in a cellular multi-hop scenario (SDM, omni antennas, LOS)

BS's situation has not changed, so its coverage is equal to the one in the TDM approach. The outer areas are covered by the relays, but compared to the TDM case, the relays' coverage has seriously shrunk. Some parts of the relay subcell have to be covered by the BS, others are not covered at all. In SDM, relays operate simultaneously and interfere with each other. The number of interferers has more than tripled and the two relays of the same cell are quite close. This reduces the CINR and thus the coverage area.

Figure 12.16 plots the DL and UL CINR versus the BS radius of the SDM scenario. For the DL case shown in Figure 12.16a only cluster order twelve is valid. The maximum cell radius is 650 m. For all other cluster orders, the CINR never reaches the minimum threshold of 6.4 dB.

Furthermore it can be seen that the curves for cluster order seven and twelve change their shapes around the radius of 750 m. With small radii, the receiving SS is covered by the BS although the distance to the relay is much shorter than the distance to the BS. The interference during the relay phase is so high that the perceived CINR from the BS surpasses the CINR from the relay. Passing a certain distance, the relay becomes the best server for the SS of interest.

In UL, the situation is similar. Only cluster order twelve allows valid cell radii, which range up to 663m. The shape of the curves indicates that the transmitting SS is always covered best by the BS. The inter- and intra-cell interference degrades the CINR during the relay phase. In UL the interference situation is even worse, since the sources of interference, i.e., the co-channel SSs might be closer to the receiving BS. Thus, the average received interference is increased by the correction factors of Tables 12.4, 12.5, and 12.6.

Figure 12.17 plots the SDM scenario under NLOS conditions. The high pathloss degrades inter and intra-cell interference so that even cluster order three results in a valid deployment. As in the previous scenarios, the NLOS case is noise-limited. In DL, all cluster orders allow nearly the same maximum cell radius of approximately 175 m. The UL interference is higher and the valid radius ranges only up to 95 m.

12.2.5 Space Division Multiplex in Combined LOS-NLOS Scenarios

Section 12.2.4 showed that a simultaneous operation of multiple relays within one cell is not advantageous due to heavy mutual interference. However, in special environments the scenario itself limits the mutual interference.

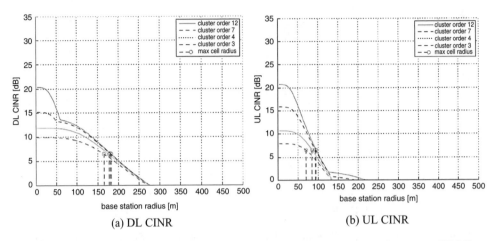

(a) DL CINR   (b) UL CINR

Figure 12.17   DL and UL CINR in a cellular multi-hop scenario (SDM, omni antennas, NLOS)

In urban Manhattan-like scenarios, the source and the destination have a direct LOS connection along the streets. In contrast, the first tier of interferers is shadowed behind buildings, thus a NLOS pathloss results. The same effect occurs in wide-area scenarios when the BSs are deployed with an antenna tilt. Like this, the SSs of the cell have a LOS connection while the SSs of distant co-channel cells do suffer from NLOS attenuation. In both deployments, the carrier signal is attenuated with the LOS pathloss coefficient, while the interfering signals are attenuated with the NLOS coefficient.

Figure 12.18 plots the best server of such a combined LOS-NLOS scenario. Omni-directional antennas are used. The NLOS propagation reduces the co-channel interference of distant cells

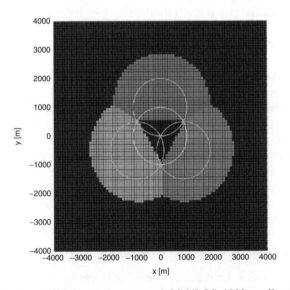

Figure 12.18   DL best server (SDM, omni-antennas, LOS-NLOS, 1000m cell radius, cluster order 7)

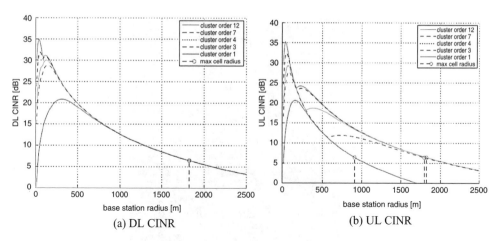

(a) DL CINR   (b) UL CINR

Figure 12.19   DL and UL CINR in a cellular multi-hop scenario (SDM, omni antennas, LOS-NLOS)

and the mutual intra-cell interference of relays of the same cell. Thus, the coverage of the relay-enhanced cell is extended far beyond the cell border in the shown example. The corresponding relays' coverage areas of the same cell adjoin.

Figure 12.19 plots the DL and UL CINR versus the BS radius of the LOS-NLOS scenario. The shape of the curve differs from the previous ones. Having small radii, the interferers are close and the system is interference-limited. With an increasing radius, the interference attenuates faster than the carrier signal, because they are following different pathloss coefficients. Due to the relative increase in the carrier signal compared to the interference, the CINR increases.

By increasing the radius further, the level of interference becomes small compared to the noise level and the behavior of the system switches to noise-limited. From this point, the attenuation of the carrier signal results in a decreasing CINR. The resulting maximum cell radius lies far beyond that switching point in the noise-limited part. For all cluster orders the maximum radius in DL is 1825 m. Note that, cluster order one is valid and it results in the same coverage area of all the other cluster orders.

The UL case in Figure 12.19(b) shows the same behavior, which results in the same maximum cell radius of 1825 m. Only cluster order one restricts the radius to 910 m. The shape of the curves indicates that by increasing the radius, the best server transits from the BS to the relay for some cluster orders. With small radii the BS provides access while the relays provide access with large radii.

12.2.6 Space Division Multiplex with Directive Antennas

In order to control the interference in SDM operation in pure LOS and NLOS scenarios, directive antennas can be used. As shown in Figure 12.10, the directive antenna covers the entire subcell area but the relay does not interfere with the two other subcells of its own cell. Like this, the two closest sources of interference are suppressed and the number of distant co-channel SSs is reduced to twelve.

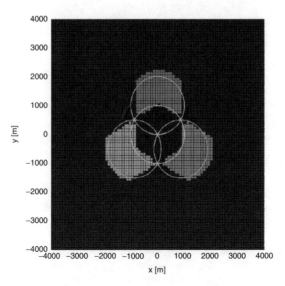

Figure 12.20   DL best server (SDM, directional antennas, LOS, 1000m cell radius, cluster order 7) [15]

Figure 12.20 plots the best server of a scenario utilizing directive antennas. Since the BS still uses omni-directional antennas and its operation is separated in time, the BS coverage does not change compared to the scenarios using omni-directional antennas. However, the relay coverage looks much better. It can be seen that the SS of interest at the distant subcell border can be covered assuming an example radius of 1000 m. Furthermore, the antenna angle of 240° is visible. Due to the ideal antenna shape, the relay does not provide a sufficient CINR level in the original (single-hop) BS cell border.

Figure 12.21 plots the CINR versus the BS radius for the LOS scenario. Compared to the SDM case with omni-antennas, the situation is improved. Cluster orders seven and twelve

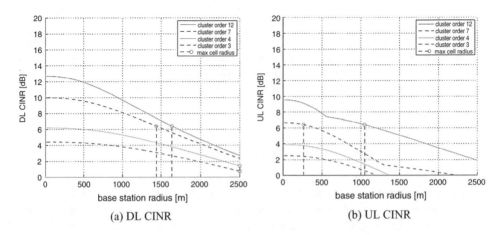

(a) DL CINR   (b) UL CINR

Figure 12.21   DL and UL CINR in a cellular multi-hop scenario (SDM, directional antennas, LOS)

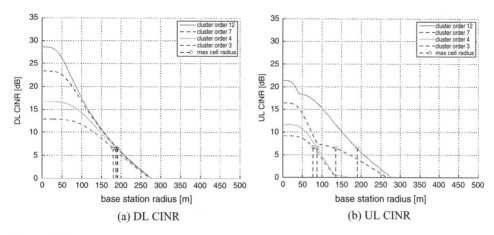

Figure 12.22   DL and UL CINR in a cellular multi-hop scenario (SDM, directional antennas, NLOS)

allow for radii up to 1640 m. This range is close to the radii of the TDM case shown in Figure 12.13(a). However, cluster orders below seven are not sufficient for this setup.

In UL, the same cluster orders are valid, but the achievable range is much smaller (refer to Figure 12.21b). Again, the random positions of interfering SSs cause this decrease. However, the UL CINR is not the limiting factor, because it can be improved by the advanced features, described in Section 12.1.5.

The NLOS scenario is plotted in Figure 12.22. Due to the noise-limited nature of that scenario, the reduced interference has nearly no effect on the cell size. Valid cluster orders between three and twelve allow for radii of up to 960 m. Compared to the TDM operation in Section 12.2.3, cell sizes are approximately the same. Only a single-frequency network, i.e., cluster order one, is not possible.

Beside cluster order twelve, the UL scenario shown in Figure 12.21(b) shows minor performance change from the DL. The transition of the serving station from the BS to the relay causes the change in the shape of the curve and therewith the large variance in the valid UL cell radii.

In Manhattan-like LOS-NLOS scenarios, the usage of directional antennas causes only a marginal improvement in the coverage area. In these environments, the mutual interference is already so low that a further reduction by means of directional antennas is not necessary.

12.2.7 Summarized Coverage Areas of Cellular Single-hop and Multi-hop Scenarios

Table 12.7 summarizes the resulting coverage areas in cellular multi-hop and single-hop deployments. The table contains results for different scenarios using cluster order seven. In order to calculate the coverage area by Equations (12.5) and (12.12) the maximum valid cell radii have been taken from the analysis above. Note that, the coverage of the multi-hop throughput scenario equals the single-hop case. The SDM operation using omni-directional antennas in a

Table 12.7 Maximum BS coverage area in cellular singlehop and multihop networks (DL, cluster order 7)

pathloss	singlehop coverage		multihop coverage		
	1 sector*	3 sectors	TDM	SDM omni anten	SDM dir. anten
NLOS	$0.0889\ km^2$	$0.0938\ km^2$	$0.287\ km^2$	$0.253\ km^2$	$0.281\ km^2$
LOS	$2.5981\ km^2$	$7.2892\ km^2$	$21.868\ km^2$	**	$16.117\ km^2$
LOS-NLOS	$8.6533\ km^2$	$8.6533\ km^2$	$25.960\ km^2$	$25.960\ km^2$	$25.960\ km^2$

*Applicable to multihop *throughput scenario*.
**no valid cell radius.

LOS environment did not result in a valid cell radius. However, Section 12.4 explains that by combining SDM and TDM, the same coverage area of the TDM mode can be achieved.

12.2.8 Capacity of Cellular 802.16 Networks

Having calculated the CINR in a cellular scenario, it can be converted into the resulting cell capacity. The link capacity at a certain position $Cap(x,y)$ is a function of the perceived CINR. Reference [6] lists gross PHY data rates for each available PHY mode in the given scenario (20 MHz bandwidth, Cyclic Prefix (CP) of 1/4). Minimum receiver requirements (in terms of required CINR) for each PHY mode are given in [9]. If perfect link adaptation is assumed, the perceived signal quality can be converted into an achievable data rate. Figure 12.23 shows this correlation. In order to convert PHY data rates into MAC data rates, retransmissions due to packet errors and MAC protocol overhead have to be taken into account. Since the MAC overhead mainly depends on the implemented multi-hop protocol, the following analysis focuses on PHY layer capacity.

SSs that are served by relays perceive a data rate that is influenced by the link capacity of both hops involved. The overall capacity can be calculated by Equation (12.14), in which

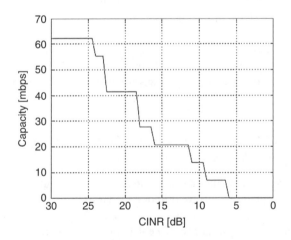

Figure 12.23 Link capacity vs. CINR

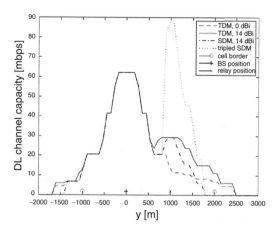

Figure 12.24   DL capacity while traversing the scenario (omni antennas, LOS) [15]

Cap_{hop1} is the capacity of the BS-to-relay link and Cap_{hop2} is the relay-to-SS link capacity [11]. In order to increase the link capacity on the first hop, the relay may apply receive antenna gain [2]. This advancement of the first hop converts into an enhanced overall channel capacity.

$$\frac{1}{Cap_{overall}} = \frac{1}{Cap_{hop1}} + \frac{1}{Cap_{hop2}} \qquad (12.14)$$

By means of the CINR-to-capacity conversion shown in Figure 12.23 and Equation (12.14), the capacity of a SS that is traversing the scenario can be calculated. For instance, Figure 12.24 shows the instantaneous DL capacity while traversing the coverage scenario using omni-directional antennas. The positions of the BS, the relay and the cell edge are indicated. The dashed line shows the TDM case, which corresponds to Figure 12.11b. It can be seen that the main cell is covered by the BS and the relay extends the coverage area to the right-hand side.

If the relay station is equipped with an antenna that provides receive antenna gain, it can improve its received CINR on the first hop. This directly converts into an increased link capacity, i.e., Cap_{hop1} in (12.14). The solid line in Figure 12.24 shows that a receive antenna gain of 14 dBi can significantly improve the capacity in the relay subcell.

Section 12.2.4 outlined the possibility of operating the relays simultaneously in SDM. The dashed-dotted line in Figure 12.24 shows (again) that a simultaneous operation of relays decreases the instantaneous channel capacity in a LOS scenario using omni-directional antennas. However, since three relays operate simultaneously, the dotted line indicates the tripled SDM capacity. It can be seen that in some regions the SDM mode is quite beneficial. In the following it will be shown that the overall cell capacity can be increased in some scenarios due to the SDM operation.

Another effect is visible in the LOS scenario shown in Figure 12.24. The additional interference in SDM mode affects the coverage so that the cell area cannot be covered completely. To overcome this drawback, the mode of operation could be switched from SDM back to TDM once the coverage is lost. This would result in a capacity that follows the envelope of both curves and the entire cell area can be covered again. However, a coordinated switching between TDM and SDM mode requires central coordination by the BS. This is not possible with distributed scheduling approaches.

By means of the instantaneous capacity of a single SS at various positions, the mean cell capacity can be calculated. Habetha [4] derives Equation (12.15), which calculates the reciprocal cell capacity as the integral of the reciprocal transmission capacities, which depends on the distance r to the BS. The cell edge is R meters away from the BS. The dependency on the azimuth angle α is neglected, because a (single-hop) cell nearly covers a circular area.

$$\frac{1}{\text{Cap}_{\text{cell}}} = \int_0^1 \frac{1}{\text{Cap}(\rho)} 2\rho \, d\rho \quad \text{with} \quad \rho = \frac{r}{R} \tag{12.15}$$

The coverage area of an ideal single-hop cell covers a hexagon and a relay enhanced cell in the coverage scenario spans over three hexagons (see Figure 12.9a). Thus (12.15) has been generalized to calculate the cell capacity of various coverage areas. By means of Equation (12.16) the mean cell capacity has been calculated for single-hop, multi-hop-throughput, and multi-hop-coverage scenarios.

$$\frac{1}{\text{Cap}_{\text{cell}}} = \int_{\text{cell area}} \frac{1}{\text{Cap}(x, y)} \, dx \, dy \tag{12.16}$$

Note that the DL CINR values and therewith the actual link capacity Cap(x,y) have been calculated based on a worst case assumption (refer to Section 12.1). Optional features of the IEEE 802.16 standard, such as subchannelization and Advanced Antenna System (see Sections 12.1.4 and 12.1.5) are able to increase the CINR and therewith the link capacity during the data transmission phase.

The following Tables 12.8, 12.9, and 12.10 contain the mean capacity of a cell and the area spectral efficiency, which shows the capacity per bandwidth per square kilometer coverage area. The channel bandwidth of the scenario is 20 MHz. The coverage area depends on the propagation conditions and on the scenario. According to the previous dimensioning approach, the cell radius in LOS and LOS-NLOS environments is assumed to be 1000 m. Under NLOS propagation conditions the cell radius is set to 150 m. In single-hop and multi-hop throughput scenarios, coverage areas of 2.598 km^2 and 0.0585 km^2 result (refer to Equation (12.5)). Compared to single-hop scenarios, multi-hop coverage scenarios triple the coverage area of a single BS so that values of 7.7942 km^2 and 0.1754 km^2 follow (refer to Equation (12.12)). All entries were calculated for cluster order seven.

Table 12.8 lists the mean cell capacity and the area spectral efficiency for the single-hop case. Table 12.9 contains the results for the multi-hop throughput scenario, in which the relays are

Table 12.8 Mean capacity and area spectral efficiency of a singlehop cell (cluster order 7, LOS/LOS-NLOS radius = 1000 m, NLOS: radius = 150 m)

scenario	mean cell capacity [Mbps]		area spectral efficiency [bps/Hz km ^{2}]	
	1 sector	3 sectors	1 sector	3 sectors
LOS	12.9747	21.0642	0.2497	0.4054
LOS-NLOS	28.9193	28.9193	0.5566	0.5566
NLOS	44.2868	47.9382	37.879	41.003

Table 12.9 Mean capacity and area spectral efficiency of a relay enhanced cell (throughput scenario, cluster order 7, LOS/ LOS-NLOS: radius = 1000 m, NLOS: radius = 150 m)

scenario			mean cell capacity [Mbps]		area spectral efficiency [bps/Hz km ^{2}]	
			gain 0 dBi	gain 14 dBi	gain 0 dBi	gain 14 dBi
LOS	TDM	omni	15.4676	20.0534	0.2977	0.3859
LOS	SDM	omni	13.0587	13.6253	0.2513	0.2622
LOS	SDM	directive	17.0894	20.4717	0.3289	0.3940
LOS-NLOS	TDM	omni	29.6229	32.1805	0.5701	0.6193
LOS-NLOS	SDM	omni	39.9708	45.8409	0.7692	0.8822
LOS-NLOS	SDM	directive	39.9888	45.8647	0.7696	0.8827
NLOS	TDM	omni	31.6841	35.0398	27.1005	29.9707
NLOS	SDM	omni	32.0656	32.7795	27.4268	28.0374
NLOS	TDM	directive	38.5744	42.1917	32.9940	36.0880

placed within the original cell. Relays operate either in TDM or in SDM. In order to increase the link capacity, the relay may apply a receive antenna gain of 14 dBi.

Figure 12.25 visualizes the mean cell capacity of exemplary single-hop and multi-hop throughput scenarios. In single-hop deployments under LOS conditions, sectorization significantly increases the cell capacity, because the system is interference-limited. Under NLOS conditions, the benefit of sectorization decreases. In LOS-NLOS scenarios, which are noise-limited, sectorization is no longer effective because the interference is limited by the propagation conditions. In NLOS environments the cell capacity reaches a maximum, but it has to be considered that the coverage area of the cell is much smaller.

Table 12.10 Mean capacity and area spectral efficiency of a relay enhanced cell (coverage scenario, cluster order 7, LOS/ LOS-NLOS: radius = 1000 m, NLOS: radius = 150 m)

scenario			mean cell capacity [Mbps]		area spectral efficiency [bps/Hz km^2]	
			gain 0 dBi	gain 14 dBi	gain 0 dBi	gain 14 dBi
LOS	TDM	omni	11.8409	19.1668	0.0760	0.1230
*LOS	SDM	omni	15.1573	17.3118	0.0972	0.1110
**LOS	SDM	omni	11.3091	14.1873	0.0725	0.0910
LOS	SDM	directive	21.1643	29.4440	0.1358	0.1889
LOS-NLOS	TDM	omni	16.7554	22.9660	0.1075	0.1473
LOS-NLOS	SDM	omni	27.0160	44.6449	0.1733	0.2864
LOS-NLOS	SDM	directive	27.0160	39.2832	0.1733	0.2520
NLOS	TDM	omni	13.4483	24.1832	3.8342	6.8949
NLOS	SDM	omni	23.0167	34.6783	6.5623	9.8872
NLOS	TDM	directive	25.4360	41.2509	7.2521	11.7611

*15.2% uncovered cell area
**uncovered area served in TDM, omni mode

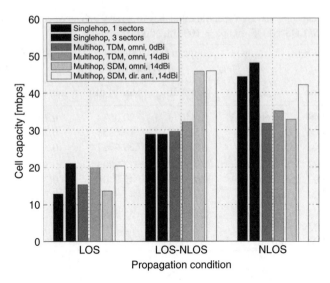

Figure 12.25   Mean cell capacity of the single-hop and multi-hop throughput scenarios [15]

Looking at TDM modes of multi-hop throughput deployments, it can be seen that receive antenna gain at the relay can improve the cell capacity under all propagation conditions. SDM operation of relays using omni-directional antennas is only beneficial when the interference can be limited under LOS-NLOS propagation conditions. In LOS and NLOS environments, the mean cell capacity during SDM operation falls below the TDM case.

In order to control inter- and especially intra-cell interference in SDM mode, directive antennas were introduced. In NLOS scenarios they increase the cell capacity beyond the TDM case. Under LOS conditions, directive antennas are only able to increase the cell capacity up to the original TDM mode. Here a further reduction of the inter-cell interference by means of Spatial Filtering for Interference Reduction (SFIR) or coordination across BSs and relays could be beneficial.

Comparing single-hop and multi-hop throughput scenarios, it can be seen that under LOS and NLOS conditions, the multi-hop capacity is less or equal to the single-hop capacity. If one considers the additional MAC overhead that is needed to control the multi-hop transmission, a single-hop deployment should be preferred. In scenarios where the relay subcells are highly shadowed from each other, i.e., under LOS-NLOS conditions, a multi-hop deployment can increase the cell capacity, especially in SDM mode. However, the additional MAC overhead and the cost of deploying and operating the relay devices have to be taken into account.

Table 12.10 lists the mean cell capacity and the area spectral efficiency of the multi-hop coverage scenario. Figure 12.26 visualizes the results of some scenarios. Like in the multi-hop throughput scenarios, the introduction of receive antenna gain increases the cell capacity of the multi-hop coverage scenarios under all propagation conditions. Since the relays are located further away from the BSs in the coverage scenarios, the benefit of receive antenna gain on the first hop is greater than in the previous throughput scenarios.

SDM operation of relays using omni-directional antennas can further increase the cell capacity in LOS-NLOS and even in NLOS scenarios. Here, the interference is limited by the propagation conditions. Additionally, the distance between simultaneously operating relay

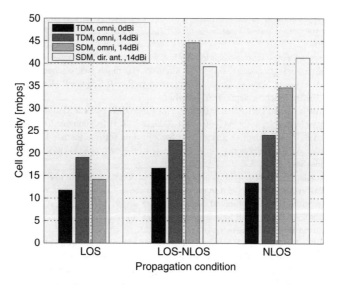

Figure 12.26   Mean cell capacity of multi-hop coverage scenarios [15]

stations is larger in the coverage scenarios compared to the throughput scenarios. Again, omni-directional antennas at the relays generate too much mutual interference in the LOS environment so that the mean cell capacity during SDM operation drops below the TDM case.

SDM operation of relays with directive antennas outperforms the other modes of operation under LOS and NLOS conditions. A significant increase is visible. In LOS-NLOS environments, the mutual interference of concurrently operating relays is negligible even with omni-directional antennas. Thus, the larger area (360° instead of 240°) that is covered by the antenna increases the mean cell capacity.

In order to finally compare single-hop and multi-hop coverage deployments, the different sizes of their coverage areas have to be considered. By simply comparing their area spectral efficiencies, single-hop deployments seem to be beneficial because their mean cell capacities are higher and their coverage areas are smaller. Under all propagation conditions, the maximum area spectral efficiency of multi-hop coverage deployments (LOS: 0.1889 bps/Hz/km^2, LOS-NLOS: 0.2864 bps/Hz/km^2, NLOS: 11.761 bps/Hz/km^2) is lower than the efficiency of single-hop deployments even without sectorization (LOS: 0.2497 bps/Hz/km^2, LOS-NLOS: 0.5566 bps/Hz/km^2, NLOS: 37.879 bps/Hz/km^2). The real benefit of multi-hop deployments is the cost-efficient roll-out and operation of such networks. Thus, a fair comparison would have to take Capital Expenditures (CAPEXs) and Operational Expenditures (OPEXs) into account, which is beyond the scope of this analysis.

References

[1] J. Eberspächer and H.J. Vögel, *GSM Global System for Mobile Communication*. Stuttgart: Teubner Verlag, 1999.

[2] N. Esseling, B. Walke, and R. Pabst, 'Emerging Location Aware Broadband Wireless Adhoc Networks', in *Fixed Relays for Next Generation Wireless Systems* New York: Springer Science+Business Media, Inc., 2005, 71–91.

[3] L.C. Godara, 'Application of Antenna Arrays to Mobile Communications. II. Beam-Forming and Direction-Of-Arrival Considerations', in *Proceedings of the IEEE*, 85(8):1195–1245, 1997. ISSN 0018-9219. doi: 10.1109/5.622504.

[4] J. Habetha, 'Entwurf eines cluster-basierten ad hoc Funknetzes', PhD thesis, RWTH Aachen University, Lehrstuhl f¨ur Kommunikationsnetze, July 2002. http://www.comnets.rwth-aachen.de.

[5] J. Habetha and J. Wiegert, 'Network Capacity Optimisation, Part 1: Cellular Radio Networks', in *10th Symposium on Signal Theory*, Aachen, Sept. 2001, 125–132.

[6] C. Hoymann, 'Analysis and Performance Evaluation of the OFDM-based Metropolitan Area Network IEEE 802.16', *Computer Networks*, 49(3):341–363, Oct 2005. doi: 10.1016/j.comnet.2005.05.008.

[7] C. Hoymann and B. Wolz, 'Adaptive Space-Time Sectorization for Interference Reduction in Smart Antenna Enhanced Cellular WiMAX Networks', in *Proceedings of the 64th IEEE Vehicular Technology Conference*, Montreal, Canada, Sept. 2006, 5.

[8] C. Hoymann, K. Klagges, and M. Schinnenburg, 'Multi-hop Communication in Relay Enhanced IEEE 802.16 Networks', in *Proceedings of the 17th Annual IEEE International Symposium on Personal, Indoor and Mobile Radio Communications*, Helsinki, Finland, Sept. 2006, 4.

[9] IEEE. IEEE Std 802.16-2004, IEEE Standard for Local and metropolitan area networks, Part 16: Air Interface for Fixed Broadband Wireless Access Systems, Oct. 2004.

[10] IEEE. IEEE Std 802.16e-2005, IEEE Standard for Local and metropolitan Area Networks, Part 16: Air Interface for Fixed and Mobile Broadband Wireless Access Systems Amendment for Physical and Medium Access Control Layers for Combined Fixed and Mobile Operation in Licensed Bands, Feb. 2006.

[11] T. Irnich, D. Schultz, R. Pabst, and P. Wienert, 'Capacity of a Relaying Infrastructure for Broadband Radio Coverage of Urban Areas', in *Proceedings of Vehicular Technology Conference* (VTC fall), Orlando, FL, USA, Oct. 2003. http://www.comnets.rwth-aachen.de.

[12] IST-WINNER. D5.4, Final Report on Link Level and System Level Channel Models. Deliverable, IST-4-027756 WINNER II, September 2006.

[13] Bernhard Walke, *Mobile Radio Networks: Networking and Protocols*. Chichester: John Wiley & Sons, Ltd, 2001.

[14] H. Wijaya, 'Broadband Multi-hop Communication in Homogeneous and Heterogeneous Wireless LAN Networks', PhD thesis, RWTH Aachen University, Lehrstuhl für Kommunikationsnetze, Feb. 2005. http://www.comnets.rwth-aachen.de.

[15] C. Hoymann, M. Dittrich, and S. Göbbels, 'Dimensioning Cellular Multihop WiMAX Networks'. *Proceeding IEEE Mobile WiMAX Symposium*, Orlando, 2007.

Part Four

Multimedia Applications, Services, and Deployment

13

Cross-Layer End-to-End QoS for Scalable Video over Mobile WiMAX

Jenq-Neng Hwang, Chih-Wei Huang, and Chih-Wei Chang

13.1 Introduction

There has been a gradual paradigm shift from analog to digital media, from push-based media broadcasting to pull-based media streaming, and from wired interconnectivity to wireless interconnectivity. Wireless broadband access with provisioned quality of service (QoS) for digital multimedia applications to mobile end users over wide area networks is the new frontier of telecommunications industry. The shift from wired to wireless internet is also coming as a strong wave. The wireless LAN (WLAN or the so-called Wi-Fi standards) technologies, IEEE 802.11 a/b/g, and the next generation very high data rate (> 200 Mbps) IEEE 802.11n WLAN are being deployed everywhere with very affordable installation cost. Almost all newly shipped computer products and more and more consumer electronics come with WLAN receivers for internet access. Cellular-based technologies such as the 3G networking are being aggressively deployed by the traditional telecommunication carriers to provide mobility support to internet access, with more and more multimedia application services being offered. On the other hand, mobile wireless microwave access (WiMAX) serves as another powerful alternative for mobile internet access by data communication carriers. The fixed or mobile WiMAX (IEEE 802.16d and 802.16e) can also serve as an effective backhaul for WLAN whenever it is not easily available, such as remote areas or moving vehicles with compatible IP protocols.

Thanks to the perfect synergy between WLAN and the mobile WiMAX, and cost-effective chipset designs based on volume, new scalable wireless distribution system architecture without large investment is envisaged. This all IP-based system architecture requires a few WiMAX

Mobile WiMAX Edited by Kwang-Cheng Chen and J. Roberto B. de Marca.
© 2008 John Wiley & Sons, Ltd

base-stations for broadband access coverage as well as acting as the backhaul for the WLAN networks in its coverage footprint. Quality of Service (QoS) plays the most critical role in the success of many multimedia communication applications over IP networks. There has been a substantial amount of research regarding end-to-end QoS in both end-system centric and network centric perspectives, such as scalable multimedia coding, adaptive protection, traffic classification, and network adaptation [1].

In order to ensure effective dissemination of compressed multimedia data over wireless networks, the main challenges are:

1. *Network heterogeneity*. QoS is further degraded due to more dynamically changing end-to-end available bandwidth caused by the wireless fading/shadowing and link adaptation. Moreover, the increased occurrence of wireless radio transmission errors also results in higher bursty rate of packet loss when compared with wired IP networks. To overcome all these extra deficiencies caused by the wireless networks, several additional QoS mechanisms spanning physical, Medium Access Control (MAC), network and application layers have to be incorporated in *cross-layered* fashion.
2. *User diversity*. Wireless networks provide not only mobility but also diversity of on-line mobile devices. From powerful laptops, personal data assistants (PDAs), to cell phones, computation power differs from couple gigahertz to a couple of megahertz with memory capacity from gigabytes to megabytes. Coupled with input/output limitations, attractive multimedia contents should be delivered also in different ways. *Scalable* codecs, *layered* multicast, and *adaptive* protection are favorite techniques.
3. *Application requirements*. For applications with various purposes, loss, delay, and bandwidth demands also vary. The *priority* queuing and traffic mapping are key components that need to be put together to satisfy all the requirements.

The multimedia (more specifically, the bandwidth-demanding video) data have two properties that affect the design of a congestion/error control for its transmission. First, video data is loss-tolerant (if loss is low) since the receiver can use error-concealment techniques to mitigate the undesirable visual perception. Moreover, the source can add some redundancy to the data stream to be more robust to loss, especially in error-prone wireless environments. Second, the stability of video quality is important to the audience's overall subjective perception. Frequent quality changes within a short period can be very annoying. These two properties are not applicable to general TCP bulk data delivery. Designing congestion control for multimedia applications should take them into consideration explicitly. Within the 802.11 and 802.16 standards, there are plenty of QoS mechanisms, mainly dealing with the last-mile connection (between an access point or a base station with the client mobile node) based on MAC or physical layers approaches, which are expected to be effective yet realize their full capability. In the real world, the multimedia contents are not disseminated only with the last mile connections, thus, to achieve satisfactory QoS in end-to-end fashion, a wide range of techniques in different aspects need to be carefully studied and applied as a whole.

The chapter is organized as follows: Section 13.2 addresses critical techniques for effective end-to-end wireless multimedia dissemination. In Section 13.3, QoS provisioning at the MAC layer is discussed. Section 13.4 describes an integrated cross-layer framework for mobile WiMAX networks, followed by the conclusion in Section 13.5.

13.2 Critical End-System Techniques

The end-systems (servers and clients) must be aware of network condition and perform proper source/transmission adaptation for quality multimedia dissemination. Packet loss and delay are the most damaging factors in multimedia networking. They can result from wired network congestion or wireless fading/shadowing, etc. These two degradation sources should be handled differently to achieve effective QoS performance. We will start with discussions of several techniques associated with congestion control, followed by those for error control.

13.2.1 Advances in Scalable Video Coding

In order to meet the increasingly heterogeneous clients' different requirements of video quality in the environment of dynamically varying available channel bandwidth, bit stream scalability in multimedia delivery is an inevitable trend. Receiver-driven layered multicasting [2, 3], which assigns each scalable video layer as one multicast group and enables every receiver to determine how many layers to subscribe, with efficient scalable video coding can provide an effective distribution of large-scale real-time multimedia streaming over heterogeneous networks.

MPEG-4 fine granularity scalability (FGS) [4] was proposed for streaming scalable video over the Internet. The basic concept of MPEG-4 FGS is to use the layer structure to encode bit-stream. The sender can send the appropriate layer bit-stream to fit the bandwidth. Although the SNR, temporal and spatial scalabilities are provided in MPEG-4 FGS, the video quality of MPEG-4 FGS is more than 2 dB worse than non-scalable MPEG-4 in high bit-rate [4]. Moreover, the H.264/AVC coding standard [5] has been proved to offer higher coding efficiency, and its coding performance is better than most existing video coding standards. Therefore the scalable H.264/AVC extension [6] (H.264/SVC), proposed by the image communication group of HHI, was thus chosen to be the starting point of MPEG's scalable video coding (SVC) standard (see Figure 13.1).

Key features to make H.264/SVC successful are:

- *Hierarchical prediction structure.* The motion-compensated temporal filtering (MCTF) improves the coding efficiency while provide temporal scalability.
- *Layered coding scheme and inter-layer prediction mechanisms.* For spatial scalability, original video is down-sampled into small resolution versions and stand-alone coders allowing inter-coder prediction. The design preserves coding efficiency for each layer.
- *Base-layer compatibility of H.264/AVC.* The base layer has high coding efficiency as H.264/AVC and is very decodable in various conditions.
- *Fine granular scalability (FGS).* This also supports SNR scalability using progressive refinement slices.
- *Usage and extension of network abstraction layer (NAL) unit concept of H.264/AVC.* Therefore bit-streams for a reduced spatial/temporal resolution can be obtained by discarding NAL units. Progressive refinement units can also be arbitrarily truncated to further adjust the bit-rate.

13.2.2 End-to-End Congestion Control

For end-to-end QoS, congestion control is one of the most important tasks required at end-systems. Basically, end-systems adjust data rates according to observed network conditions,

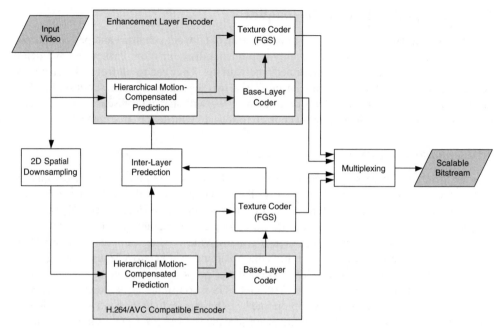

Figure 13.1   The basic coder structure for the scalable extension of H.264/SVC. Source: [6]

i.e., packet loss and delay statistics. TCP transport protocol is a typical example. However, the long delay created by retransmission used in TCP is not practical in multimedia applications, such as real-time streaming. Unlike TCP, most UDP control schemes widely used in multimedia communication try not to grab more channel shares than a TCP session under the same environment. TCP-friendly and available bandwidth estimation methods are two main concepts commonly adopted in real-time multimedia communication.

13.2.2.1 TCP-Friendly Methods

Many existing rate-based protocols for multimedia (mainly UDP) traffic try to be TCP-friendly. Some protocols, such as TCP Friendly Rate Control (TFRC) [7] and Smooth Multirate Multicast Congestion Control (SMCC) [8], use the explicit TCP throughput equation for rate adjustment:

$$T_{TCP} = \frac{s_{TCP}}{R_{TCP}\sqrt{\frac{2P_{TCP}}{3}} + t_{RTO}\left(3\sqrt{\frac{3P_{TCP}}{8}}\right)p\left(1 + 32P_{TCP}^2\right)} \tag{13.1}$$

where T_{TCP}, s_{TCP}, R_{TCP}, p_{TCP}, and t_{RTO} are the upper bound of the sending rate, packet size, round trip time (RTT), loss rate, and retransmission time out (RTO) respectively. The receiver measures the loss event rate and feeds it back to the sender. The sender measures RTT using the feedback message and then evaluates the acceptable transmit rate by Equation (13.1). Thus the sender can adjust the transmit rate to match the evaluated one.

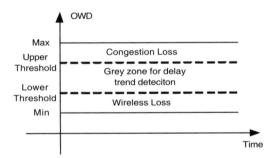

Figure 13.2 OWD zones for packet loss classification. Source: [22]

For better efficiency in wireless networks, TFRC Wireless [9], MULTFRC [10], and VTP [11] have also been proposed as extended solutions. Packet loss classification (PLC) has been proposed and added to TFRC to deduce random losses over wireless links from congestion related decisions. We will provide more details of PLC in next sub-section.

Overall, in order to be TCP-friendly, a lot of effort (especially in the multicast case) is required to acquire the RTT information and packet loss rate to estimate the proper rate or implement the additive increase and multiplicative decrease (AIMD) mechanism for rate control. Several studies in the literature [12, 13] have reported that TCP-based congestion control has fairness problems and difficulty in utilizing the network resource efficiently over the high per-flow bandwidth-delay product networks. For fairness issues, one example is that even with two receivers sharing the same bottleneck of a network, the steady throughputs are different if their round trip times are different. It is debatable whether the rate should be the dominant factor to determine the bandwidth share for rate-based protocols. In the TCP or TCP-friendly congestion control algorithms, packet drop is also an important index to indicate network congestion.

13.2.2.2 Using Available Bandwidth Estimation

Available bandwidth estimation tools also play an important role in layered multicast congestion control. Generally, receivers perform bandwidth detection before subscribing to a proper layer according to its estimated available bandwidth. PLM [2] uses packet pairs for layered video streaming over fair scheduler networks. BIC [3] modified delay trend detection from Pathload [14] for a receiver-driven layered multicast protocol, and proposed an effective bandwidth inference scheme with minimum additional network traffic for each receiver to subscribe to the appropriate multicast group.

Delay Trend Model and Detection

Conventionally, assuming there are N links from the sender to the receiver, $l = 1, \ldots, N$. The capacity of link l is C_l. The sender sends out a probe of multiple packets, $h = 1, \ldots, M$, at a known rate (pr) from time a until time b. The size of packet m is S_m. Denote the cross-traffic (excluding the probe packets) through each link between time a and b as $B_l[a, b]$. The path's end-to-end available bandwidth in time interval $[a, b]$ can be defined as:

$$A[a, b] = \min_l \left(C_l - \frac{B_l[a, b]}{b - a} \right) \qquad (13.2)$$

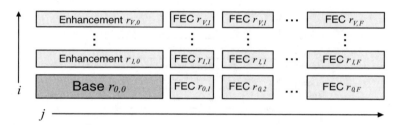

Figure 13.3   Rates of (i, j) in layers. Source: [31]

There has been active research in available bandwidth estimation recently, such as Delphi [15], TOPP [16], Pathload [14], IGI [17], pathChirp [18], and Spruce [19]. However, these tools have special requirements for their probes, such as fixed packet size and exact inter-packet spacing, which may not be available for most practical multimedia applications. Moreover, existing tools generate hundreds of kilobytes of probe traffic (or more) and take a long time to give the available bandwidth estimation. It has been observed that there is no need to estimate the absolute available bandwidth for congestion control purposes. It is good enough to know whether the path has enough bandwidth of a specific rate (e.g., the cumulative rate of the next layer) for congestion control purposes.

A delay trend model [3, 20], which uses tens of variable-size packets, has been proposed to check whether the path has the available bandwidth of a specific rate. The model relies on correctly detecting whether there are trends among the measured one-way delays (OWDs) and packet sizes. A reliable and consistent trend detection algorithm is the core function in the model. Our proposed trend detection algorithm, called Fullsearch, uses statistical tests to check the existence of a trend. Define $I(X)$ is 1 if X holds, and 0 otherwise. If the measured OWDs of packet h are D_h, $h = 1, ..., M$, the test used is:

$$S_{Fullsearch} = \frac{\sum_{h=2}^{M} \sum_{u=1}^{h-1} I(D_h > D_u)}{\frac{M(M-1)}{2}} \tag{13.3}$$

where OWD is defined as the relative one-way trip time, measured by the receiver as the time difference between the receiving time and the packet sending timestamp recorded in the field of multimedia UDP packet header, plus a fixed bias.

As Equation (13.3) shows, the Fullsearch algorithm reflects the statistical relationship among all pairs of measurements. If there is no trend among the measurements, the test result is around 0.5; if there is a strong increasing trend, the test result approaches one. A trend threshold th can be used to check whether there is an increasing trend. If the test result is larger than th, the measurements have an increasing trend, otherwise no increasing trend.

Packet Loss Classification (PLC)

In a wireless network environment, common channel errors due to multi-path fading, shadowing, and attenuation may cause bit errors and packet losses, which are quite different from the packet loss caused by network congestion. In congestion control, the packet loss information can serve as an index of network congestion for effective rate adjustment; therefore wireless packet loss can mistakenly lead to dramatic performance degradation.

Generally, the classification algorithms of packet loss depend on the analysis of statistical behavior of some observed values: packet timestamp and packet serial number in the packet header. Spike-train [21] and ZigZag [9] investigate the OWD difference between two classes of packet loss. Unfortunately, these methods may produce unreliable classification performance when OWD is around the threshold, which depends on the network topology. We proposed a PLC algorithm [22] based on trend detection of OWD when it falls in the ambiguous zone where the PLC is not straightforward. The algorithm can greatly benefit rate-based congestion control algorithms for multimedia over IP networks.

shows OWD zones in wired/wireless heterogeneous networks along the time. Assume there are maximum OWD and minimum OWD values observed for this specific end-to-end link in steady state. There are also upper and lower thresholds that divide the range into reliable zones of congestion/wireless loss and a gray zone. The delay trend detection algorithm is performed to detect the congestion in this special area. When a packet loss is observed at arbitrary time, it should be considered as a congestion loss if the evaluated delay trend is in an ascending phase or in upper reliable zone; otherwise, it is categorized as wireless loss.

13.2.3 Layered Coding and FEC Structure for Error Control

For multicast video streaming, our scalable video is created by the MCTF scalable extension of H.264/AVC [6], which is an emerging compression technology with coding efficiency comparable to original H.264/AVC standard. With MCTF using the lifting framework, temporal decomposition can be achieved nicely for SNR, temporal and spatial scalabilities.

To provide maximum flexibility in both quality and protection, video data and error protection codes are formatted in layers [23], where each layer is assigned to a multicast group as, with rate $r_{i,j}$, i and j are indexes of video and FEC layers respectively. Layers with $j = 0$ contain only video streams, otherwise j indicates level of protection for a specific i. There are V enhancement video layers and F protection.

There are many ways to construct FEC packets, e.g., packet repeating, or block erasure codes such as the Reed-Solomon code. We decided on corresponding feasible block parameters $(n_{i,j}, k_i)$ for video layer i through network analysis under the structure illustrated in Figure 13.4 . Elliot's model [24] is used to simulate the bit error rate resulting in the packet loss. This method

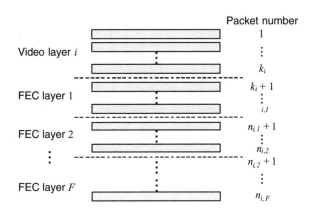

Figure 13.4   Layered FEC structure. Source: [31]

can be approximated as a two-state Markov chain with parameters p and q representing transition probabilities from (packet) loss state to received state and from received state to loss state respectively. During the process, target decoding error rates, $e_{FEC(n,k)}$, can be derived accordingly [25]. We set k_i first then apply the estimated (p, q) value sets representing wireless channel conditions to evaluate minimum $n_{i,j}$ satisfying:

$$e_{FEC(n,k)} < e_{FEC,i} \tag{13.4}$$

where $e_{FEC,i}$ is the preset loss requirement for each video layer i. If the popular Reed-Solomon code is our choice, $e_{FEC(n,k)}$ can be derived as following steps [25]. First, we come out the steady state average packet loss rate P_L:

$$p_L = \frac{q}{p+q} \tag{13.5}$$

and then we have

$$e_{FEC,(n,k)} = \sum_{m=n-k+1}^{n} P(m,n) \tag{13.6}$$

$$P(m,n) = \sum_{s=1}^{n-m+1} p_L G(s) H(m, n-s+1) \tag{13.7}$$

with

$$G(s) = \begin{cases} 1, & \text{for } s = 1 \\ p(1-q)^{s-2}, & \text{for } s > 1 \end{cases} \tag{13.8}$$

$$H(x,y) = \begin{cases} G(y), & \text{for } x = 1 \\ \sum_{s=1}^{y-x+1} g(s) H(x-1, y-s), & \text{for } 2 \leq x \leq y \end{cases} \tag{13.9}$$

$$g(s) = \begin{cases} 1-p, & \text{for } s = 1 \\ p(1-q)^{s-2}, & \text{for } s > 1 \end{cases} \tag{13.10}$$

where $P(m, n)$ is the probability of m lost packets out of n consecutive packets. Conclusively, the right amount of FEC packets can thus be decided assuming (p, q) estimated at receivers for a known FEC scheme.

In other words, by modeling the network conditions using F number of (p, q) sets, we can derive the corresponding $n_{i,j}$, $j = 1 \sim F$ in ascending order to generate F protection layers. If we want to judge how many FEC layers to subscribe after the whole scenario has been constructed, we can again choose a data layer i, plug in newly estimated (p, q), find minimum n_{new} through the same process, and then pick j with $n_{i,j}$ closest to but larger than n_{new}.

Regarding inter-layer dependency and protection levels for every layer, the overall data rate is:

$$R_{v,f} = \sum_{i=0}^{v} \sum_{j=0}^{f} r_{i,j} \tag{13.11}$$

where v is the number of enhancement layers *subscribed* and f, same for every i, is the FEC layers requested. Unequal protection to different video layers is supported by different $e_{FEC,i}$, not by f. For instance, under channel condition $(p, q) = (0.8, 0.2)$, we preset $e_{FEC,0} = 0.001$, $e_{FEC,5} = 0.01$, and $k_0 = k_5 = 8$ resulting in $n_{0,j} = 17$ and $n_{5,j} = 15$ if we choose R-S block code as FEC. It also implies that a video layer requires all lower ones to be available for decoding (due to the cumulative layer structure of the adopted scalable codec), while not every sub-layer FEC is needed. Therefore, receivers acquire distinct video quality and amount of protection by subscribing to proper groups.

13.2.4 Embedded Layered Probing and Join Decision

In order to increase the data rate, either for more video data or more loss recovery, available bandwidth estimation has to be performed in advance to prevent congestion. In contrast to extra probing packets, we embed probing streams in regular ones through effective scheduling of packet transmission and take advantage of the fact that streaming systems usually have a decoding buffer to tolerate some amount of delays [23, 26].

As shown in Figure 13.5 on our embedded probing, the stream is periodically separated into probing and regular intervals alternatively with period T. The length of probing interval, t_p, is further divided into certain uniform probing regions according to number of possible layers to go, e.g., r_{p1} to r_{p4}. In each region, previously generated packets are delayed in transmission creating temporarily higher sending rate within it.

On the receiver side, the interval and region a packet belongs to can be distinguished from its RTP timestamp; then Fullsearch delay trend detection, as shown in Equation (13.3) is applied to packets in time slots at objective rates. The duration of each probing region is set to be the time to send 50 packets in regular intervals of base layer $(i, j) = (0, 0)$ and remains the same for every layer.

Other important facts are that the probing rate we are looking for is an aggregation rate $R_{v,f}$ not $r_{i,j}$. The aggregation rates of all possible target subscriptions should be included in a probing interval. Given that the layered streams are ready at the server, what layers to subscribe is based on three types of information: channel estimation (p, q), probing results (available bandwidth), and observed packet loss rates. Relying on PLC, we are able to continuously monitor $(\hat{p}, \hat{q})$ (update every T seconds for enough samples) and congestion packet loss (update every second). Table 13.1 addresses all possible cases and reactions in our proposed architecture where five moves are allowed in the system. '$-$' means do not care because the rate is getting lower or low congestion loss is a must for a positive probing. Every period T, receivers come out a (p, q) pair, and map it to demand FEC, j, for current video layer i conditional on Equation (13.4). If the new j is larger than current j, we categorize it as 'worse' in wireless condition.

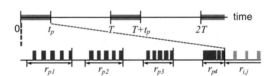

Figure 13.5   Embedded probing. Source: [31]

Table 13.1 Adaptation rules

Events	Channel estimation	Probing result	Congestion loss	Action path
Increase FEC	Worse	Positive	–	I
		Negative	(Yes)	II
Decrease FEC	Better	–	–	III
Change quality	Same	Positive	–	IV
		Negative	Yes	V

'Better' and 'same' are defined analogously. For each condition, we increase FEC, decrease FEC, and change video quality accordingly taking into account the observations. For example, if the wireless channel is worse and probing for more FEC in the same video layer has failed, we go along path II with more FEC but less quality; if channel does not change and probing for higher rate at the same protection level is positive, it is time to subscribe to more video contents. Figure 13.6 shows a partial view of with five action paths notated.

The aggregation rates of probing regions will match the target rates. Depending on channel quality, end users select region 1 and 2 or 3 and 4 for delay trend detection and make a decision to either stay or take one of five moves accordingly. The only information needed for feedback is the resulting change of layer subscription.

13.3 Mobile WiMAX QoS Provisioning

WiMAX is based on 802.16d [27] and 802.16e [28] standards published in 2004 and 2006 respectively. 802.16d supports fixed terminals only using OFDM while 802.16e is able to provide services to mobile terminals by OFDMA technologies.

13.3.1 Internet Protocols

The Internet Engineering Task Force (IETF) developed two QoS protocols, IntServ [29] and DiffServ [30], for internet QoS provisioning in addition to the best-effort service.

IntServ or Integrated Services is a QoS architecture with the underlying reservation mechanisms called Resource ReSerVation Protocol (RSVP). An IntServ-enabled network is expected to provide fine-grained guaranteed services to each traffic flow. All machines on the network capable of sending QoS data send every 30 seconds a PATH message, which spreads out

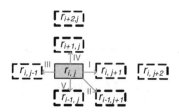

Figure 13.6 Optimization paths. Source: [23]

through the networks. Those who want to listen to them send a corresponding RESV (short for 'Reserve') message which then traces the path backwards to the sender. The RESV message contains the flow specs. However, the protocol is not popular since the number of states to store in routers is too large and limits the scalability.

DiffServ or Differentiated Services, in contrast, is a class-based protocol which aggregates flows with similar QoS requirements into a class providing a simpler and more scalable mechanism. DiffServ operates on the principle of *traffic classification*, where each data packet is placed into a limited number of traffic classes, rather than differentiating network traffic based on the requirement of an individual flow. Each router on the network is configured to differentiate traffic based on its class. Each traffic class can be managed differently, ensuring preferential treatment for higher-priority traffic on the network. The priority is indicated by encoding a 6-bit value, called the Differentiated Services Code Point (DSCP), in the 8-bit Differentiated Services (DS) field of the IP packet header. In practice, most networks follow some commonly defined priorities, such as Default, Expedited Forwarding (EF), Assured Forwarding (AF), etc.

Nowadays, with advances in network backhaul capacity and optic fiber technology, wired networks infrastructure is likely to provide sufficient bandwidth most of the time while wireless networks can be the primary link suffering from packet losses. Networks with large coverage, lots of users, and diversity in conditions, especially mobile WiMAX, need extra care. The power of DiffServ, which is to drop relatively unimportant packets before critical packets, is essential for end-to-end QoS under this situation. The transition of its usage from wired to wireless networks also raises its importance.

13.3.2 WiMAX QoS Support

With QoS in mind, the design of 802.16 MAC supports traffics with a wide range of demand based on scheduling services. More specifically, the data handling mechanism, which is supported by the MAC scheduler, is determined by a set of QoS parameters that quantify aspects of its behavior. Outbound transmission scheduling selects the data for transmission in a particular frame/bandwidth allocation and this is performed by the base station (BS) for downlink, and subscriber station (SS) for uplink. The following details are taken into account for each active service flow:

- the scheduling service specified for the service flow.
- the values assigned to the service flow's QoS parameters.
- the availability of data for transmission.
- the capacity of the granted bandwidth.

13.3.2.1 Scheduling Services

Uplink request/grant scheduling is performed by the BS with the intent of providing each subordinate SS with bandwidth for uplink transmissions or opportunities to request bandwidth. BS scheduler can then anticipate the throughput and latency needs of the *uplink* and provide polls and/or grants at the appropriate times. Table 13.2 details the five types of services.

Download services are defined in the same way as upload services under different names due to no need for polling. Instead of showing corresponding applications, available parameters

Table 13.2   Uplink request/grant scheduling services

Class	Abbreviation	Description
Unsolicited grant service	UGS	Real-time uplink service flows that transport fixed-size data packets on a periodic basis, such as T1/E1 and Voice over IP without silence suppression
Real-time polling service	rtPS	Real-time uplink service flows that transport variable size data packets on a periodic basis, such as moving pictures experts group (MPEG) video
Extended real-time polling service	ertPS	Real-time service flows that generate variable size data packets on a periodic basis, such as Voice over IP services with silence suppression
Non-real-time polling service	nrtPS	Delay-tolerant data streams consisting of variable-sized data packets for which a minimum data rate is required, such as FTP
Best effort	BE	Data streams for which no minimum service level is required and therefore may be handled on a space-available basis

to request are listed for download types in Table 13.3. Depending on the assigned classes and attached parameters, BS is responsible for making decisions on how much resource to grant to each SS.

Unsolicited grant service (UGS) supports real-time applications generating fixed-rate data provided by fixed or variable length protocol data units (PDUs). The transmission opportunities are granted by periodic basis and associated parameters, such as tolerated jitter, service data unit (SDU) size (in the case of fixed length SDU), minimum reserved traffic rate, maximum latency, request/transmission policy, and unsolicited grant interval.

Table 13.3   Data delivery services

Class	Abbreviation	Parameters
Unsolicited grant service	UGS	Tolerated jitter, service data unit (SDU) size (in case of fixed length SDU), minimum reserved traffic rate, maximum latency, request/transmission policy, and unsolicited grant interval
Real-time variable rate	RT-VR	Maximum latency, minimum reserved traffic rate, maximum sustained traffic rate, traffic priority, request/transmission policy, and unsolicited polling interval
Extended real-time variable rate	ERT-VR	Maximum latency, tolerated jitter, minimum reserved traffic rate, maximum sustained traffic rate, traffic priority, request/transmission policy, and unsolicited grant interval
Non-real-time variable rate	NRT-VR	Minimum reserved traffic rate, maximum sustained traffic rate, traffic priority, and request/transmission policy
Best effort	BE	Maximum sustained traffic rate, traffic priority, and request/transmission policy

Extended real-time variable rate (ERT-VR) service supports applications with variable data-rates, which require guaranteed data rate and delay while sensitive to *jitter*, for example, VoIP with silence suppression and interactive conferencing. QoS Parameters for this type are maximum latency, tolerated jitter, minimum reserved traffic rate, maximum sustained traffic rate, traffic priority, request/transmission policy, and unsolicited grant interval.

Real-time variable rate (RT-VR) service supports real-time applications with variable bit rates which require guaranteed data rate and delay. The BS is supposed to allocate sufficient resources to the connection for at least min$\{S, R * T\}$, where S denote the amount data arriving at the transmitter's queue during time interval T with $R = minimum\_reserved\_traffic\_rate$. Any SDU should be delivered within the latency requirement. When $S > R * T$, delivery of each specific SDU is not guaranteed. The associated QoS parameters include maximum latency, minimum reserved traffic rate, maximum sustained traffic rate, traffic priority, request/transmission policy, and unsolicited polling interval.

Non-real-time variable rate (NRT-VR) service supports applications that require a guaranteed data rate but insensitive to delays. Delivery of each specific SDU is not guaranteed if $S > maximum\_sustained\_traffic\_rate * T$. The associated QoS parameters include minimum reserved traffic rate, maximum sustained traffic rate, traffic priority, and request/transmission policy.

Best effort (BE) service is for applications with no rate or delay requirements. The associated QoS parameters include maximum sustained traffic rate, traffic priority, and request/transmission policy.

For the layered multicast application, ERT-VR, RT-VR and NRT-VR are three services to be used in the integrated multicast streaming system.

13.3.2.2 QoS Operation and Service Flow Management

802.16/WiMAX MAC is connection-oriented. For each direction (uplink or downlink), a 16-bit connection identification (CID) is assigned to each established connection. After that, a service flow identification (SFID) is associated to the CID to provide a set of QoS parameters such as latency, jitter, and throughput assurances.

When a service flow is formed, it can be deleted or modified by management message sets: Dynamic Service Change (DSC) and Dynamic Service Deletion (DSD); the operation can be initialized by either BS or SS. So we can change the existing service flow specification when it is needed. BS will make all decisions. The processes include:

- a configuration and registration function for pre-configuring SS-based flows and traffic parameters;
- a signaling function to dynamically establish QoS-enabled service flows and traffic parameters;
- utilization of MAC scheduling and QoS traffic parameters for uplink service flows;
- utilization of QoS traffic parameters for downlink service flows;
- grouping of service flow properties into named Service Classes (identifiers for a specific set of QoS parameter set values).

The principal mechanism for providing QoS is to associate packets traversing the MAC interface into a service flow appropriately as identified by the transport CID.

13.4 The Integrated Cross-Layer System

The best news for the QoS demanding service introduced by WiMAX is to explicitly specify QoS needs, such as minimum reserved traffic rate, maximum sustained traffic rate, maximum latency, and sometimes tolerated jitter. The scheduler will guarantee minimum rate, try to achieve maximum rate and hold other constraints by vendor-specific tweaks. For uplink transmission, requesting processes are also involved. Contrary to conventional distributed coordination standards such as Ethernet and 802.11, the centralized coordination of WiMAX can result in better efficiency, guaranteed QoS support, and few collisions while necessary mechanisms for robust wireless communication like ARQ and link adaptation are still playing their roles.

13.4.1 System Overview

An end-system driven solution featuring embedded probing for layered multicast of scalable video [23] can thus be adopted for WiMAX systems as an example. Figure 13.7 illustrates our proposed WiMAX scalable video layered multicast system, where video quality degradation resulting from wireless packet loss is protected by adequate FEC erasure codes with embedded probing performed first to assure enough available bandwidth for redundancy and fair share with other sessions. On the other hand, video quality degradation resulting from wired congestion loss can be adequately mitigated through less video/FEC layers subscriptions (i.e., less multicast group joining). The fundamental spirit of this proposed system is to decouple several important modules (scalable video layer creation, packet loss classification, bandwidth probing, and adaptive FEC insertion) and conduct an effective integrated tradeoff analysis to reach optimal number of video layers and FEC protection levels under all the resource constraints.

13.4.2 Priority Service Flow Mapping

In terms of layered streaming (downloading), the minimum rate constraint is to have base layers video or relatively more important packets delivered so as to guarantee the acceptable quality. When it comes to multicast design, there are several possibilities involving multiple scheduling services [31].

- Guarantee the base video/protection layers only, RT-VR only (see Figure 13.8(a)). By setting minimum rate at the base-layer rate and maximum rate at the best quality/protection rate, the application layer probing will force BS to temporarily get extra capacity if available. If the result of end-to-end available bandwidth estimation is positive, application clients (at SS) can then subscribe to more video/FEC layers.
- Adaptive base rates (minimum reserved rates), RT-VR only (seeFigure 13.8(b)). Similar to the first method in (a), except through channel estimation, it adaptively sets the minimum rate at certain level of protection depending on current channel quality. Competitions with other traffic flows are needed for extra quality only. The adaptive algorithm may need careful design for not setting minimum reserved rate too high so as not to grab too much bandwidth and thus affect inter-connection fairness.

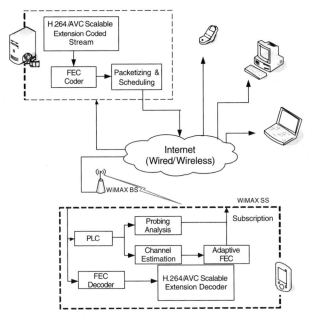

Figure 13.7   WiMAX scalable video streaming system. Source: [31]

- Use both RT-VR and NRT-VR (see Figure 13.8(c)). Transmit base-layers in RT-VR but enhancement layers through the NRT-VR provide more flexibility to drop late packets and tolerate more varying receiving quality of clients in a WiMAX network.
- Use ERT-VR only (see Figure 13.8(d)) if the content is jitter-sensitive and does not allow much buffering at receivers.

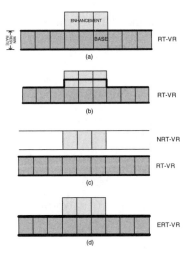

Figure 13.8   Layered multicast packets arrangement in service flows. Source: [31]

Implementation of the proposed schemes can be done by setting the proper DSCP in IP header and multiple MAC layer queues. Base-layer and enhancement-layers have different priority values in the type of service column of IP header or values compatible with DiffServ framework. Besides, service flow ID (SFID) and connection ID (CID) tags are also specified for streams to be assigned to the desired service type in WiMAX MAC. There are also corresponding priority queues for each service type. For example, base and enhancement layers in Figure 13.8(a) method have different IP priorities, identical SFID and CID. Therefore they end up with different queues in the same service type (RT-VR). Due to higher IP priority, base-layer packets will be sent out first with guaranteed minimum rate and the amount of transmitted enhancement packets will depend more on contention among connections.

When broken down into its components, end-to-end available bandwidth estimation and PLC could both benefit from QoS features offered by WiMAX, e.g., knowing the minimum and maximum rate, extreme bandwidth estimation errors can be avoided. In PLC, less collision loss could make it more reliable in differentiating wired congestion loss from wireless transmission loss since the wireless loss is now more reflected in the signal quality.

In brief, the advanced QoS design in WiMAX can offer a much more reliable wireless multimedia transmission performance and more flexibility for service providers to customize both software and hardware in the path.

13.4.3 Performance

With priority assignment, the end-to-end video quality can be improved a lot when the wireless network condition is suffering insufficient available bandwidth. Using NS2 [32] and WiMAX module [33], we compared the video streaming (download) performance of 4 schemes in Figure 13.8. The 'foreman' CIF video clip at 30 fps is encoded by H.264/SVC [6] to a stream with 500 Kbps base layer and 200 Kbps (total 700 Kbps) SNR enhancement layer. The topology of Figure 13.9 is used with nine 500 Kbps upload or download cross traffic flows (2 in each service type except only 1 in UGS) and one variable rate download link to control the available bandwidth. So there are in total 10 mobile WiMAX nodes (one for each traffic flow). Available bandwidth is measured at the BS from the NS2 trace ranging from 0 to 1000 Kbps. Downloading (BS to SS) video stream is applied to the network to record packet loss rates and PSNRs.

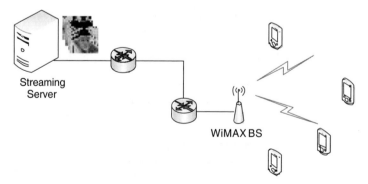

Figure 13.9 The simulation topology. Source: [31]

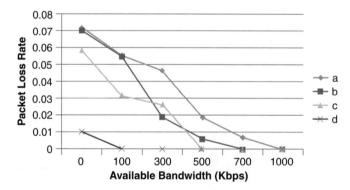

Figure 13.10   Packet loss rates of mapping schemes. Source: [31]

Figures 13.10 and 13.11 show packet loss and PSNR results for four cross-layer mapped schemes and an original scheme with no distinguishing between base and enhancement layers. The loss rates are higher when there is less available bandwidth. Also with larger portion of the stream in higher priority scheduling type, we can see fewer losses. An interesting observation is that scheme (c) using two service types outperforms (a) and (b), schemes which implies due to setting the parameter of requested minimum bandwidth, packet loss can also be significantly reduced at the enhancement layer. In Figure 13.11, all four custom schemes receive much better quality with about 5 dB gain of quality. This results from the protection of base layer through minimum rate reservation.

The decoded video frames are visualized in Figure 13.12. Four frames of original method at 31.7 dB are shown in the upper row. Obvious error blocks and noise can be seen on the foreman's face and background with homogeneous color. If we properly map the layered information to the MAC layer priority classes, we can see the lower four frames from scheme (a) have much better quality at 36.2 dB of PSNR.

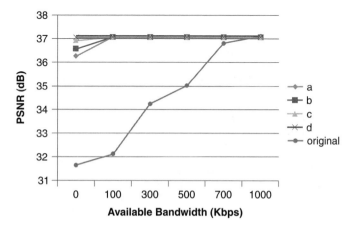

Figure 13.11   PSNR of mapping schemes. Source: [31]

Figure 13.12   Decoded video frames from original transmission (upper row) at 31.7 dB and scheme (a) (lower row) at 36.2 dB

13.5 Conclusion

With increasing deployment of wireless broadband infrastructure, large-scale real-time video streaming, such as IPTV, will be the killer application for such an infrastructure. It is never enough to count on the MAC layer QoS mechanism provided in the last-mile wireless connection. Therefore we have to resort to the end-to-end QoS from the cross-layer perspective. By incorporating the DiffServ mechanism with the MAC QoS features offered by one of the four proposed WiMAX service flow arrangements, our proposed layered scalable video multicast over WiMAX can achieve more flexible layer construction and subscription while still remaining reliable in diverse channel conditions and suiting users' demands. The system optimality comes from the best tradeoff between the number of video layers subscription and the number of additional FEC packets insertion to simultaneously satisfy the estimated available bandwidth and the estimated wireless channel error condition.

References

[1] Q. Zhang, W. Zhu, and Y. Zhang, 'End-to-end QoS for video delivery over wireless Internet', in *Proceedings of the IEEE,* Jan. 2005.

[2] A. Legout and E.W. Biersack, 'PLM: Fast Convergence for Cumulative Layered Multicast Transmission Schemes', in *Proceedings of ACM SIGMETRICS,* 2000.

[3] Q. Liu, J. Yoo, B.-T. Jang, K. Choi, and J.-N. Hwang, 'A scalable VideoGIS system for GPS-guided vehicles', *Signal Processing: Image Communication,* **20**, Mar. 2005, 205–208.

[4] W. Li, 'Overview of Fine Granularity Scalability in MPEG-4 Video Standard', *IEEE Trans. on Circuits and System for Video Technology,* **11**, Mar. 2001, 301–317.

[5] 'Draft ITU-T Recommendation and Final Draft International Standard of Joint Video Specification (ITU-T Rec. H.264 | ISO/IEC 14496-10 AVC)', Joint Video Team (JVT) of ISO/IEC MPEG and ITU-T VCEG Mar. 2003.

[6] H. Schwarz, D. Marpe, and T. Wiegand, 'Overview of the Scalable H.264/MPEG4-AVC Extension', in *IEEE International Conference on Image Processing* Atlanta, GA, 2006.

[7] M. Handley, S. Floyd, J. Padhye, and J. Widmer, 'TCP Friendly Rate Control (TFRC): Protocol Specification', RFC Editor RFC 3448, 2003.

[8] G.I. Kwon and J.W. Byers, 'Smooth Multirate Multicast Congestion Control', in *IEEE INFOCOM,* 2003.

[9]   S. Cen, P. Cosman, and G. Voelker, 'End-to-end Differentiation of Congestion and Wireless Losses', *IEEE/ACM Transactions on Networking,* Oct. 2003.

[10] M. Chen and A. Zakhor, 'Rate Control for Streaming Video Over Wireless', in *IEEE InfoCom,* 2004.

[11] G. Yang, L.-J. Chen, T. Sun, M. Gerla, and M. Y. Sanadidi, 'Smooth and Efficient Real-time Video Transport in Presence of Wireless Errors', *ACM Transactions on Multimedia Computing, Communications and Applications,* 2006.

[12] G. Hasegawa and M. Murata, 'Survey on Fairness Issues in TCP Congestion Control Mechanisms', *IEICE Transactions on Communication,* Jan. 2001.

[13] D. Katabi, M. Handley, and C. Rohrs, 'Congestion Control for High Bandwidth-Delay Product Networks', in *SIGCOMM,* 2002.

[14] M. Jain and C. Dovrolis, 'Pathload: A Measurement Tool for End-to-End Available Bandwidth', in *Passive and Active Measurements* Fort Collins, CO, USA, 2002.

[15] V. J. Ribeiro, M. Coates, R. H. Riedi, S. Sarvotham, and R. G. Baraniuk, 'Multifractal Cross-Traffic Estimation', in *Proc. of ITC specialist seminar on IP traffic Measurement,* 2000.

[16] B. Melander, M. Bjorkman, and P. Gunningberg, 'A New End-To-End Probing and Analysis Method for Estimating Bandwidth Bottlenecks', in *IEEE Globecom,* San Francisco, USA, 2000.

[17] N. Hu and P. Steenkiste, 'Evaluation and Characterization of Available Bandwidth Techniques', *IEEE JSAC Special Issue in Internet and WWW Measurement, Mapping, and Modeling,* 2003.

[18] J. Ribeiro, R.H. Riedi, R.G. Baraniuk, J. Navratil, and L. Cottrell, 'PathChirp: Efficient Available Bandwidth Estimation for Network Paths', in *Passive and Active Measurement Workshop,* 2003.

[19] J. Strauss, D. Katabi, and F. Kaashoek, 'A Measurement Study of Available Bandwidth Estimation Tools', in *Proceedings of the 3rd ACM SIGCOMM Conference on Internet Measurement* Miami Beach, FL, USA, 2003.

[20] Q. Liu and J.N. Hwang, 'A Scalable Video Transmission System using Bandwidth Inference in Congestion Control', in *IEEE ISCAS* Vancouver, 2004.

[21] Y. Tobe, Y. Tamura, A. Molano, S. Ghost, and H. Tokuda, 'Achieving Moderate Fairness for UDP Flows by Path-Status Classification', in *IEEE LCN* Tampa, FL, 2000.

[22] H.-F. Hsiao, A. Chindapol, J. Ritcey, Y.-C. Chen, and J.-N. Hwang, 'A New Multimedia Packet Loss Classification Algorithm for Congestion Control over Wired/Wireless Channels', in *IEEE ICASSP,* 2005.

[23] C.-W. Huang and J.-N. Hwang, 'An Embedded Packet Train and Adaptive FEC Scheme for Effective Video Adaptation over Wireless Broadband Networks', in *IEEE International Packet Video Workshop* Hangzhou China, 2006.

[24] E. Elliot, 'Estimates of Error Rates for Codes on Burst-Noise Channels', *Bell System Technique Journal,* 1963.

[25] Q. Zhang, G. Wang, Z. Xiong, J. Zhou, and W. Zhu, 'Error Robust Scalable Audio Streaming Over Wireless IP Networks', *IEEE Transactions on Multimedia,* Dec. 2004.

[26] C.-W. Huang, S. Sukittanon, J.A. Ritcey, A. Chindapol, and J.-N. Hwang, 'An Embedded Packet Train and Adaptive FEC Scheme for VoIP over Wired/Wireless IP Networks', in *IEEE ICASSP,* 2006.

[27] IEEE802.16-2004, 'IEEE Standard for Local and Metropolitan Area Networks Part 16: Air Interface for Fixed Broadband Wireless Access Systems', Oct. 2004.

[28] IEEE802.16e-2005, 'IEEE Standard for Local and metropolitan area networks Part 16: Air Interface for Fixed and Mobile Broadband Wireless Access Systems Amendment 2: Physical and Medium Access Control Layers for Combined Fixed and Mobile Operation in Licensed Bands and Corrigendum 1', 2006.

[29] 'Resource ReSerVation Protocol (RSVP)', IETF Network Working Group RFC 2205, 1997.

[30] 'An Architecture for Differentiated Services', IETF Network Working Group RFC 2475, 1998.

[31] C.-W. Huang, J.-N. Hwang, and D.C.-W. Chang, 'Congestion and Error Control for Layered Scalable Video Multicast over WiMAX', in *IEEE Mobile WiMAX Symposium,* Orlando, FL, USA, 2007.

[32] 'The Network Simulator - ns-2', http://www.isi.edu/nsnam/ns/.

[33] J. Chen, C.-C. Wang, F.C.-D. Tsai, C.-W. Chang, S.-S. Liu, J. Guo, W.-J. Lien, J.-H. Sum, and C.-H. Hung, 'The Design and Implementation of WiMAX Module for ns-2 Simulator', in *Proceeding from the 2006 Workshop on Ns-2: The IP Network Simulator,* Pisa: ACM Press, 2006.

14

WiBro – A 2.3 GHz Mobile WiMAX: System Design, Network Deployment, and Services

Hyunpyo Kim, Jaekon Lee, and Byeong Gi Lee

14.1 Introduction

The history of WiBro (abbreviation of *Wireless Broadband*) dates back to 2002 when the Korean government re-allocated the 100 MHz frequency band at the 2.3 GHz spectrum from fixed to portable Internet services. Triggered by the new opportunity, some Korean local research organizations including ETRI, Samsung Electronics and KT started developing the WiBro system. In 2004, the Korean government and the *Telecommunication Technology Association* (TTA), the Korean domestic standardization body, issued the three basic requirements on WiBro system in 2004: (1) 2.3 GHz frequency band; (2) 9 MHz of channel spacing (with an effective bandwidth of 8.75 MHz); and (3) *time-division duplexing* (TDD).

The WiBro development group participated in the IEEE802.16e standardization activity and made an active contribution to help them incorporate the WiBro technical features in the IEEE802.16e-2005 standard [1]. Since then, WiBro has been fully harmonized with Mobile WiMAX based on IEEE 802.16e-2005 standard and the Mobile WiMAX system profile [2], which contains a comprehensive collection of features that the equipment has [3]. As a result, WiBro could provide the same capabilities and features as the standard Mobile WiMAX system, so was regarded as the Mobile WiMAX of Korea [4].

In 2005, the Korean government issued WiBro licenses to three operators including KT, to which the B band in Figure 14.1 is allocated. Based on this 27 MHz band, KT began to prepare to offer commercial services, including network deployment and application service

Mobile WiMAX   Edited by Kwang-Cheng Chen and J. Roberto B. de Marca.
© 2008 John Wiley & Sons, Ltd

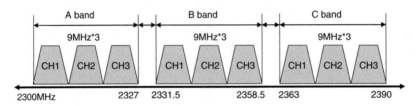

Figure 14.1   Frequency allocation for WiBro services in Korea

development. In June 2006, KT did a large-scale trial of the 2.3 GHz-based Mobile WiMAX network deployed in the metropolitan Seoul area, and then it began to start full commercial services in April 2007. Now its coverage has been expanded to the entire city of Seoul and suburban cities, encompassing 12 million people.

Figure 14.2 shows the overall KT WiBro network architecture in connection with *public-switched telephone network* (PSTN), Internet and CDMA cellular network. Samsung Electronics designed and developed *mobile stations* (MSs) (or *subscriber stations* (SSs)), *base stations* (BSs) (or *radio access stations* (RASs)) and *access service network* (ASN) *gateways* (G/W) (or *access control routers* (ACRs)). KT deployed the *access network* (AN), and implemented AAA, billing system, server farm, back-end platform and other necessary network elements [5]. The technical aspects of the BSs (or RASs) and ASN G/W (or ACR) of the WiBro system will be discussed in Sections 14.2, 14.3 and 14.4: and the network deployment will be discussed in Sections 14.5 and 14.6.

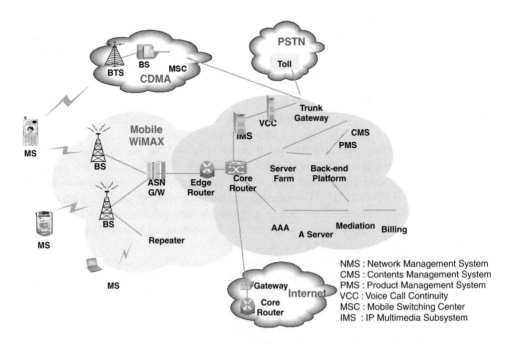

Figure 14.2   Overall WiBro network architecture

14.2 Mobile WiMAX Network

The Mobile WiMAX network, overall, is composed of the *access service network* (ASN) and the *connectivity service network* (CSN). BS (or ACR) and ASN G/W (or RAS) are the major network elements in ASN, and mobile *WiMAX system manager* (WSM) is the network element that manages them. CSN is composed of the *authentication, authorization and accounting* (AAA) servers, *home agent* (HA), *dynamic host configuration protocol* (DHCP) server [6], *domain name service* (DNS) server, and *policy and charging rules function* (PCRF) server. ASN is connected to CSN via router and switch.

14.2.1 Network Configuration

Figure 14.3 shows the configuration and the network elements of the Mobile WiMAX network. We examine the network elements one by one in the following:

Radio Access Station (RAS) (or Base Station (BS))
RAS is the system that connects ACR and MS. It provides wireless connection to MS under IEEE 802.16d/16e standards to support wireless communication services to subscribers. RAS conducts various functions including wireless signal exchange with MS, modulation/demodulation signal processing for packet traffic signal, efficient use of wireless resources, packet scheduling for *quality of service* (QoS) assurance, assignment of wireless bandwidth, *automatic repeat request* (ARQ) processing and ranging function. In addition, RAS controls the connection for packet calls and handover.

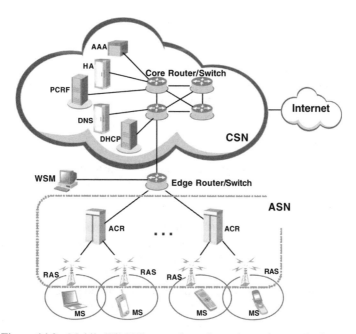

Figure 14.3   Mobile WiMAX network configuration and network elements

Access Control Router (ACR) (or ASN Gateway (ASN G/W))

ACR is the system connecting CSN and RAS. It enables multiple RASs to work with IP network, sends/receives traffic between an external network and MS, and controls QoS. ACR is connected to AAA server in the Diameter protocol method and works with DNS server in the DNS protocol method.

Mobile WiMAX System Manager (WSM)

WSM provides the management environment for the operator to operate and maintain ACR and RAS.

Home Agent (HA)

HA accesses other networks or private networks and enables *mobile IP* (MIP) users to access Internet. HA works with ACR that performs *foreign agent* (FA) function for mobile IPv4 and works with MS to exchange data for mobile IPv6.

Authentication, Authorization and Accounting (AAA) Server

AAA server interfaces with ACR and carries out subscriber authentication and accounting functions. It interfaces with ACR via the Diameter protocol and provides *extensible authentication protocol* (EAP) certification [7].

Domain Name Service (DNS) Server

The DNS server manages the domain names. It interprets the domain or the host names to the IP addresses that are composed of digits.

Dynamic Host Configuration Protocol (DHCP) Server

The DHCP server manages the setup and the IP address of the MS. It performs the management and allocation of the IP addresses and other setup information for the MS. Since ACR includes the DHCP server and relay agent functions, ACR carries out the DHCP server function when external DHCP server does not exist.

Policy and Charging Rules Function (PCRF) Server

The PCRF server manages the service policy and sends QoS setting information for each user session and accounting rule information to ACR.

14.2.2 System Functions

Figure 14.4 shows the functions of the ASN systems – ACR and RAS. Each block name complies with the standards of the mobile WiMAX NWG (*Network Working Group*).

ACR (or ASN G/W)

ACR supports the *convergence sublayer* (CS) and performs the packet classification and *packet header suppression* (PHS) functions. It carries out the header compression function, and supports *robust header compression* (ROHC) defined in the NWG standard. In addition, it performs the paging controller and location register functions for an MS in idle mode [8].

In authentication, ACR performs the authenticator function and performs the key distributor function to manage the higher security key by working with the AAA server as an AAA client. At this time, RAS performs the key receiver function to receive the security key from the key distributor and manages it. ACR works with the AAA server of the CSN for authentication

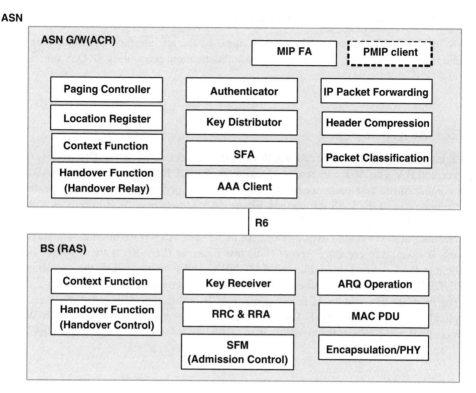

Figure 14.4   Configuration of mobile WiMAX system functions (based on profile C)

and charging services and with the HA of CSN for *Mobile IP* (MIP) service. ACR, as the FA of MIP, supports both *Proxy MIP* (PMIP) and *Client MIP* (CMIP).

RAS (or BS)

RAS performs the *service flow management* (SFM) function to create/change/release connections for each *service flow* (SF) and the admission control function while creating/changing the connections. In regard to the SFM function of RAS, ACR carries out the *SF authentication* (SFA) and *SF identification* (SFID) management functions. ACR carries out the SFA function to obtain QoS information from *policy function* (PF) and apply it in the SF creation. It also performs the SFID management function to create/change/release SFID and map the SF according to the packet classification.

In handover, RAS performs the handover control function to determine the execution of the handover and processes the corresponding handover signaling. The ACR confirms the neighbor BS list and relays the handover signaling message to the target system. At this time, ACR and RAS carry out the context function to exchange the context information between the target system and the serving system. RAS performs the *radio resource control* (RRC) and *radio resource agent* (RRA) functions to collect/manage the radio resource information from MSs and the RAS itself.

14.3 ACR (ASN-GW) System Design

ACR (or ASN-GW) processes the control signal of the air interface and the bearer traffic. When working with CSN, ACR provides the authentication, accounting, IP QoS function, and others.

14.3.1 ACR Architecture

ACR interfaces with RAS, other ACRs, WSM, DNS servers, AAA server, DHCP server, PCRF server, and HA. The ACR can access a maximum of 1,000 RASs on the basis of RAS ID, and the number of the maximum acceptable RASs varies depending on the call model, the RAS configuration, and the RAS throughput. Figure 14.5 depicts how ACR interfaces with RAS, other ACR, WSM, DNS, AAA, and DHCP.

The interface between an ACR and an RAS in the same ASN is R6 interface and its physical access is done over *gigabit Ethernet* (GE)/*fast Ethernet* (FE). R6 is the interface between ACR and RAS defined in Mobile WiMAX NWG and is composed of the signaling plane (IP/UDP/R6) and the bearer plane (IP/*generic routing encapsulation* (GRE)) [9].

The interface between an ACR and another ACR in different ASN is R4 interface and its physical access method is GE/FE. R4 is the interface between ACRs defined in Mobile WiMAX NWG and is composed of the signaling plane (IP/UDP/R4) and the bearer plane (IP/GRE).

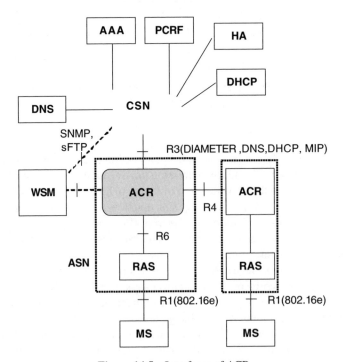

Figure 14.5   Interfaces of ACR

The interface between an ACR and a WSM complies with the *simple network management protocol* (SNMP)v2c/SNMPv3 (which are the IETF standards), *secure file transfer protocol* (sFTP), or Samsung's proprietary standard and its physical access method is GE/FE.

The interface between an ACR and a HA complies with the IETF standard MIP specification and its physical access method is GE/FE.

The interface between an ACR and an AAA server complies with the IETF standard Diameter specification and its physical access method is GE/FE.

The interface between an ACR and a PCRF server complies with the IETF standard Diameter specification and the 3GPP standard *policy and charging control* (PCC) specification and its physical access method is GE/FE.

The interface between an ACR and a DNS server complies with the IETF standard DNS specification and its physical access method is GE/FE.

The interface between an ACR and a DHCP server complies with the IETF standard DHCP specification and its physical access method is GE/FE.

14.3.2 ACR Functions

ASN performs various functions, including mobility support, call processing, bearer processing, and working.

14.3.2.1 Mobility Support Function

ASN can contain several ACRs and each ACR divides several MSs into one or more groups and manages each group as one IP subnet. So the CSN communicating with MS considers MS as an end host connected to an IP subnet.

ACR provides the following functions to maintain the connection between MS and the network wherever the MS moves in such an IP network structure at any status.

Optimized Hard Handover
ACR performs the L2 handover with no re-authentication, keeping the anchor function of MIP FA. So ACR can carry out the hard handover optimized up to the highest level defined in IEEE 802.16.e standards by minimizing the break time caused by the handover.

MIP Function
ACR supports Simple IP and MIP and provides both *Proxy MIP* (PMIP) and *Client MIP* (CMIP) as the FA of MIP. In PMIP, when an MS does not support the MIP stack, ACR performs the FA function of MIP and the MIP client function replaces the MS. Thus, ACR enables continuous services even when the MS supports only Simple IP. In CMIP, the MS supports MIP stack and the ACR acts only as FA of the MIP.

Handover Processing under L2 Layer
When an MS in Awake Mode or Sleep Mode moves, ACR makes the handover process only under L2 layer to perform the handover quickly. Although an MS moves to another IP subnet, the L3 layer between the ACR, which is the current session anchor, and the MS does not change and only the L2 layer is extended to transmit traffic. In these handover methods, the

R3 relocation procedure to change L3 layer from a serving ACR to a target ACR is carried out when the MS status is transited to Idle Mode after handover.

In general, if an MS supporting MIP moves in a subnet area, MIP handover is carried out and FA is changed. However, Samsung's ACR extends L2 path and deals with handover quickly when an MS in Awake Mode/Sleep Mode moves in a subnet area.

Handover Processing Including L3 Layer
When an Idle Mode MS not being served moves to another ACR area or an MS in Awake Mode or Sleep Mode moves to an ASN composed of another provider's devices, the handover function including the L3 layer is performed. The handover function, including the L3 layer, relocates the anchor point of R3, which is the interface with the CSN, to a target ASN when an MS is moved to another ASN area. In the handover, including the L3 layer, the R3 relocation procedure is carried out after performing the handover under L2 layer to minimize the break time.

Context Transmission Function
ACR stores and updates various types of context information to manage all the status of MS, such as Awake Mode, Sleep Mode and Idle Mode. The context information includes MS ID information, service flow information, security information, paging information, and other MS information. ACR transmits the context information of an MS to a target RAS or a target ACR when the MS performs handover, location update, and *quick connection setup* (QCS) and enables MS to access the target RAS/target ACR quickly.

Paging Controller
ACR as a paging controller performs paging and MS's location management functions. ACR transmits paging message to overall paging group area to enable the Idle Mode MS to enter the network and changes the status of the MS from Idle Mode to Awake Mode. ACR can perform paging function by composing the paging groups diversely.

In addition, ACR manages the location of Idle Mode MS in a paging group. If an MS in Idle Mode receives a paging message from the paging controller, the MS acquires the current location information, performs the location update procedure, if necessary, and notifies the paging group where the MS is involved to the ACR.

Fast BS Switching (FBSS)
ACR manages the active set defined in IEEE 802.16e standards and can quickly provide anchor BS switching between RASs by supporting *fast BS switching* (FBSS) effectively. The FBSS deals with the signaling for handover in advance and sends Indication message via the dedicated channel, *channel quality indicator channel* (CQICH), so that it can raise the handover success rate and reduce the break time more than the reduction made by the optimized hard handover.

14.3.2.2 Call Processing Function

Service Flow Authorization (SFA) Function
When an MS accesses the Mobile WiMAX network initially or requests the creation/change/deletion of service flow during the access, ACR can create, change or delete the connection corresponding to the service flow via the process of *dynamic service addition* (DSA)/*dynamic service change* (DSC)/*dynamic service deletion* (DSD).

MS Authentication Function

ACR performs the MS authentication function using *extensible authentication protocol* (EAP) by working with the AAA server to determine the legitimacy of an MS. In the MS authentication using EAP, the ACR carries out the authenticator function and transfers EAP payload between the MS and the AAA server by acting as a 'pass-through' agent independently of the EAP-method on the upper EAP. The MS transmits the EAP payload to RAS via the *privacy key management* (PKM) message. The EAP payload transmitted to the RAS is transmitted to the AAA server through R6 interface (between RAS and ACR) and Diameter interface (between ACR and AAA server).

Since the ACR sustains the anchor authenticator function, the re-authentication procedure of an MS may be omitted if the ACR has not been changed in handover or in re-entry of an MS in Idle Mode. At this time, the EAP-*transport layer security* (EAP-TLS) method based on X.509 certificate is supported for MS authentication. The detailed EAP method may vary depending on the provider's policy.

Subscriber Authentication Function

ACR performs the subscriber authentication function by using EAP after working with the AAA server. At this time, the EAP authentication and key agreement method or EAP tunnel TLS method is supported for subscriber authentication. The detailed EAP method may vary depending on the provider's policy.

Security Key Management Function

If MS certification or subscriber authentication is successfully completed, the ACR receives the upper security key, *master session key* (MSK) from the AAA server and then it creates and manages the security key for the MAC management message authentication and traffic encryption.

ACR manages such security information as *security association* (SA) and shares the SA information between MS and ACR via the authentication process.

Accounting Information Collection and Report Functions

ACR collects *call detail record* (CDR) in preparation for imposing the Mobile WiMAX service fee to subscribers. ACR collects the accounting information on the used time, used data packet, service level and QoS. In addition, ACR sends the collected accounting information to the AAA server via the Diameter protocol. Since the accounting information is collected in the unit of service flow, some differentiated accounting policy can be provided in terms of service flow.

14.3.2.3 Bearer Processing Function

Packet Classification

Since the IEEE 802.16e MAC standard adopts the connection-oriented method, all uplink/downlink packets are mapped to a specific connection for the packet exchange. The function to classify traffic packets to map with the MAC connection depending on the service flow is called packet classification. The IEEE 802.16e-2005 standard defines the packet classification rule including ATM, IP, Ethernet/*virtual local area network* (VLAN) and ROHC. Among them, Mobile WiMAX prescribes only IP and ROHC as compulsory requirements. So ACR provides packet classification for IP and ROHC.

Packet Header Suppression (PHS)

The IEEE 802.16e standard defines PHS and enables suppression of the repeated part of a packet header for an efficient use of radio resources after classifying packets. MS and ACR set PHA parameter and determine the part to be deleted from the packet header, and this is called the PHS rule. ACR exchanges the PHS rule with the MS via the DSA procedure while setting a connection. ACR and MS suppress or restore the packet header according to the PHS rule during the packet exchange.

Robust Header Compression (ROHC)

ROHC is the way to compress a packet header including the IP header and the algorithm defined in IETF RFC3095. ROHC can manage the packet status dynamically and has a high level of efficiency in header compression, while the PHS simply suppresses and restores a part of packet. In addition, PHS is defined in the IEEE 802.16e-2005 standard and applies to Mobile WiMAX, but the ROHC applies to various technologies including WCDMA and has high compatibility. However, ROHC is more complicated in algorithm than PHS. In addition, its feedback path is managed to enhance the robustness of the protocol.

Data Path Function (DPF)

ACR interfaces with multiple RASs or ACRs and the interface function on these bearer planes are called *data path function* (DPF). The NWG standard divides the type of DPFs into Type 1 and Type 2. For Type 1, IP packets are exchanged but, for Type 2, MAC SDU is exchanged. ACR supports DPF Type 1. The communication with RAS and another ACR observes the WiMAX NWG R6 interface and the WiMAX NWG R4 interface, respectively.

IP QoS Function

ACR provides *differentiated service* (DiffServ)-based QoS. DiffServ is the way to apply a differentiated scheduling by varying the *differentiated services code point* (DSCP) value in accordance with the QoS level. ACR provides IP QoS by varying the DSCP value in accordance with the QoS level of the service flow. As such, in ASN, the DSCP value is used to provide QoS.

End-to-End QoS Structure

For subscribers to feel actual QoS, end-to-end QoS, not just the QoS in a particular section, should be ensured. The section between MS and RAS is an air section and service is provided to the section according to the QoS defined in the IEEE 802.16e-2005 standard. The section between ACR and RAS is the MAC section and the QoS to be applied to each service flow is set after classifying the service flow. The QoS set in the MAC section is mapped with the QoS in ASN.

The information on QoS classes and QoS parameters of the service flow for the end-to-end QoS is stored in the AAA server or the PCRF server. ACR receives the QoS information from the AAA server or the PCRF server when the relevant service flow is created. Based on this QoS information, ACR transmits the QoS information corresponding to the service class of the air section defined to RAS and sets the DSCP value for the QoS of the MAC section.

Simultaneous IPv4/IPv6 Support

ACR supports the dual stack of IP (i.e., IPv4 and IPv6) simultaneously. The dual stack function of ACR is roughly divided into the dual stack function for MS access and the dual stack function for the working between *network elements* (NEs). The dual stack function for the MS access and the dual stack function for the connection between NEs can be performed, separately. For example, when an MS accesses the IPv6 network, normal service can be provided even if ACR and RAS are connected via IPv4 network. Since ACR is connected with RAS via a tunnel, the IP method for the tunnel can be independently selected from the protocol version used in MS.

IP Routing Function
Since ACR provides several Ethernet interfaces, it stores the information on the Ethernet interface to route IP packets to the routing table. The operator organizes the routing table for the ACR operation. The way to organize and set the routing table is similar to the standard setting of the router. The routing table of ACR is configured depending on the operator's setting and configuration. The settings of the routing table are similar to the standard setting of the router.

ACR can set the routing table to support the static and the dynamic routing protocols, such as *open shortest path first* (OSPF), *intermediate system to intermediate system* (IS-IS) and *border gateway protocol* (BGP). In addition, ACR supports the IP packet routing function to transmit the packets that is handled inside the system via the interface specified by the system routing table. ACR also supports the function to forward the IP packets received from the outside according to the information of the routing table.

Ethernet/VLAN Interface Function
ACR provides the Ethernet interface and supports the link grouping function, VLAN function and Ethernet *class of service* (CoS)function under IEEE 802.3ad for the Ethernet interface. At this time, the MAC bridge function defined in IEEE 802.1d is excluded. ACR enables several VLAN IDs to be set in one Ethernet interface and maps the DSCP value of IP header with the CoS value of the Ethernet header in the transmit packet to support the Ethernet CoS.

14.3.2.4 Connecting Function

HA Connecting Function
To support the MIP service, the ACR supports CMIP, which is the IETF standard MIP, and PMIP, which is defined in the WiMAX NWG standard, for the interface between ACR and HA. As an FA, ACR transmits the *care-of-address* (CoA) by MS and the binding information between HAs to HA and exchanges MS traffic via the tunneling interaction with HA. In addition, ACR additionally performs the PMIP client function for the MS not providing MIP stack.

ACR can connect with multiple HAs and specify the HA connection for each MS. At this time, the information on the HA connected with each MS is informed from the AAA server to the ACR in the initial MS authentication stage.

AAA Server Connecting Function
ACR connects with the AAA server under the IETF standard Diameter specification and performs the MS authentication and subscriber authentication functions according to the EAP method by connecting with the AAA server. The EAP-method is implemented in MS and the AAA server. ACR relays the EAP payload to the AAA server. In addition, ACR collects the accounting information on subscriber access and the used time, the used data packets and

the service level and QoS and transmits the information to the AAA server via the Diameter protocol.

PCRF Server Connecting Function

The PCRF server determines the accounting rule on the basis of the QoS policy and the service flow and sends it to ACR. According to the policy received from the PCRF server, ACR can create/change/release the service flow dynamically and controls the QoS complying with the service flow. In addition, ACR creates accounting data by using the accounting rule received from the PCRF server. The interface between the ACR and the PCRF server is based on the Gx interface of 3GPP Rel 7 and the Gx interface is determined by using the Diameter. The Gx interface is for provisioning the service data flow based charging rules between the *traffic plane function* (TPF) and the *charging rules function* (CRF), also known as the service data flow based charging rules function.

IP Address Allocation Function

ACR allocates an IP address to MS in the DHCP or MIP method. The way that ACR allocates an IP address to MS is subject to the service type provided to the MS. When Simple IP service is provided to MS, ACR allocates an IP address to the MS by using DHCP. At this time, ACR acts as a DHCP server or a DHCP relay agent. When PMIP service is provided to MS, ACR allocates an IP address to the MS by using DHCP and uses MIP for HA. When CMIP service is provided to MS, ACR transmits a home IP address to the MS by using MIP.

14.4 RAS (BS) System Design

The main function of RAS is to perform air-interface processing based on the Mobile WiMAX profiles of IEEE 802.16e specification. Table 14.1 lists some characteristics of the WiBro system that meets the Mobile WiMAX profile. Note that the data rates of 45 Mbps *downlink* (DL) and 12 Mbps *uplink* (UL) can be obtained in the WiMAX Wave 2 system using 10 MHz bandwidth (based on 64 QAM with 5/6 rate in DL and 16 QAM with 3/4 rate in UL; utilizing 26 symbols in DL and 12 symbols in UL). The system supports 2×2 *multi-input multi-output* (MIMO) technology, 120 km/h mobility, and 1–15 km of coverage. It adopts the *time-division duplexing* (TDD) technology for duplexing and the *orthogonal frequency division multiple access* (OFDMA) technology for multiple access, taking the *cyclic prefix*

Table 14.1 WiBro system specification

Parameters	Value or technology
Multiple Access	OFDMA
Duplexig	TDD
Channel Bandwidth	9 MHz
MIMO technology	2×2
Peak data rate	45 Mbps DL, 12 Mbps UL @10 MHz
Mobility	120 km/h
Service coverage	1~15 km
Cyclic prefix	12.8 us (1/8)

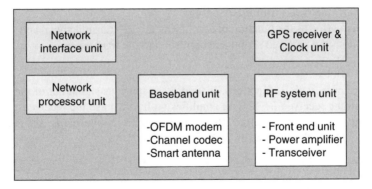

Figure 14.6 Functional architecture of RAS

(CP) of 1/8 size. RAS is controlled by ACR, and it processes functions to link ACR and MSs. The functions include modulation/demodulation of packets, wireless resources management, packet scheduling to guarantee QoS, and connecting with ACR for handover.

14.4.1 RAS Architecture

RAS, in general, is composed of five functional units, namely *global positioning system* (GPS) receiver and clock unit, RF system unit, baseband unit, network process unit, and network interface unit. Figure 14.6 illustrates this.

14.4.1.1 GPS Receiver and Clock Unit

The GPS receiver and clock unit receives the GPS signal and generates timing reference to maintain system synchronization.

14.4.1.2 RF System Unit

The main components of the *radio frequency* (RF) system include front-end unit, power amplifier/low-noise amplifier, and transceiver.

Front-End Unit
Front-end unit sends out RF signals to antenna, where the received signals are filtered for the frequency band of interest and then amplified. Switch is used for TDD and duplexer is used for FDD to separate the transmitted RF signal and the received RF signal at the front end unit. In addition, it may support the diagnosis function on the RF transmit/receive paths.

Specifically, the key functions of *front-end board* (FEB) include the followings: Transmit RF signal by antenna; suppression of spurious out-of-band signals, which is emitted from the received RF signal; low-noise amplification of the received pass band RF signal; distribution of the down-converted RF signals to several channel cards; and TDD switching function for the RF.

Power Amplifier/Low-Noise Amplifier (LNA)

Power amplifier performs the function of amplifying the RF transmit signal and the low-noise amplifier amplifies the received signal and delivers it to the transceiver.

Transceiver

Transceiver performs various functions. It suppresses spurious signal out of the band, which is emitted from the received RF signal; amplifies with low noise the received pass band RF signal; distributes the down-converted RF signals to several channel cards; and performs TDD switching function for the RF transmit/receive path.

14.4.1.3 Baseband Unit

Baseband unit includes OFDM modem, channel codec, and smart antenna/*space time coding* (STC) unit.

OFDM Modem

OFDM *modulator/demodulator* (modem) performs the following functions: Modulation and demodulation of OFDM signal, synchronization for packet traffic burst, link control (such as power control, frequency offset control, timing offset control), and inter-carrier interference cancellation.

Channel Codec

Channel *coder/decoder* (codec) is responsible for coding/decoding the duo-binary *convolutional turbo code* (CTC) and memory management for *hybrid ARQ* (HARQ) support.

Smart Antenna/STC

Smart antenna performs TDD RF calibration and beam-forming for supporting the *space division multiple access* (SDMA) scheduler. STC and frequency hopping diversity coding are supported for 2 and 4 transmit antennas.

14.4.1.4 Network Processor Unit

Network processor unit of RAS is responsible for such functions as scheduling multi-user traffic with QoS support, radio resource management, and handover supports in cooperation with ACR.

14.4.1.5 Network Interface Unit

Network interface unit supports the interface between ACR and RAS.

14.4.2 RAS Functions

The main function of RAS is to perform air-interface processing based on the IEEE 802.16e specifications. RAS is controlled by ACR, and it processes functions to link ACR and MSs. The functions include modulation/demodulation of packets, call processing, wireless resources management, packet scheduling to support QoS, connecting with ACR for handover, operation and maintenance, and other additional functions.

14.4.2.1 Call Processing Function

The *RAS main block* (RMB) is responsible for call processing. It provides an initial access to MSs, and allocates the *connection identifier* (CID) to MSs and supports handover. Also, RAS supports the location update and registration between MSs and ACR, and transmits subscriber data between MSs and ACR.

14.4.2.2 Handover Optimization Function

RAS may support the handover between sectors (i.e., *inter-sector HO*), the handover between ACR (i.e., *inter-ACR HO*), and the handover between *frequency allocation* (FA) (i.e., *inter-FA HO*). To maintain the call quality, the packet loss rate and the handover delay time are to be optimized.

Maintaining Call Quality during Handover
RAS may use a flexible frequency management to obtain higher *signal to interference and noise ratio* (SINR) for MSs in the area. Also, it can raise the link performance during handover by supporting soft handover between sectors in the upper link.

Minimizing Delay Time
To minimize the handover delay, the target RAS may reuse the session data setup between the serving RAS and the MSs, and minimize the time for re-entry of the MSs to the target RAS in the mobile area.

14.4.2.3 System Resource Management Function

RAS can detect the service status of MS for each FA/sector, and may not assign additional calls to FA/sector where the service is not available, but assigns calls to the service available FA/sector. RAS manages the MS Awake/Sleep Mode and the MS connection status, and controls the receive traffic volume received from the ACR by limiting the assigned call counts for certain time period based on the specified overload grades.

Wireless Resource Management Function
RAS can manage the failure status, MS connection status, and the Awake/Sleep Mode status of MS for each FA/sector.

Scheduling Function
To support QoS, RAS provides the QoS scheduling, where the scheduler manages QoS parameter for each subscriber. QoS parameter can be set up separately.

Overload Control
If the system gets overloaded, RAS can limit the new calls per certain hours not to exceed the threshold value defined on the overload level for certain period of time, based on the specified overload grades.

14.4.2.4 Convenient Operation and Maintenance Function

In the event of failure, cancellation, board status change, link status change, board switching, and link switching occurs, the corresponding data are reported to the network manage system (i.e., WSM) in real time. Through the WSM, operator can inquire the configuration data of RAS and support the growth/reduction of the network.

Configuration Management
While operating the system, the operator can change or delete the system operating parameters, and also change the configuration data.

Status Management
Upper processors of RAS may manage the status (such as operation status, duplex status) of lower processors. Operator can inquire about the status of the system, sector, FA, and repeater services, and the network management system can categorize and select the commands that can be executed in the ACR and the RAS.

Measuring and Statistics
RAS counts the events in the system to create the statistics data by checking the status, maintenance and the performance of the system, and sends the data to network management system, so that operator can check the statistics data as necessary.

14.4.2.5 Additional Functions

System Test Function
RAS can provide the diagnosis and failure detection functions for the *inter-processor communication* (IPC) and RF path of the boards, which are related to other system services except for the power supply module.

Operation Test Function
RAS can support the diagnostic function for the call path in the link and the system. The diagnostic results and the quality measurements are output, through the network management system, in the format that can be analyzed.

Failure Diagnosis and Processing
RAS may generate alarms in the case of system failure or operation failure, and report the status to the network management system in real time according to the severity criterion that operator had preset. If any mechanical failure that affects the service, it switches the boards and the links, and executes the overload control function for the device to stop the service. In addition, if there is any failure in the duplex boards or links, it switches the failed board or link, and reports the switching status to the network management system. In the case of software failure, it restarts the failed block or recovers it by restarting or auto-loading.

Alarm
The alarm is categorized into multiple levels depending on the level of critical effects on the system, for example, critical alarm, major alarm, minor alarm and warning alarm. The alarm is switched off if the system has recovered and is back to normal operation.

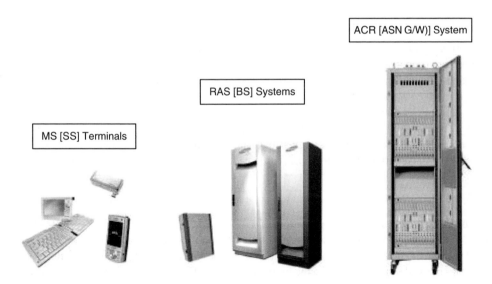

Figure 14.7 WiBro systems developed by Samsung Electronics

Auto Switching

If a failure occurs, duplex boards and devices, except for the channel cards, are designed to switch to each other automatically, without affecting the services that the system is providing. In channel card switching, the service is dropped since the standby channel card does not back up the data of the present active channel card. Nevertheless, it does not affect new services.

Remote Access

The operator can perform maintenance and debugging remotely using the remote access function of the RAS.

Figure 14.7 shows the MS terminals, RAS (or BS) systems, and ACR (or ASN G/W) system developed by Samsung Electronics.

14.5 Access Network Deployment

Deployment of the access network is a very important process for operator as it is the dominant portion of the total investment for providing services and, in addition, it is highly correlated with the service quality. In order to achieve cost-effective network deployment, it is necessary to determine the optimal location and height, the type of the equipment (such as indoor/outdoor, BS, repeaters, FA, sector) and the sector configuration [10]. Also, it is necessary to conduct medium-scale tuning processes by determining the position (feeder type), type, azimuth, and down-tilt of the antenna. Then fine-scale tuning can be done by adjusting the engineering parameters such as the transmit power allocation and hand-over parameters. Good service quality and maximum coverage with optimal investment can be ensured by arranging such an optimal cell planning.

14.5.1 Radio Network Planning (RNP)

In designing the radio network, the following two factors should be considered: first, it is important to understand the system and the radio environment. To do this, we need to analyze the propagation characteristics of the 2.3 GHz radio wave and determine the service quality based on the *received signal strength indicator* (RSSI) and CINR values. Secondly, it is important to do efficient cell planning of the ground. In support of this, we need to choose the potential BS sites by considering the center of high data traffic areas, relatively high buildings to insure the *line-of-sight* (LOS) site as much as possible, and good locations for indoor services of huge buildings adjacent to the main roads.

14.5.1.1 Process of RNP

Figure 14.8 describes the overall procedure of radio network planning. This whole RNP process, in general, is composed of three stages, which are dimensioning, preliminary planning, and final planning [11]. If there is a network already, all the steps may not be required. In addition, the steps marked by dashed lines in the figure may be omitted depending on the specific project.

14.5.1.2 Dimensioning

The dimensioning process is the first stage of the RNP. Its goal is to determine the BS counts and the network configuration based on the analysis of the coverage and capacity. In the dimensioning stage, the first procedure is the coverage planning which is done based on the link budget. It is composed of the following four steps:

1. Analyze the link budget parameters such as slow fading margin, handover gain, interference margin, receiver sensitivity, and noise figure.
2. Calculate the *maximum allowable path loss* (MAPL).
3. Analyze the prediction model.
4. Determine the BS counts and configuration.

Capacity planning is another key process in the dimensioning stage, which can be done based on the user traffic. It is composed of the following four steps:

1. Generate the user traffic model.
2. Calculate the system capacity.
3. Determine the traffic loading.
4. Determine the BS counts and configuration. The BS counts and configuration process may be terminated by examining the balance of the coverage and capacity planning.

14.5.1.3 Preliminary Planning

Preliminary planning is the second stage of the RNP. It is the preparation process before doing on-site cell planning based on the BS counts and the configuration obtained in the dimensioning stage. The following five steps are performed in this stage:

1. Field survey.
2. CW test and prediction model tuning.

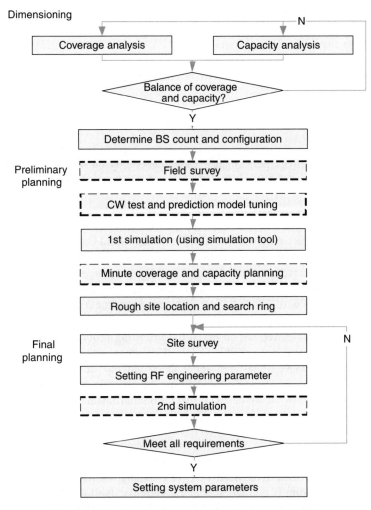

Figure 14.8 Overall procedure of radio network planning

3. The 1st simulation (using the cell planning tools).
4. Fine coverage and capacity planning.
5. Coarse site location and search ring.

14.5.1.4 Final Planning

Final planning is performed based on the results of the preliminary planning. In this stage, site survey is performed and all the system parameters are determined. Specifically, this stage is composed of the following four steps:

1. Site survey: First, site position is determined, and equipment type is determined as well, by considering all the aspects including the coverage, capacity, and interference. Throughout the overall RNP procedures and network optimization process, site location and height are

the most important factors that influence on the network performance. Second, antenna position, as well as antenna type and configuration (azimuth and down-tilt) values, are determined to ensure the desired coverage.

2. Setting all the RF engineering parameters.
3. The 2nd simulation (using the detailed RF engineering parameters).
4. Setting system parameters.

14.5.1.5 Case Studies

As discussed earlier, network performance varies depending on the network planning. We introduce some of the real RNP results [7] obtained during the Mobile WiMAX network deployment by KT.

Case 1 Yeoksam 5 Area

Yeoksam 5 area covers the crossroads near Gangnam station and Gangnam street. The Mobile WiMAX systems installed are the wall-mounted type and the environment-friendly type. The design point was that this area is one of the most thriving and congested areas in metropolitan Seoul, so decentralizing of the traffic was important. We installed antennas with a tilt over 20°, with 15dB gain, and with a wide horizontal beam angle. The α sector was designed to be partially covered by in-building repeaters.

While installing the antennas, we observed the following problems: Due to the building structure in that area, it was not possible to install antennas at the edge of the building. In addition, there existed weak signal area under the γ antenna. It turned out that coverage adjustment or additional repeater was needed during the network optimization process.

Tables 14.2 and 14.3 show the RF design forms for BS and antenna, respectively, and Figure 14.9 shows the RF design map for the case study of the Yeoksam 5 area.

Case 2 Pangyo IC Area

The Pangyo IC area includes the Gyeongboo Expressway which has about 10 lanes and an *interchange* (IC) that connects to the nearby city of Bundang. We installed antennas on a 45 m high pole. The design point is that the area is a very noisy due to high-speed traffic, so we needed to install high gain antennas to avoid the *pseudo-random* (PN) noise generated on the expressway and to achieve a wide coverage. We used antennas with a narrow horizontal beam angle to reduce the interference among sectors.

While installing the antennas, we observed that there was a slope road in the $\beta1$ direction. So we would have to adopt an antenna with wide vertical beam angle if there was a weak signal area on that road.

Table 14.2 RF design form (BS) of Case 1

BS ID	SL0511W	Type	Outdoor, standard
BS name	Yeoksam 5	FA/sec	1FA/3S
BS location	18FL(rooftop), Gangnam Bldg	Latitude	37-29-50.702
		Longitude	127-01-42.427
Coverage	Teheran Road, Gangnam Road, Yeoksam-dong, Taegeukdang		

Table 14.3 RF design form (Antenna) of Case 1

Antenna information	α			β			γ		
	$\alpha0$	$\alpha1$	$\alpha2$	$\beta0$	$\beta1$	$\beta2$	$\gamma0$	$\gamma1$	$\gamma2$
Type	**65-15-TA(5)**	**65-15-TA(5)**		**65-15-TA(5)**			**65-15-TA(5)**		
Azimuth	80	140		165			300		
M_Tilt	20	20		15			15		
Quantity	2			2			2		

Tables 14.4 and 14.5 show the RF design forms for BS and antenna, respectively, and Figure 14.10 shows the RF design map for the case study of the Pangyo IC area.

14.5.2 Network Implementation and Optimization

KT installed BSs, different types of repeaters (optical or RF), feeder line to cover the entire metropolitan Seoul area and its vicinities, as shown in Figure 14.11.

In order to achieve the target performance of the network, we conducted a network optimization process. In network optimization, there are two different types – RF optimization and system optimization.

RF optimization is the overall process of improving the RF parameters (such as CINR and RSSI). Specifically, the process includes the adjustment of antenna such as gain, tilt (mechanical and electrical) and direction; relocation of equipments; and implementation of additional antennas.

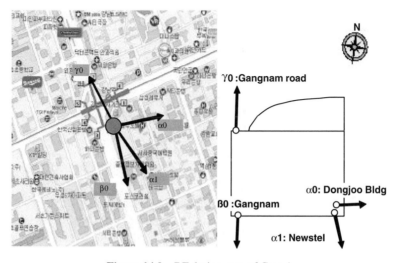

Figure 14.9 RF design map of Case 1

Table 14.4 RF design form (BS) of Case 2

BS ID	KG0023W	Type	Outdoor, standard
BS name	Pangyo IC	FA/sec	1FA/2S
BS location	337-1 Geumto-dong,	Latitude	37-24-20.533
	Seongnam City	Longitude	127-06-28.777
Coverage	Pangyo IC, Gyeongboo Expressway, Outer Beltway		

Table 14.5 RF design form (Antenna) of Case 2

Antenna information	α			β			γ		
	α0	α1	α2	β0	β1	β2	γ0	γ1	γ2
Type	65-17-TA(5)	65-17-TA(2)		33-19-TA(4)	65-17-TA(4)				
Azimuth	70	150		240	330				
M_Tilt	50	10		0	0				
Quantity	2			2	2				

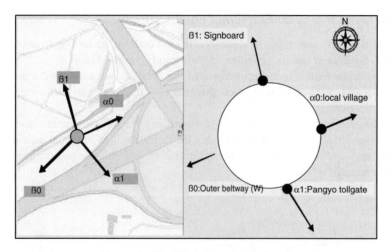

Figure 14.10 RF design map of Case 2

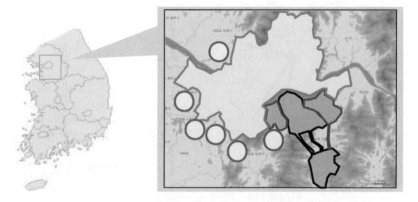

Figure 14.11 WiBro services coverage map in metropolitan Seoul and its vicinity

Table 14.6 Target performances of selected items

	Items		Target	Remarks
Data	Success rate of network connection		> 98%	
	Transmission completion rate		> 97%	FTP file transfer
	Throughput per user (minimum)	Downlink	512 Kbps	FTP file transfer at cell edge
		Uplink	128 Kbps	
	Throughput per user (average)	Downlink	3 Mbps	FTP file transfer at random place
		Uplink	1 Mbps	
	HO latency		< 150 ms	
Service	Streaming service	Completion rate	> 95%	

System optimization is the process of upgrading the software package in the system. It adjusts the system parameters such as output, timer, and so on.

While conducting a large-scale trial, we were able to define the target performances of some major items as listed in Table 14.6. For commercial service, target performance was set to be much higher than given on the table. We performed the network optimization process iteratively to achieve this target performance, finally meeting the target performance for commercial services.

14.6 Core Network Deployment

In addition to BS (or RAS) in the access network, we need to install other network elements in the core network such as ASN G/W (or ACR), *elementary management system* (EMS), AAA, HA, DNS server, DHCP server, NMS, aggregation switch, and router. Besides, we need to deploy the transmission lines connecting those network elements and the connection to backbone network for commercial services.

14.6.1 Core Network Planning

In order to deploy efficient core network and to provide quality services, we need to establish suitable design criteria. The design criteria KT adopted for the core network deployment were targeted at the following goals: Accommodation of high capacity of data traffic according to traffic prediction, efficient mobility support with minimized handover traffic between ASN G/W, economical implementation of the network, reliable network for ensuring good quality services, and flexible and scalable network architecture for easy addition, removal and substitution.

In general, the core network design process is divided into three stages, namely, network design, equipment planning, and implementation planning.

Table 14.7 Design standard of network elements

	Item	Number of BSs accommodated	
		Maximum capacity	Standard capacity
ASN G/W	Small (1.2Gbps)	60	50
	Medium (2.4Gbps)	120	100
	Large (3.6Gbps)	180	150
Aggregation switch		30	24

14.6.1.1 Network Design

Network design stage is composed of the following four steps:

1. Analyze the market forecast data.
2. Predict the data traffic. As a means, first, classify terminal type; second, calculate data traffic of each terminal type in each year; and calculate the average monthly data traffic, the hourly data traffic at the busiest hour, and the peak data traffic.
3. Establish a design standard for each network element.
4. Determine the network topology and the routing policy.

For example, the design standard of ASN G/W and the aggregation switch for WiBro services were established as listed in Table 14.7.

To decentralize the traffic and avoid service discontinuity caused by network failure, we protected each node by doubling it. In KT network, the main and local nodes were doubled and the center nodes were located in two different places. It was possible to easily expand the network due to the layered network architecture that we had nationwide. We had 217 branches, 31 local nodes of IP premium network, which was used as a backbone for WiBro services.

The connection between the Mobile WiMAX network and the Internet backbone was done in the following two ways:

1. *Direct connection type*: It is the basic approach adopted by KT. Mobile WiMAX network is directly connected to the Internet backbone. So it is adequate to handle high data traffic and to support broad coverage. The connection follows the pattern: BS – aggregation switch – ASN G/W – PE router (local node).
2. *Interconnection type*: BS is connected to the local nodes of IP premium network. So it is adequate to save network implementation cost and support small coverage at remote site. The connection follows the pattern: BS – aggregation switch – PE router (local node) – ASN G/W.

14.6.1.2 Equipment Planning

Equipment planning stage is the process of making basic equipment planning (i.e., network element planning) and is composed of the following three steps:

1. Determine the target coverage.
2. Gather the information to plan, such as the predicted traffic data and the on-site data at the installation place.

3. Decide the specification and the quantity needed at each location, including the general requirement (or equipment specification), installation plan, the number of lines required, and the detailed list of the needed equipment quantity.

14.6.1.3 Implementation Planning

Implementation planning is the final stage of the core network planning to prepare for the detailed layout of the equipment and the construction plan. It is composed of the following three steps:

1. Do on-site investigation.
2. Draw the equipment layout.
3. Make the construction plan.

14.6.2 Authentication, Authorization and Accounting (AAA)

AAA and supplementary servers are needed to manage subscriber's access to the network and services and to generate the billing data. KT WiBro service adopted the *single sign-on* (SSO) concept by the interconnecting network and by service authentication for specific value-added application services. In addition, KT designed the system such that it generates different billing rates according to the contents type and the service class. Such functionality was made possible by getting support of several different servers as follows: (1) Authentication server for authentication of the user; (2) Session server for session DB; (3) *Operation and maintenance platform* (OMP) server for operation and management of AAA; (4) *Authentication center* (AuC) server for management of authentication key; (5) *Statistics server* for authentication and account statistics; and (6) *Billing server* for the generation of packet data record.

14.6.3 Aggregation Switch (L2 switch)

Taking into account the installation cost of *wavelength division multiplexing* (WDM), aggregation switch (L2 switch) was deployed to accommodate multiple BSs. In the case of one BS, it is directly connected to WDM without an aggregation switch.

14.6.4 Transmission Line Connection

The transmission line between ASN G/W and the aggregation switch was deployed on WDM to ensure network reliability. Both primary and backup lines are installed. The transmission line between the aggregation switch and BS is deployed on metro-Ethernet. It was opted because its per-line cost was much cheaper than that of the synchronous optical transmission system and it was more flexible to accommodate BS traffic over 10 Mbps.

14.7 WiBro Services

From the network performance perspective, Mobile WiMAX network performs superior to any other existing mobile networks, as described in several documents [12], [13]. Such superiority may be characterized by the keywords – mobility, broadband, all IP, always-on, low-cost, and so on.

WiBro services pursued mobile *triple play service* (TPS), that is, the convergence of communication, Internet, and broadcasting services by keeping up with the market demand and by utilizing the advantage of the Mobile WiMAX network. To provide differentiated services to the users, KT took advantage of the distinctive capabilities of WiBro network and services as follows:

1. Supporting open network and services based on all IP network, which is effective in supporting the managed PDA, interactive e-learning, mobile commerce, charge per sale, etc.
2. Offering Web 2.0-based service in mobile environment, which is effective in supporting personal mobile media, location based community service, customized web contents, etc.
3. Offering larger upload throughput, which is effective in supporting multi-party video conferencing, integrated communicator, m-IP channel, online game, etc.

In order to provide differentiated WiBro services, KT developed efficient service platform and software architecture as well, as will be addressed below.

14.7.1 Service Platform

In order to provide a diversified set of services to users, a client software at user devices and an application service platform are needed in addition to the access network. Figure 14.12 shows the overall architecture of the WiBro service platform. Presence server and call control manager belong to core part of the service platform. Servers related to messaging function and other applications are also implemented in the service platform.

14.7.1.1 Software Architecture of Smart Phone

User devices for WiBro services are divided into two groups: One is the communication module-only type such as PCMCIA card or USB dongle. The other group is the user devices with built-in communication module such as smart phone, PDA, PMP and embedded laptop computer. To support various application services in the second group, KT considered an efficient software architecture including OS, middle ware, and applications. For example, a dual mode smart phone (WiBro + CDMA) was introduced onto the Korean market, which can find versatile applications in WiBro and CDMA services. In order to manage those applications effectively, KT adopted the software architecture shown in Figure 14.13.

14.7.1.2 Connection Manager

Connection manager is to control network entry based on user's configuration (automatic, manual network entry, and so on). Besides, the Connection manager can display and manage various useful information on the network status including signal strength, connection time and transmission rate.

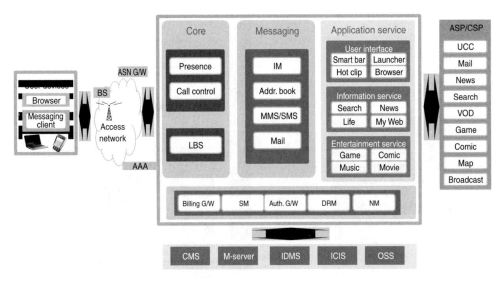

Figure 14.12   Service platform architecture

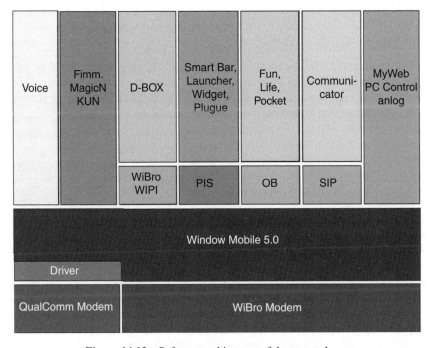

Figure 14.13   Software architecture of the smart phone

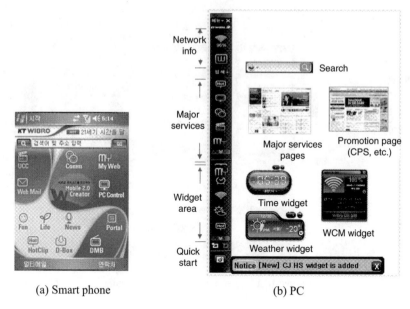

(a) Smart phone   (b) PC

Figure 14.14   Display image of the launcher

14.7.1.3 Launcher

Launcher is an integrated UI platform to accommodate flexible requirements of operators, application service providers and users. Figure 14.14 shows the launcher image of the smart phone and PC. Major application services can quickly be started on the launcher screen. Users can use the search function by entering the keyword on the launcher without opening a web browser.

14.7.2 Major Application Services

KT WiBro services are based on the mobile TPS concept. It enables to offer varieties of application services in addition to the basic Internet connection. The WiBro services are categorized into three groups, namely, core, differentiated, and competitive service groups, as shown in Figure 14.15. The core service group contains the most fundamental services, including the following five attractive services: Web Mail, Multi-Board, My Web, PC Control, and mobile UCC. The differentiated service group contains the WiBro differentiated services, including entertainment contents, on-line game, e-learning, and *location-based service* (LBS). The competitive service group utilizes the high-performance feature of the Mobile WiMAX network, including full web browsing and m-IP channel services.

Among the three service categories, KT will concentrate on the core service in the initial stage and will gradually expand to the differentiated and competitive services.

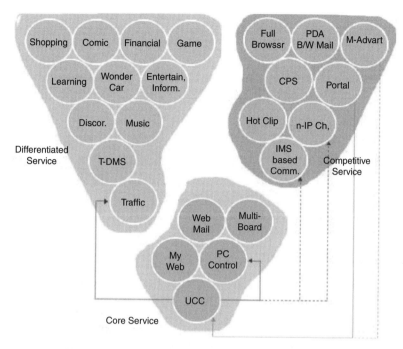

Figure 14.15   Grouping of KT WiBro application services

14.7.2.1 My Web

My Web is an RSS service providing the user's favorite information without visiting each websites unnecessarily. By My Web service, users can get the desired information, check the blogs of friends and can even get the entire posts without logging on and worrying about costly packets.

Figure 14.16 shows the concept and display image of My Web service. In My Web service, users can define the preferred group and the channel belongs to each group. Channel refers the web sites providing the posts, such as blogs, news feeds, or podcasts. To use this service, users need to register specific web sites by entering the address of the blog, the xml address or the RSS feeding address in the URL. Once registration is done, the most recent posts are provided to the subscriber. When a new page of the channel is loaded, a number appears that indicates the amount of new information.

There is a search function in My Web services. There are two types of searching – online search (or Internet search) and offline search (in the user device). In general, the information provided by the RSS may not load the entire post. However, when My Web searches the load data, the entire post is loaded without accessing or visiting the original post. Consequently, there is no waste of packets. The service is now being expanded to show the entire post. There is an auto-search function which enables reserved search. If the user subscribes to a keyword, then the post with the keyword will be saved automatically.

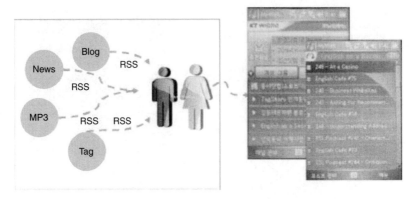

Figure 14.16   The concept and display image of My Web service

14.7.2.2 Web Mail

Web Mail is an integrated e-mail managing service on the smart phone or PDA that manages multiple POP3 e-mail or web mail that users usually use, such as Yahoo mail, Gmail or Hotmail. To use Web Mail service, the user needs to register his/her e-mail accounts on the registration page of the Web Mail service. Multiple e-mail accounts from different mail services can be registered and managed by single client software of the Web Mail service.

There are Inbox, Outbox, Draft, Temp, Storage and Recycled directories in the mailbox. In Inbox directory, retrieved mail list is shown, including e-mail service name, sender's name, the subject of the e-mail, the received date and the attached file icon. By clicking the desired e-mail to read, user can read the whole e-mail contents on the pop-up window after the contents are received. Basically, the contents are shown in text form, but optionally user can see the original web page in html form also. Attached file with e-mail can also be read in the Web Mail service. After the attached file is downloaded, the proper application of the file is executed. When sending, the default e-mail account that the user has chosen when registering the e-mail account is automatically selected as the sender. The receiver can be chosen from the *personal information management system* (PIMS) contact list of the device.

14.7.2.3 Multi-Board

Multi-Board is the next-generation rich-media communication platform that enables users to access any time and anywhere. Multi-Board provides business people a virtual space where people can collaborate with others beyond the geographical restriction. Multi-Board provides multi-way conference environment from anywhere. Even if users are away, they still can join the conference through the Multi-Board service. People at remote sites can get real-time training and education effectively through the Multi-Board service. The trainer can use various materials such as office documents, pictures and even DVD for the trainees away from the classroom as if both parties are taking a class sitting in the same place.

As shown in Figure 14.17(a), multi-party video conferencing up to 12 can be done in the Multi-Board service. During video conferencing, a clear sound and video are provided and more than 500 audiences can join a single session. Multi-Board supports diverse screen

(a) 1:n video conferencing   (b) Video conferencing with file sharing

Figure 14.17   Screen image of the Multi-Board service

layout modes. Depending on the conference and meeting needs, screen layout may be chosen. The rich media communication features of the Multi-Board are useful especially for business collaboration. Meeting participants can share the applications on their desktop computer and co-edit spreadsheet and word documents. It can remotely access a third party's PC by getting authorization with application sharing and can easily manage and control the workspace. The participants can share streaming media, DVD and multimedia contents without any buffering. One of the most distinguished features that Multi-Board provides is the triple service in a single session. People can communicate talking and looking at each other and can also share movies and data, as Figure 14.17(b) shows.

14.7.2.4 Mobile UCC

User created content (UCC), generally called *user generated content* (UGC), refers to the media content produced by the end-users. One of the major trends in current Internet industry is highly related to UCC. KT developed the Mobile UCC service to keep up with this market trend. Users can experience video recording, live broadcasting, Hot UCC, My Album and other services, and access other popular UCC providing web sites. Figure 14.18 shows the overall flow of mobile UCC.

After recording video or taking photos, users can easily upload the contents to the associated UCC portals including SeeU or MBox. UCC live broadcasting function is also available in the Mobile UCC service. If the user sets the video to start broadcasting after completing the creation of the title, content, and the setting of the category and the number of viewers, then the viewers will be able to see the user's station on the on-air list of the associated UCC portals. Users can check the number of viewers and end the broadcasting. Users can enjoy diverse contents via Hot UCC menu and can easily check the contents in the My Album menu. Users can easily upload the contents from My Album to the associated UCC portals as well.

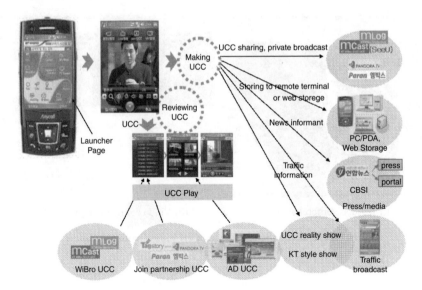

Figure 14.18 Overall flow of Mobile UCC service

14.7.2.5 PC Control

PC Control service provides managing functionalities of the various contents in remote terminal. Figure 14.19 shows the basic concept of the PC Control service. When PC Control is running, the local WiBro mobile device, home PC and office PC are linked together. By opening mobile window explorer at local mobile device, users can see the files stored in the office PC. Users can copy or move specific files between the linked devices by the PC Control service. By operating a local mobile device, users can edit office files or play multimedia files stored at remote devices. It is also possible to share multimedia files with friends by the PC Control function over the WiBro network.

14.7.3 Communicator

Voice call is still a fundamental communication service even today and VoIP is now popular at the IP data network. For an efficient VoIP service, Mobile WiMAX network supports different service classes, including *unsolicited grant service* (UGS) and *extended real-time polling service* (ertPS).

In addition to VoIP service, different types of services including voice, message, and e-mail are integrated into single communicator application in the WiBro services. They include the call type services such as mobile VoIP, *push-to-talk* (PTT), *push-to-view* (PTV); the messenger type services such as chatting, file/folder transfer; the message type services such as *short message service* (SMS), *multimedia message service* (MMS), e-mail; and the value-added services such as file and application sharing.

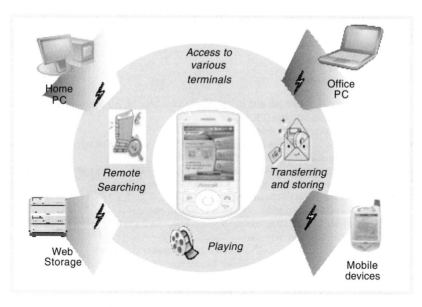

Figure 14.19   Concept of PC Control services

14.7.4 m-IP Channel Service

Broadcast or multicast information that needs to be delivered to multiple users in a single
or multiple cells can share the radio resources dedicated for this service in Mobile WiMAX
network. This is the *multicast-broadcast service* (MBS), which is an efficient way of utilizing
the radio resources without allocating radio resources to each user.

MBS is viewed as an attractive service that can differentiate Mobile WiMAX network from
all other wireless networks by providing real-time, high-quality and interactive multimedia
contents to the users. Push type broadcasting service based on zone, as shown in Figure 14.20,
is one of the examples of MBS. KT develops the m-IP channel services that support multiple

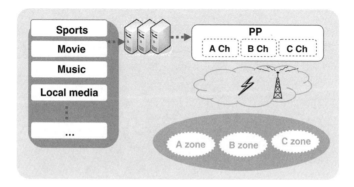

Figure 14.20   Push type broadcasting service based on zone

channels of broadcasting contents (30 fps, 320 × 240, H.264/AAC+ video) at the rate of 512 Kbps/channel.

References

[1] IEEE Std 802.16e-2005 and IEEE Std 802.16-2004/Cor1-2005, 'Part 16: Air Interface for Fixed and Mobile Broadband Wireless Access SDystems', Feb. 2006.

[2] WiMAX Forum™, 'Mobile System Profile', v1.0.0, May 2006.

[3] H. Kim et al, 'Mobile WiMAX standardization activity and WiBro globalization strategy', *The Journal of the Korean Institute of Communication Sciences*, **23**, April 2006, 4.

[4] WiMAX Forum™, 'Relationship between WiBro and Mobile WiMAX', white paper, October 2006.

[5] KT, 'Mobile Internet Business Project Report', KT Corporation, May 2006.

[6] IETF RFC 2131, Dynamic Host Configuration Protocol (DHCP), R. Droms, Mar. 1997.

[7] IETF RFC 3748, Extensible Authentication Protocol (EAP), B. Aboba et al., June 2004.

[8] IETF RFC 3095, Robust Header Compression (ROHC): Framework and Four Profiles: RTP, UDP, ESP, and Uncompressed, C. Bormann et al., July 2001.

[9] IETF RFC 1701, Generic Encapsulation Protocol (GRE), S. Hanks et al., Oct. 1994.

[10] S. Kim, 'WCDMA RNP Special Topic Guidance Engineering Parameter Analysis', Huawei Tech., October 2004.

[11] KT, 'WiBro Engineering Project Report', KT Corporation, June 2007.

[12] WiMAX Forum™, 'Mobile WiMAX – Part I: A Technical Overview and Performance Evaluation', White Paper, June, 2006.

[13] WiMAX Forum™, 'A Comparative Analysis of Mobile WiMAX Deployment Alternatives in the Access Network', May 2007.

15

A New WiMAX Profile for DTV Return Channel and Wireless Access

Luís Geraldo Pedroso Meloni

15.1 Introduction

Brazil has recently defined its Digital TV system based on the Japanese ISDB-T system. Digital television systems use many different modules compliant with many standards that can be represented by many block layers as shown in. The Brazilian system (SBTVD-T *Sistema Brasileiro de Televisão Digital Terrestre*) uses the same modulation technique of the ISDB-T system, also known as BST-OFDM – Band Segmented Transmission – Orthogonal Frequency Division Multiplexing, which consists of a series of frequency blocks called OFDM segments that provide DTV transmission to fixed and mobile receivers simultaneously, by means of OFDM segment hierarchical transmission. The modulation schemes vary according to the channel quality from the more robust DQPSK to 64QAM which offers higher bit rates. The transport stream standard used by all DTV systems is based on MPEG-2, defined in the ITU-T H.222 standard.

Several improvements have been included in the ISDB system, such as better video and audio codecs. The ISDB uses the MPEG-2 HDTV and MPEG-2 AAC for video and audio coding respectively. The video coding of the SBTVD-T uses the H.264, level 4.0, offering a high resolution of 1080i and audio coding uses the HE-AAC standard.

Concerning the Return Channel (RC), among the technologies currently previewed in ISDB are dial-up modems, ISDN for wired lines and PDC – Personal Digital Cellular and PHS – Personal Handy-Phone System for wireless access. PDC is a standard developed and used exclusively in Japan. The SBTVD-T has complemented the above return channel technologies by including WiMAX/WiMAX-700, GSM/GPRS, and CDMA2000/1xRTT, among others.

Mobile WiMAX   Edited by Kwang-Cheng Chen and J. Roberto B. de Marca.
© 2008 John Wiley & Sons, Ltd

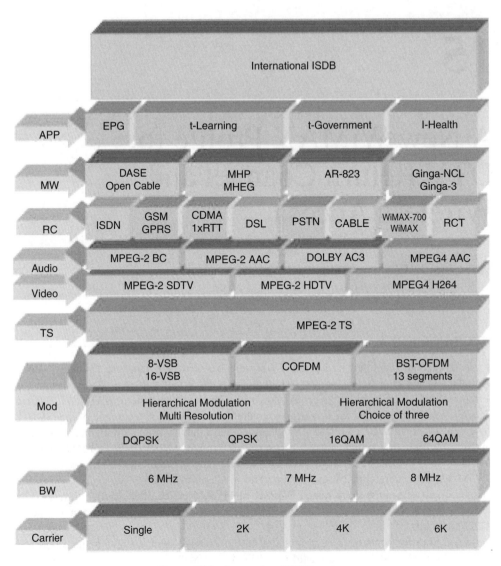

Figure 15.1 International ISDB system

The SBTVD-T specification defines a new middleware based on the NCL language, that is a declarative language for hypermedia documents authoring and object synchronizing. Another important innovation regards the proposal of a new WiMAX profile that covers the UHF and VHF television bands which is the main purpose of this chapter.

The main motivation to include wireless standards as the return channel in the SBTVD-T is based on the increasing rate of mobile phone users in Brazil, having surpassed the number of fixed lines by more than double. It is expected that the same rising rate will be observed in wireless Internet access in this decade.

The WiMAX-700 is a new WiMAX profile that covers 400 MHz to 960 MHz band, including all the UHF TV channels and also optionally includes the VHF band, by operating in a secondary band from 54 MHz to 400 MHz. The WiMAX-700 presents several advantages over the current profiles as will be seen.

This new DTV system assembling modern technologies is very well suited to applications crafted on the digital convergence, such as t-learning, t-health and t-government, which are very attractive in developing countries. Technical groups in Brazil and Japan are working on harmonizing both specifications as an International ISDB body. It is expected that other countries, mainly in Latin America, will choose this system for its DTV.

This chapter is structured as follow: Section 15.2 presents a brief history of the SBTVD-T, Section 15.3 presents the use of WiMAX as return channel for DTV, the advantages of WiMAX-700 are summarized on Section 15.4, Section 15.5 presents the network architecture, Section 15.6 presents the WiMAX-700 channeling, Section 15.7 presents results of a simulation environment including several traffics estimates; and finally Section 15.8 concludes the chapter.

15.2 A Brief History of the SBTVD-T

In 1998, through an initiative of the Brazilian Television Engineering Society (SET) and the Brazilian Association of Radio and Television Broadcasters (ABERT), studies aiming to compare the main DTV systems (ATSC, DVB and ISDB) were conducted in the city of São Paulo in Brazil, under the responsibility of Mackenzie University. These studies delivered detailed reports comparing the main systems [1].

The *Sistema Brasileiro de Televisão Digital* was launched by the Brazilian Presidential Decree 4.901 of November 2003. In May 2004, FINEP-Studies and Projects Funding Agency delivered several Request for Proposals (RFP) for the studies and development of the main modules of the SBTVD, in a total of 18 RFPs. At the beginning of 2005, 22 R&D consortia started the development of the Brazilian DTV system.

A second Brazilian Presidential Decree 5820 of June 2006 defined the ISDB-T as basis for the signal pattern for the SBTVD-T, i.e. the modulation scheme, also incorporating technological innovations approved by the Developing Committee. In November 2006, the SBTVD forum was created, with representatives of the broadcasting, industry and scientific communities. The SBTVD forum delivered the first version of the ISDTV specifications in April 2007. The first deployment of the SBTVD-T is scheduled to happen in December 2007.

Some of the objectives of the Presidential Decree 4.901, well focused on the mitigation of the digital divide and promoting technological convergence, are:

- Mitigate the social divide, and promote cultural diversity of the Nation and the mother language by means of the digital technology access, aimed at the democratization of information.
- Promote the creation of a universal network for distance learning.
- Stimulate research and development and promote the expansion of Brazilian industries concerning the telecommunications and computing technologies.
- Plan the transition from analog to digital television, in such a way as to guarantee the gradual user adoption while being compatible with population income.
- Improve the radiofrequency use of spectrum.
- Contribute to technological and business convergence of the communications services.

Although some of these objectives are difficult to attain mainly due to conflicts of interest among the several players, the SBTVD-T is presently one of the best performing DTV systems within the International ISDB body.

15.3 WiMAX as Return Channel for DTV

In Digital Television, the interactivity allows data communications between receiver stations with applications and services eventually available on the broadcaster's signal. Data communications to set-top boxes is implemented by means of data carrousel sent by the broadcaster transmitter station. In the reverse link data communications are provided by a subsystem called return channel, which allows in some cases only half the duplex communications and offers low bit rate transmissions, such as the Return Channel Terrestrial from DVB-T. In the SBTVD-T the return channel is full duplex and the middleware explores this functionality.

One of the available technologies for interactivity channel is WiMAX which offers great communication capability for the DTV return channel subsystem. Figure 15.2 shows the return channel scheme supported by the SBTVD-T. The figure shows the three subsystems of the DTV. The broadcast subsystem generates the transport stream for broadcasting transmission. The transport stream multiplexes video, audio, and data. The receiver implements the inverse operation disposing data for users. Finally, the return channel allows the communications between users and interactive applications at the broadcaster or at any server connected to the Internet, outside the broadcast facilities. In this scheme access to the Internet is independent of the return channel technology.

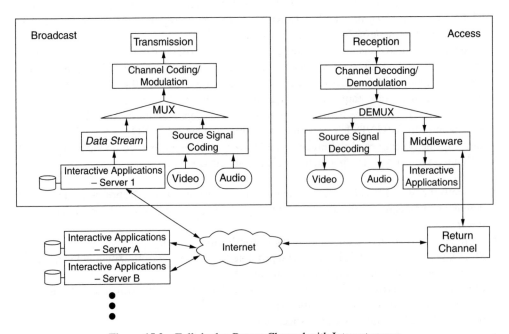

Figure 15.2 Full-duplex Return Channel with Internet access

WiMAX has the advantage of full duplex transmission, offering high bit rate to users, besides the compatibility with packets networks also offering access to the Internet.

15.4 WiMAX-700 Advantages and RC Application

The use of WiMAX at the 700 MHz profile offers several advantages [2], [3]:

- The profile offers a signal reach of up to 65 km, which is a very attractive characteristic in low density population areas. In these cases the service may be deployed with only one base station, which drastically reduces capital expenditures.
- At this frequency profile the signal indoor penetration is much better than the current profiles; as is well known, this characteristic reduces with increasing frequency [4].
- WiMAX offers a low cost compared to other wireless data communications, e.g. mobile services. The number of BSs necessary to cover the same area is about 10% of those at 3.5GHz profile.
- Even the use of lower frequencies presents advantages of using entertainment components, which are less expensive than those of communications and information technologies.
- WiMAX offers QoS and secure communications, and this is very attractive to rural areas.
- A frequency allocation scheme uses TV channels as secondary service and the channel allocation is dynamic; once the regulatory agency has allocated the primary service, the system allows easy operation in other TV channels available.
- The channel allocation scheme includes a Return Channel Provider entity that controls the channel allocation as opposed to the scheme described in IEEE 802.22.
- The simulation analysis presented in this study shows that for the bit rates appropriate for interactive DTV, up to four TV channels are enough for RC interactivity.
- The system operates harmonically with DTV standards, being compatible with the International ISDB.
- The system is flexible, allowing from hundreds to thousands users per BS.
- It is the only solution to provide universal access in continental countries like Brazil where the only exception is in dense forests.

WiMAX-700 specifications are already agreed.

These advantages justify the interest of defining a new frequency profile of WiMAX-700 which includes the UHF band and optionally the VHF band.

Figure 15.3 illustrates wireless transmissions of the proposed return channel for the SBTVD-T. The WiMAX-700 radiofrequency solution uses a wireless modem for the defined frequency band. The modem communicates with the base station offering Internet access. The modem connects the return channel to the set-top box (STB) by means of the USB or Ethernet interfaces. The modem may also include firewall or router functionalities for Internet access to other devices such as PCs, PDAs or others. The WiMAX modem can also be integrated inside the STB.

Several important aspects concerning the Figure 15.3 scheme are relevant. The return channel is not necessarily linked to the broadcast station, i.e., the broadcast station and the return channel provider may be physically separated. A common scenario in a big city would be the use of a single broadcast antenna in an appropriately high site and several base stations spread over

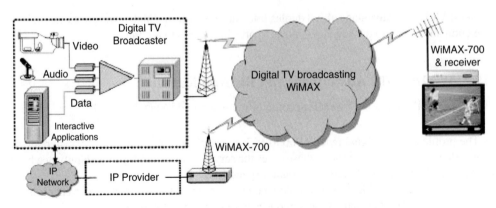

Figure 15.3   Return channel architecture with WiMAX-700

the city in a distribution similar to the cells as in mobile services. In this manner, the system deployment depends on the existence of an entity denominated Return Channel Provider. This entity may be private or public, and it has fundamental importance, it is responsible for planning services, system deployment, users' registry, and base station operations, among other responsibilities.

Some WiMAX-700 requirements are presented in for the SBTVD-T. The proposed WiMAX profile is different from the IEEE802.22, also known as cognitive radio in several aspects. In the proposed profile the channel allocation is by means of a second service allocation and it assumes a centralized control by a Return Channel Provider. On the contrary, in the IEEE802.22, links are provided by means of an opportunistic communications using a spectrum sensing scheme without a centralized control. The proposed profile uses the WiMAX standards as its basis [5], [6].

Table 15.1   WiMAX-700 Requirements for the SBTVD-T

Functional requirements	Description
Physical layer (PHY)	According to the IEEE802.16 in and [6]
	Coverage radius may reach approximately 65 km, appropriated to rural areas;
	NLOS - Non Line of Sight;
	Operation in 400 MHz to 960 MHz, and optionally from 54 MHz to 400 MHz.
MAC layer	According to the IEEE802.16 in and [6]
	Compatible with system architectures based on packets, such as TCP-IP, IP protocols, Ethernet/IEEE 802.3, etc.
	Connection oriented services;
	PMP - Point to Multipoint;
	Manageable QoS.
Interface with other modules	USB preferably;
	Ethernet IEEE 802.3
	Compatibility with upper layer network and transport protocols.
Mobility	Allows receiver mobility – IEEE802.16e-2005.
Identification	Each modem has a single identifier (ID).

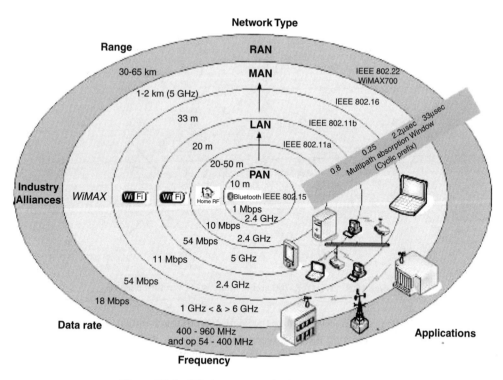

Figure 15.4 Wireless communications IEEE standards

The new profile may play an important role in offering broadband Internet access mainly in rural areas as can be shown in illustrating the common IEEE wireless standards. The different standards have been conceived focused on the coverage area. Bluetooth is dedicated to the Pico Area Network offering communications between peripheral devices up to 10 m. The IEEE802.11 (Wi-Fi) is conceived to Local Area Network with reach up to 33 m at version IEEE802.11b. The WiMAX uses different profiles (2.4 GHz, 3.5 GHz and 5.8 GHz); the profile of 5.8 GHz has a reach up to 2 km. The main advantage of the WiMAX-700 is the reach up to 65 km very attractive for rural areas. As mentioned above, WiMAX-700 has a different approach from IEEE802.22 and some harmonization of part of the specifications is also possible.

15.5 Network Architecture

The proposed return channel provides a wireless communications to a nearby base station. This scheme allows the coexistence of DTV broadcast networks with a wireless telecommunications networks. The networks architecture is depicted in. The base station indicated by letter A is responsible by TV signal broadcast for covering the service area. The return channel is implemented by means of a transceiver (sender and receiver) at the user set-top box, which besides the digital TV reception; it has a low-power transceiver that allows data communications to a nearby return node representing a base station. Letter B in represents users and letter C the return nodes. This scheme is similar to mobile telephony where the coverage cells are defined

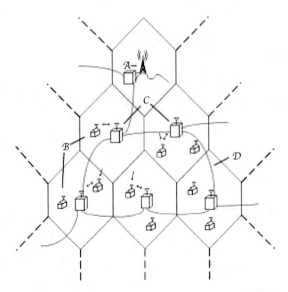

Figure 15.5 WiMAX-700 architecture as DTV return channel

by the radio bases. It is also possible to use the same conventional antenna for VHF and UHF for WiMAX-700 when targeting low cost receivers.

In this scheme the return channel nodes have equipments for wideband Internet access, represented by letter D in. The Return Channel Provider is responsible for these connections which can be provided by different forms. In some places, depending on the availability, optical linking, Ethernet or other can be used; another alternative of lower cost is the use of point-to-point wireless links.

The STB may also use different return channel modality, the SBTVD-T middleware manages which return channel will be in use.

The IEEE802.16 standard defines a specification for air interface for wireless Metropolitan Area Networks (MAN) [5], [6]. This wireless MAN may offer wideband Internet access, an alternative for other access modalities such as cable or fiber optics, offering a less expensive solution. It is an access solution for the last mile that fulfills the requirements of the return channel of the SBTVD-T. The WiMAX-700 specifications were submitted to the SBTVD forum, which details the necessary enhancements for operating with the DTV system in the new frequency profile.

The use of IEEE802.16 as return channel for DTV offers several advantages. Upgrades and enhancements of the WiMAX which occur along the standard evolution may be promptly incorporated in the return channel, such as the recent IEEE802.16e revision that allows mobility. With this revision is possible to use WiMAX in mobile DTV terminals.

15.6 WiMAX-700 Channelling

Several improvements have been introduced in the WiMAX-700 mainly concerning the synchronization and equalization techniques[7]., [8] and [9]. This section review the scheme proposed in the SBTVD-T concerning spectrum channeling.

WiMAX-700 allows wireless communications in the frequency from 400 MHz to 960 MHz, which includes the UHF (Ultra High Frequency) band, and optionally includes the VHF (Very High Frequency) in a secondary band.

The channeling method allows simultaneously operation of the digital transmission systems by means of OFDM/OFDMA techniques or other techniques in such way to allow coherent operation with the analogue or digital television systems. The method allows the reutilization of frequencies for transmission using OFDM/OFDMA modulation in free channels at the defined frequency profile, operating harmonically with other systems, with emphasis to the television broadcasting and mobile telephone services.

The television systems around the world use the following bandwidths: 6 MHz, 7 MHz or 8 MHz. Aiming to harmonize WiMAX with the several standards, the following bandwidths and the respective masks have been defined in the WiMAX-700:

- 1.5 MHz, 2 MHz, 3 MHz, 6 MHz or 12 MHz – for television system with 6 MHz channeling,
- 1.75 MHz, 3.5 MHz, 7 MHz or 14 MHz – for television system with 7 MHz channeling,
- 2 MHz, 4 MHz, 8 MHz or 16 MHz- for television system with 8 MHz channeling.

In WiMAX-700 the following channeling is defined by the initial frequency:

$$f_i = S_f + nBT + kBW; \tag{15.1}$$

where: S_f is the starting frequency of the TV systems in MHz, according to different global systems, n and k are integers,

$$BT \in \{6, 7, 8\}[MHz] \tag{15.2}$$

represents television channel bandwidth; and

$$BW \in \{1.5, 1.75, 2, 3, 3.5, 4, 6, 7, 8, 12, 14, 16\}[MHz] \tag{15.3}$$

defines the bandwidths of WiMAX channels. For instance, for operation of WiMAX-700 in Brazil, the following values are used:

$$S_f \in \{54, 76, 174, 470\}[MHz], \tag{15.4}$$

and

$$BW \in \{1.5, 2, 3, 6, 12\}[MHz]. \tag{15.5}$$

Figure 15.6 illustrates the use of three WiMAX channels inside the 6 MHz used in TV systems. The WiMAX-700 masks must fit inside the 6 MHz DTV mask.

Generally the WiMAX systems employ frequency reuse factor of 3, or even 1, in which case is implemented an internal frequency reuse, by means of a sub-channeling or frequency hopping. Such reuse factors of 3 or 1 may be difficult in the 600 MHz or 700 MHz bands. In this case, the signal propagation has different characteristics, and the reuse patterns may be similar to those of mobile phones operating at frequencies bellow 1 GHz.

Several reuse patterns are possible; some of them are illustrated in Figures 15.7 and 15.8. shows reuse pattern of 4x1 and 4x3, for only one sector cells or three sectors of 120°. In these

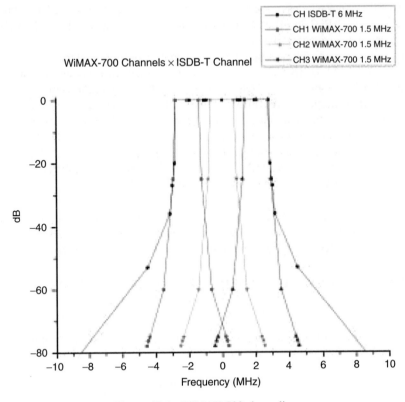

Figure 15.6   WiMAX-700 channeling

cases, 4 and 12 frequencies are used respectively. The base station antennas are placed in the center of each hexagon. In the first case, using 1.5 MHz WiMAX, it needs one 6 MHz TV channel, and in the second case, using 2 MHz WiMAX, it is necessary to use four 6 MHz TV channels.

Figure 15.8 shows the reuse pattern of 7x1 and 7x3, for only one sector cells or three sectors of 120°. In these cases 7 and 21 frequencies are used respectively. In the first case, using 1.5 MHz WiMAX it is necessary two 6 MHz TV channel, and in the second case, using 2 MHz WiMAX, it is necessary to use seven 6 MHz TV channels. Naturally, other reuse frequency patterns are possible within WiMAX-700 systems.

15.7 WiMAX-700 Capacity Simulation for Interactive DTV

For the purposes of performance analysis, several simulation models of the IEEE 802.16 wireless network have been implemented to simulate air interface traffic of the return channel of DTV [3]. The model reproduces the mechanism of IP packets transmission between a base station and the subscriber stations (SS) of a sector controlled by the base station (BS).

The simulator uses the NS-2 environment [10], which has a set of elements to simulate data transmission networks, for example, the IP protocol, wire and wireless transmissions systems,

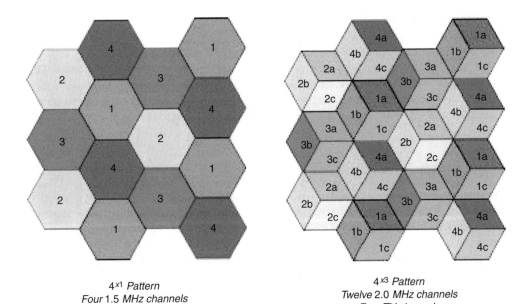

4^{x1} Pattern
Four 1.5 MHz channels
One TV channel

4^{x3} Pattern
Twelve 2.0 MHz channels
Four TV channels

Figure 15.7 Reuse frequency patterns

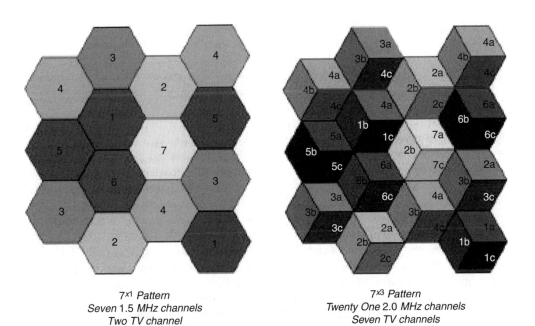

7^{x1} Pattern
Seven 1.5 MHz channels
Two TV channel

7^{x3} Pattern
Twenty One 2.0 MHz channels
Seven TV channels

Figure 15.8 Reuse frequency patterns

and MAC protocols (IEEE 802.11 and TDMA). A new element was developed to simulate the IEEE 802.16 protocol which was employed over several rounds of simulations to analyze the IEEE 802.16 network capability performance for the DTV return channel.

The NS-2 is a discrete events simulator specialized on computers networks simulation. The NS-2 is open source a software and an object-oriented simulator, written in C++, which allows code reutilization and modularity. These characteristics make possible a fast model development of the MAC layer of the IEEE 802.16 [11], [12], and [13].

15.7.1 Simulation Scenarios

The simulation uses only a basic scenario which reproduces operation in a sector controlled by one base station. The scenario was simulated several times, using distinct parameters of propagation, distinct parameters of SS distributions, different types of SS traffic as well as the number of SS.

In general, two points of view are very important for mobile systems: signal propagation and channel traffic capacity. The signal propagation determines the major radius of cells (or sector), so that the received power intensity is appropriate for a good system performance. The traffic capacity determines the major number of SS supported by a cell (or sector). In general, in the urban areas, the cells radius is limited by the traffic capacity and in the rural areas by the propagation features. The main aim of the simulation model is to allow analysis concerning the traffic capacity point of view.

It is specified the transmission mode of each node should represent a base station in the model. The transmission mode determines the modulation scheme (BPSK, QPSK, 16-QAM and 64-QAM – for IEEE802.16d) and the error correction code scheme, both defined by the standard. These parameters determine the transmission rate, as well as the error rate for a specific signal to noise ratio level. In a real network, the transmission modes are determined from the power level, interference and error rate measurements of the received signal. These parameters are entries for the simulation model.

In order to specify the sectors traffic capacity, two scenarios were used for wireless propagation and SSs distributions. Without loss of generality, both models have a circular sector of one km radius. From the traffic point of view, the cell radius and the sectors geometries are important only to determine the number of SSs which utilize each transmission mode. For example, a circular sector (360°) that covers a certain area with a certain SS density has the same number of SS of such a sector of 120° covering an area three times bigger. From the simulation point of view, the capacities in both cases are the same. The factor that introduces any cell capacity variation is the signal propagation. The SS distribution operating in each mode varies slightly in function of signal attenuation due to path losses. The COST 231 [14] propagation model was used to analyze the cells capacity and two approaches were defined depending on loss factors. The first approach, called P26, considered a loss factor of 2.6 in this model, which is a typical factor for urban and suburban areas. The second, called P35, considered a 3.5 loss factor, which represents an area with bad propagation characteristics, such as a density urbanized areas with irregular topography.

In order to define the number of SS operating in each transmission mode, a homogeneous distribution of SS in a circular area was considered.

Table 15.2 Percent distribution of SS in the sectors

Mode		SS	
UL	DL	P26	P35
QPSK 1 / 2	QPSK 3 / 4	30	23
QPSK 3 / 4	16-QAM 1 / 2	41	37
16-QAM 1 / 2	16-QAM 3 / 4	9	9
16-QAM 3 / 4	64-QAM 2 / 3	12	15
64-QAM 2 / 3	64-QAM 3 / 4	1	2
64-QAM 3 / 4	64-QAM 3 / 4	7	14

Six circulars sectors were created in both scenarios, calculated from the reach of the transmission signal in each mode. A percentage SS distribution was defined for each sector. This percentage is equal to the rate between the sector area and the total area of the cell. The total number of SS in the cell is defined for each simulated scenario. The simulation model distributes the SS in each sector and associates them with the corresponding transmission mode.

The transmission mode in the direct and reverse path is different for a same sector. The difference between approaches P26 (Figure 15.9) and P35 (Figure 15.10) is that the approach P35 has a higher SS concentration in the higher transmission rates.

In conditions of worse transmissions, a higher transmission power is needed (or in an equivalent way, a reduction of the cell radius). One effect also related to higher signal attenuation in P35 scenario, is the small sector area corresponding to a particular modulation scheme at the cell border, as compared to P26 scenario which has a smaller attenuation factor. Consequently,

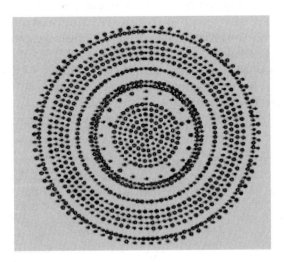

Figure 15.9 SS distribution for P26 scenario

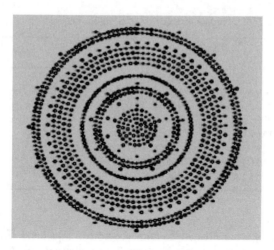

Figure 15.10   SS distribution for P35 scenario

the approach with the worst propagation factor is paradoxically the scenario more favorable regarding traffic flow.

Four traffic profiles were specified in simulations: WEB, low rate bidirectional stream, e-mail, and low rate client/server session communications. These also define five different profiles distributions which were considered for the rounds of simulation (Table 15.3). The table presents the distributions of the active SS that generate the respective traffic according to the percentage shown in the table.

The characteristics of each traffic profile are:

1. Web: This profile utilizes the connections source model using default parameters of the Packmime model (PackMimeHTTP – Bell Labs) 0 with rate of 1 page access per 2 minutes per station.
2. Low Rate client/server Session: to simulate this profile a new agent was created in the NS-2 with the following parameters:
   1. Request packets length: exponential distribution with mean of 512 bytes.
   2. Response packets length: exponential distribution with mean of 50 kbytes.
   3. Connections rate done by user: 10 per hour.
   4. Request packets number per connection: exponential distribution with mean of 8.

Table 15.3   Traffic profile distributions

Profile	WEB	Session	E-mail	Bid stream
CI-Low	0%	0%	0%	100%
CI-1	0%	0%	90%	10%
CI-2	0%	50%	45%	5%
WEB-1	25%	37%	34%	4%
WEB-2	50%	25%	22%	3%

    5. Time delay to send response packets after reception of request packets: 0.
    6. Time delay to send the next request packets after reception of response packets: negative exponential distribution with mean of 20 seconds.
3. E-mail: the profile utilizes the On-Off source model (Application/Traffic/Exponential) with the following parameters:

Receiving E-mails:
1. Transmission rate in the active state: 240 kbits/s.
2. Messages length: 15 kbytes.
3. Activity period mean time: 4 seconds.
4. Inactivity period mean time: 1200 seconds.

Sending E-mails:
1. Transmission rate in the active state: 16 kbits/s.
2. Messages length: 200 bytes.
3. Inactivity interval mean time: 600 seconds.
4. Low Rate Bidirectional Stream: This profile uses the model of simple variable rate (Application/Telnet), with the parameters:
    1. Transmission rate: 1 message each 10 seconds.
    2. Message Length: 100 Bytes.

The simulation used the NS-2 routing protocol DSDV (Destination Sequence Distance Vector) [10]. This protocol allows systems simulation in the radio environment. In the model, the IP layer of each SS has an agent which sends periodically packets of routing to the neighbors SS. When the agents receive the packets, they obtain the necessary information to construct the routing table, which will be used to redirect the traffic. All the rounds of simulation have both the application traffic and that traffic in the IP layer. It is worth noting that the DSDV protocol has been used for the simulation purposes, but may not be present in WiMAX environment.

Several rounds of simulation were executed for each arrangement of propagation scenario and for each traffic scenario, as varying the number of SS in the sector. The number of SS was assumed to vary between 500 and 1000, with increments of 50 SS for each simulation step.

15.7.2 Configuration Model

A series of simulations were done using a traffic source of constant rate. The traffic volume produced by traffic sources was greater than system capacity, allowing analysis of the MAC layer at maximum dataflow. In this way, the simulated system behavior is deterministic, and the comparison between simulation results and calculations analysis is simple and present these results. They show MAC and PHY layers (including the protocol IEEE 802.16 overhead), dataflow (bytes per second). The MAC layer dataflow amount had been compared with transmission nominal rate (operation using 3.5 MHz bandwidth) for efficiency computing.

The reverse path had an efficiency of about 80% and the direct path of about 74%. These values are due to protocol packaging, which needs additional bits transmission to control the preamble transmission times, guard time, information transmission about system management and allocation band mechanism used in simulation.

Table 15.4 Model validation results – reverse path (dataflow in bytes/s, nominal rate in Mbps and efficiency)

Mode	Reverse path			
	PHY	MAC	Rate	Efficiency
BPSK 1 / 2	151,200	140,800	1.47	77%
QPSK 1 / 2	302,400	291,200	2.94	79%
QPSK 3 / 4	453,600	441,600	4.41	80%
16-QAM 1 / 2	604,800	592,800	5.88	81%
16-QAM 3 / 4	907,200	893,600	8.82	81%
64-QAM 2 / 3	1,209,600	1,181,700	11.76	80%
64-QAM 3 / 4	1,360,800	1,331,350	13.24	80%

15.7.3 Simulation Models

The following simulation models were developed:

1. *TDD Model:* A MAC layer model was developed according to the IEEE 802.16 system operating in the OFDM/TDD mode. The model represents mainly the most essential protocol elements necessary to obtain dataflow and delay times, which are: mechanism for band allocation in direct and reverse path, mechanism for band request from SS to base station, adaptive modulation and data packaging in the MAC and PHY layers. We have looked to represent these elements according to the IEEE 802.16 specification. The band allocation mechanism is not included in this specification. So, a simple system is modeled, using the Round Robin type, which allocate an equal opportunity period for each SS for transmission. The opportunity period is an entry parameter of the model (quota). As the SS use different bit rates, the mechanism allocates higher band to the SS with higher bit rates. The developed model is quite flexible and allows configuration of several parameters. Next,

Table 15.5 Model validation result – direct path (dataflow in bytes/s, nominal rate in Mbps and efficiency)

Mode	Direct path			
	PHY	MAC	Rate	Efficiency
BPSK 1 / 2	136,800	136,200	1.47	74%
QPSK 1 / 2	273,600	273,000	2.94	74%
QPSK 3 / 4	410,400	409,800	4.41	74%
16-QAM 1 / 2	547,200	546,600	5.88	74%
16-QAM 3 / 4	820,800	820,200	8.82	74%
64-QAM 2 / 3	1,094,400	1,093,800	11.76	74%
64-QAM 3 / 4	1,231,200	1,230,600	13.24	74%

a list of parameters and used values in simulations are presented. These values are related to IEEE 802.16 system using 3.5 MHz channel band, frames of 20ms in TDD transmission mode. There are 16 sub-channels in the reverse path.

The model parameters are:

- Backoff window minimum length (2).
- Backoff window maximum length (16).
- MAC layer packets overhead (12 bytes).
- Frame length (20 ms).
- Symbol duration (68 μs).
- Frame guard time (300 μs).
- Number of direct path sub-channels (1).
- Number of reverse path sub-channels (16).
- Number of bytes per symbol per transmission mode (12, 24, 36, 48, 72, 96, 108).
- Multiplexing type (TDD)
- Quota per station used in band allocation algorithm (direct path: 4 slots, reverse path: 64 slots).
- Sub-frame minimum duration (number of slots) of reverse path in TDD mode (8).
- Burst minimum length (number of slots) to each transmission mode (direct path: 1 to all modules; reverse path: 8, 4, 4, 2, 2, 1, 1).
- Number of channels for band allocation request (48).
- Number of codes for band allocation request (8).
- Number of opportunities for band allocation request per frame (2).
- Ranging phase duration (number of symbols) in reverse path (3).
- First burst (preamble) duration (number of slots) in direct path (5).
- Periodicity of statistical logging (1 second).
- Trace level (1).

The level trace 1 generates a registry file to store and collect statistical data. These include packets total dataflow (number of bytes) and average time delay in direct and reverse paths. Dataflow is measured in enlace layer (MAC entry), which regards only the traffic in upper layers, and in Physical layer (MAC exit), which considers the additional bytes in packaging of the MAC layer. The packets' delay time considers the waiting time of transmission and its own transmission time.

1. *FDM Model*: The FDM model was developed, but not used in the Digital TV return channel study, due to the fact that it presents disadvantages in flexibility, once the available band for direct and reverse paths is constant, in contrast to TDD mode where band share is dynamic.

15.7.4 Analysis of the Results

Tables 15.6–15.8 present results for the same propagation conditions and varying number of SS for different traffic profiles. They present dataflow in the physical layer, including IEEE 802.16 protocol overhead, and dataflow in the enlace layer (IP dataflow). All rounds simulate 10

Table 15.6   Results of CI-2/P26 scenario

	CI-2 / 26					
	Dataflow UL (bytes/s)		Dataflow DL (bytes/s)		Delay (s)	
SS	Phy	Enlace	Phy	Enlace	UL	DL
500	202.340	185.206	194.794	191.101	0,124	0,026
550	265.486	247.724	245.176	239.182	2,362	0,026
600	311.100	294.019	191.865	186.745	5,877	0,022
650	258.946	241.252	251.624	245.375	3,410	0,024

Table 15.7   Results of WEB-1/P26 scenario

	WEB-1 / 26					
	Dataflow UL (bytes/s)		Dataflow DL (bytes/s)		Delay (s)	
SS	Phy	Enlace	Phy	Enlace	UL	DL
500	204.586	186.773	136.760	133.426	0,094	0,022
550	233.252	213.752	182.149	177.700	0,144	0,025
600	207.639	188.347	215.503	210.270	0,124	0,024
650	268.575	250.289	227.999	221.336	2,026	0,024
700	284.677	266.132	233.650	225.346	3,006	0,023
750	335.112	317.263	191.081	183.989	8,302	0,019
800	299.195	281.004	235.513	226.652	6,005	0,021

Table 15.8   Results of WEB-2/P26 scenario

	WEB-2 / 26					
	Dataflow UL (bytes/s)		Dataflow DL (bytes/s)		Delay (s)	
SS	Phy	Enlace	Phy	Enlace	UL	DL
500	227.484	208.056	122.622	117.760	0,094	0,020
550	247.398	226.025	182.011	176.018	0,129	0,023
600	308.059	289.310	183.974	176.308	2,166	0,022
650	286.457	267.596	205.145	197.910	1,305	0,024
700	287.001	268.079	208.733	200.645	2,690	0,022
750	323.344	305.221	189.734	180.397	7,215	0,019

minutes of network operation. To analyze the results, only the final 2 minutes were considered to ensure that the data outputs were in a statistical stable region. The obtained results correspond to confidence interval lower than 1% of average values to a confidence level of 95%. In the result analysis, all rounds of simulation whose packets' time delay were stationary around its mean value were considered. This is the reason why the maximum number of SS is different in each traffic and propagation scenario.

An important point to observe in the results is the traffic impact produced by the routing DSDV mechanism. DSDV agents investigate the neighbors, sending a broadcast message to the radio interface. When agents receive the message, they send a response informing their addresses. This traffic increases with the square of SS number. In the simulated approaches, that traffic was significant and it allowed evaluate small dataflow traffic effect originated by broadcast messages.

The maximum and minimum total averages dataflow (reverse + direct links) achieved in simulations were 567 and 292 kbytes/s (IP layer). The IEEE 802.16 protocol is optimized to broadband traffic. DSDV traffic and others simulated traffics applications were small data bursts.

The maximum and minimum rate difference shows that the system can be used in a scenario with a great number of active SS transmitting with small rates simultaneously, but also shows that it has great variation in efficiency in function of traffic profile.

It is also important to observe that the behavior of the dataflow is not monotonous with the number of subscriber stations. The main reason for this is the distribution of the distinct spatial traffic sources for each scenario. Because of the relatively small number of SSs transmitting with great rates (see Table 15.2), a distribution variation can have a significant effect in the results.

15.7.5 RF Spectrum Use

Using the spectral efficiency reported on technical literature, the number of TV channels to be used considering the use of WiMAX-700 as return channel can be calculated. In Table 15.8, 4 cells with 3 sectors in each one were considered. The following bandwidths were also considered: 1.5, 1.75, 2, 3, 3.5, 4 and 5.5 MHz. The table shows the mean bit rate per user operating simultaneously in the system (uplink and downlink). It had a user density of 500 users per sector and a spectral efficiency of 2 bps/Hz, which is a conservative value.

For a bandwidth of 3.0 MHz, 6 TV channels would be necessary, which would offer a bit rate of 12 kbps. It is important to emphasize that these numbers consider the simultaneous service to 500 users. For only 100 users, each user will take up a bit rate five times the number shown in Table 15.9. For instance, for the 3.0 bandwidth, the system will offer 60 kbps, which is the same or better bit rate than that offered by dial up modems. This bit rate is appropriate for DTV interactive applications and even for Internet access. Besides, considering the overbooked service typical in mobile system (with a 6 times factor), a base station with 3 sectors can serve 9,000 users. These conservative numbers, due to the fact that they use a spectral efficiency of 2 bps/Hz and overbook factor of 6 times, show the great potential of this wireless system with high service penetration, to achieve the main objectives of the SBTVD-T concerning the digital divide.

In fact, there is an exchange relation of frequency bandwidth and mean bit rate. The use of a channel with a bandwidth higher than 3.0 MHz would demand more TV channels, which is

Table 15.9 Number of TV channels used and mean bitrate per
user (500 users, TDD mode, 4x3 reuse)

Bandwidth (MHz)	Number of TV channels used	User bitrate (kbps) (uplink and downlink)
1.5	3	6
1.75	4	7
2	4	8
3	6	12
3.5	7	14
4	8	16
5.5	11	22

not attractive and in some areas is even unfeasible. On the contrary, channels of 1.5 MHz may
be very attractive for areas with high population density. The 2 MHz channels would allow
a mean bit rate of 8 kbps, using four 6 MHz channels. These numbers show that the more
attractive profiles uses 1.5 MHz, 2 MHz or 3 MHz, which correspond to use of 3 to 6 TV
channels for the return channel with the 4 x 3 pattern.

The numbers shown in this section are only for illustrative purposes, naturally it is possible
to offer large band Internet access with a higher investment in the network infrastructure.

15.8 Conclusion

The SBTVD-T is based on the modulation BST-OFDM of the ISDB-T system. It has introduced
several important improvements on the Japanese standard. The main improvements are a more
flexible middleware, the use of a very efficient video codec and the return channel which is
mainly focused on wireless technologies. The motivation for the search of wireless solutions
is based on the observation that mobile phone use in developing countries has been increasing
at a very high rate over the past decades. We infer that the same rising rate will be observed in
wireless Internet access in these countries.

The Return Channel (RC) technologies defined currently in ISDB are dial-up modems,
ISDN for wired lines and PDC – Personal Digital Cellular and PHS – Personal Handy-Phone
System for wireless. The SBTVD-T has complemented the above return channel technologies
by including WiMAX/WiMAX-700, GS/GPRS, and CDMA2000/1xRTT. This chapter has
reviewed some important aspects of the WiMAX-700.

The sector capacity obtained by the simulations is around 550 active stations in the simulations conditions. In all combinations of traffic profiles and propagation scenarios, stationary
results were obtained after a mean time delay below 5 seconds, for the simulations using less
than 600 subscriber stations. Also the simulations have shown that the IEEE 802.16 allows the
simultaneous treatment of a great number of stations. This is a very important aspect for DTV
return channel, even for use in areas with high SS density.

In the scenarios simulated in these conditions, we obtained a net bit rate (IP layer) above
5 kbps (uplink + downlink), and this is a typical dataflow for dial-up Internet accesses. The
simulated scenarios are adequate for return channels of DTV systems. For this purpose, we

considered that a user bit rate of few kilobits/s are enough for the majority of interactive applications and even for navigation on light Internet pages. The user bit rate may be much higher with the use of small cell sites, but these also imply a higher investment in the network infrastructure.

The development of a simulation model and the simulation results have shown that the IEEE 802.16 standard presents excellent capacity and efficiency for data transmission at variable bitrates. The system allows relatively high bitrates to be attained when compared to other wireless standards.

The simulations results have also shown that the system is very efficient for data bursts with a high volume of data and also is efficient in supporting a high number of terminals per sector. As far as the capacity is concerned, the system fits well with DTV return channels applications, even for the more demanding interactivity applications, while being a suitable solution for Internet access in emerging countries like Brazil.

References

[1] Testes em Sistemas de TV Digital, Mackenzie University, Mar. 2000, http://www.anatel.gov.br/ BIBLIOTECA/PUBLICACAO/RELATORIOS/DEFAULT.ASP, May 2007.

[2] L.M.J. Barbosa, A. Budri, J.V. Gonçalves, E. Morais, R. Moreira, R. Sonntag, and L. G. P. Meloni, 'Uma Proposta para o Canal de Interatividade para o SBTVD Através de Comunicação sem Fio em RF Intrabanda', In *XVIII Brazilian Symposium on Computer Graphics and Image Processing*, 2005, 1–6.

[3] A.K. Budri, J.V. Gonçalves, and L.G.P. Meloni, 'WiMAX Simulation Models for Return Channel in Digital Television Systems' in *VI International Telecommunications Symposium* – ITS2006, Fortaleza, 2006.

[4] Hsiao-Hwa Chen and Mohsen Guizani, *Next Generation Wireless Systems and Networks*. Chichester: John Wiley & Sons, Ltd, 2006.

[5] IEEE Standard for Local and Metropolitan Area Networks. Part 16: Air Interface for Fixed and Mobile Broadband Wireless Access Systems. Amendment 2: Physical and Medium Access Control Layers for Combined Fixed and Mobile Operation in Licensed Bands, IEEE P802.16d, 2004.

[6] IEEE Standard for Local and Metropolitan Area Networks. Part 16: Air Interface for Fixed and Mobile Broadband Wireless Access Systems. Amendment for Physical and Medium Access Control Layers for Combined Fixed and Mobile Operation in Licensed Bands, IEEE P802.16e, 2005.

[7] L.G.P. Meloni, Sistemas de TV Digital Usando Antenas Receptoras e Transmissoras de TV para Canal de Retorno e como Repetidor Digital. Required Patent PI0304.013-5, INPI.

[8] L.G.P. Meloni, Método de Canalização de Sistemas de Comunicação sem Fio sobre o Espectro de Televisão e Espectro Adjacente. Required Patent 018060114062, INPI.

[9] L.G.P. Meloni, Método de Sincronização para Receptores OFDM Através da Estimação dos Desvios Temporal e Freqüencial com Base na Análise do Sinal Recebido. Required Patent 018060120940, INPI.

[10] K. Fall, and K. Varadhan, *The NS Manual*. UC Berkeley, LBL, USC/ISI, and Xerox PARC, Apr. 2005. http://www.isi.edu/nsnam/ ns/nsdocumentation, Apr. 2007.

[11] C. Eklund, R. Marks, K. Stanwood, and S. Wang, 'IEEE Standard 802.16: A Technical Overview of the Wirelessman Air Interface for Broadband Wireless Access', *IEEE Communications Magazine*, June 2002, 98–106.

[12] A. Ahmad, C. Xin, F. He, and M. McKormic. Multimedia performance of IEEE 802.16 MAC. www.cs.nsu.edu/research/OPNET/Abstract /IEEE_802.16/ATS_2005_HF.pdf, Apr. 2007.

[13] B. Petry, 802.16 1 MAC Simulation Tools: Recommendations, Nov. 2000. http://grouper.ieee.org/ groups/802//16/tg1/mac/pres/802161mp-00_06.pdf, Apr. 2007.

[14] B. Petry, Urban Transmission Loss Models for Mobile Radio in the 900 and 1800 MHz Bands, European Corporation in the Field of Scientific and Technical Research, EURO-COST-231, Revision 2, Sep. 1991.

[15] J. Cao, W.S. Cleveland, Y. Gao, K. Jeffay, F.D. Smith, and M.C. Weigle, 'Stochastic Models for Generating Synthetic Http Source Traffic', in *Proceedings of IEEE INFOCOM*, Hong Kong, Mar. 2004.

[16] S. Ramachandran, C.W. Bostian, and S. F. Midkiff, 'Performance Evaluation of IEEE802.16 for Broadband Wireless Access', in *Proceedings of OPNETWORK 2002*, Aug. 2002.

[17] M. Pätzold, *Mobile Fading Channels*. Chichester: John Wiley & Sons, Ltd, 2002.

16

A Packetization Technique for D-Cinema Contents Multicasting over Metropolitan Wireless Networks

Paolo Micanti, Giuseppe Baruffa, and Fabrizio Frescura

16.1 Introduction

The combination of a number of recent technological developments (such as high-performance film scanners, digital image compression algorithms, high-speed data networking and storage, and advanced digital projection) has allowed many demonstrations of what is called 'Digital Cinema' (D-Cinema, DC) [1]. In this framework of technological achievements coming from industries and research organizations, Digital Cinema Initiatives (DCI) [2] is a new entity created by several studios, with the primary purpose of establishing uniform specifications for digital cinema. The DC system also defines different strategies for the distribution of the movie from studios to theatres, such as magnetic supports, satellite or Wireless MAN broadcasting [3]. In Europe, some research projects [4], [5] are addressing the whole DC chain and they are evaluating both satellite and WiMAX [6] as distribution technologies.

One of the key aspects for user enjoyment of the D-Cinema content is represented by the timely and high-quality delivery of the movies or live events, in an operating scenario where the head distribution site feeds one or more regional theatres (Multiplex), which can act as proxy to smaller, local theatres located in sub-urban area (Figure 16.1). The end user could also directly access video feeds, if he has a network connection that can offer such a high bit rate, e.g. satellite or WiMAX. Many research projects and experiments are being done, involving video broadcasting over high speed wireless connections [7], [8].

Mobile WiMAX   Edited by Kwang-Cheng Chen and J. Roberto B. de Marca.
© 2008 John Wiley & Sons, Ltd

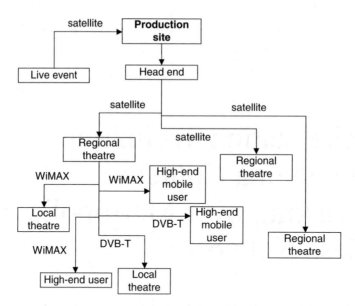

Figure 16.1   Distribution scenario for D-Cinema and live events [29]

Digital Cinema content may be delivered to the theatres by using several communication channels, and many of them could be wireless, such as, for example, DVB-T [9] and WiMAX. The destination of the video contents, in this case, can also be a mobile user: for instance, an audience located on a train or bus. Due to the very high bit rate required for the transmission of digitized contents (it can be in excess of 5-6 Gbit/s), lossy compression of the video stream is required, and the JPEG 2000 standard [10] has been chosen as source coding method, due to its higher image quality when compared to other ones. When compressed, the rate will possibly be in the 250–500 Mbit/s range.

However, due to the high bit rate still required, WiMAX may become one of the preferred wireless distribution systems. In addition to transport and content delivery to regional and local theatres, the same WiMAX network, with meshing applied [11], can be used to deliver content to high-end customers.

In addition to the file transfer approach, already used for digital cinema transfer to the theatres, the use of streaming would allow for applications such as live events projection and (Near) Movie on Demand (in this case, however, additional methods for the reduction of bit rate must be adopted) [12].

The delivery of JPEG 2000 compressed video is a well known challenge for wireless IP networks [13]. In the case of TCP, the variable delay is unacceptable for many types of real-time applications. D-Cinema will introduce even more stringent conditions, in terms of both required bandwidth and high quality demand.

As concerns the distribution, the requirements can be specified in the following terms:

1. Small bandwidth occupancy, not only of the JPEG 2000 compressed content itself, but also considering some signaling overhead;
2. Reliable transmission of the contents, by use of Forward Error Correction (FEC) techniques and/or Selective retransmission of lost data (Selective ARQ).

A possible solution can be that of adopting a multicast protocol [14], which can be adapted to the high bit rate and low packet loss rate necessary for the distribution of D-Cinema content. If using multicast transmission, a reliable protocol could be used, which guarantees the sender that every receiver has a copy equal to the original. The NORM protocol [15] may be a viable solution for such needs: it uses a NACK-oriented reliable multicast, achieving reliability by means of negative ACKs that are sent back from the receivers to the sender. In addition, it is possible to specify even a small FEC layer, which reduces retransmission rate at the expense of some added overhead.

In this chapter, we present a technique to encapsulate D-Cinema compressed video content into the existing NORM protocol packets, thus minimizing the probability of retransmission by a judicious way to split, send, and re-compose JPEG 2000 data packets. In Section 16.2 we present the technical requirements for DC target quality; in Section 16.3 we give an overview of the adopted multicast protocol, whereas in Section 16.4 the encapsulation technique is described and, eventually, in Section 16.5 we present the obtained results.

16.2 Technical Specifications for D-Cinema

16.2.1 JPEG 2000 Overview

JPEG 2000 is a standard for image and video compression [10]. The compressed images are known as *codestreams*, and they may be wrapped into a dedicated file format. This standard has many advantages, compared to its predecessor (i.e. JPEG), in particular, some of its features are:

1. higher compression ratio while keeping a comparable quality;
2. ability to encode with lossless and lossy compression;
3. progressive transmission by different spatial resolutions, components, and image qualities (*layers*);
4. random codestream access, also obtained by subdivision of the image into smaller sub-images (*tiles*), independently encoded; the basic elements of a codestream are the JPEG 2000 data *packets* (not to be confused with network packets), which represent a set of compressed data sharing some spatial, component, or quality characteristic;
5. possibility to specify the so-called *regions of interest*, that is, a part of the image which should be carefully encoded, with more detail than the rest of the image;
6. integrated error protection for noisy environments (such as wireless transmission), either by means of embedded error resilience or added redundancy (JPWL extension [16]);
7. possibility to introduce a form of integrated security (JPSEC extension [17]).

The JPEG2000 standard is completely different from its predecessor (i.e. JPEG) and other image compression standards. It uses the wavelet transform (DWT, Discrete Wavelet Transform). The main difference with the DCT (Discrete Cosine Transform) used in the JPEG standard, is that the former operates on the whole image, whereas the latter uses small parts of the image, usually 8x8 pixel blocks. This difference can be seen by observing two images compressed at high rates, one with JPEG and one with the JPEG 2000 standard: the first one will show many artifacts, i.e. blocking artifacts, due to the high compression ratio in the 8x8 pixel blocks. In the JPEG 2000 case, the image will show a uniform contour noise, but the image will still be clearly visible. The DWT consists of a digital spatial filtering, both along rows and columns,

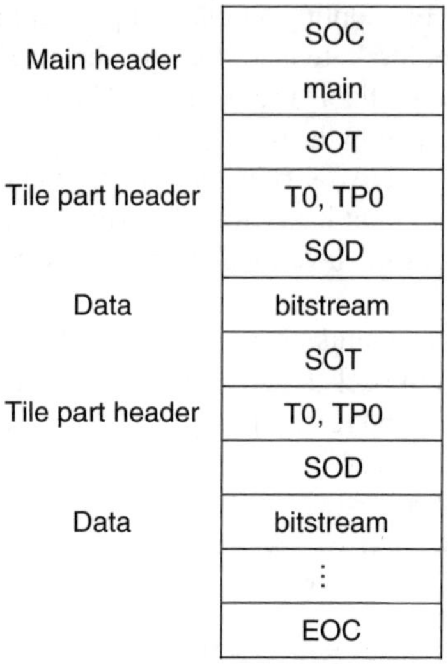

Main header	SOC
	main
Tile part header	SOT
	T0, TP0
	SOD
Data	bitstream
	SOT
Tile part header	T0, TP0
	SOD
Data	bitstream
	⋮
	EOC

Figure 16.2   JPEG 2000 codestream marker structure

which concentrates the image energy in the lower frequencies; subsequent applications of the filtering, called decompositions (the standard specifies 5 decomposition levels) will output data centered in the low and high spatial frequencies. After the quantization, data will be compressed using an arithmetic coding. First of all, data are ordered by the EBCOT (Embedded Block Coding with Optimal Truncation) algorithm, which uses information from the wavelet transform to partition data into blocks, called code-blocks, which will then feed the arithmetic coder. This is a generic compression algorithm that uses symbol probabilities to reduce the size of the symbol sequence. Compressed data will then be used in the rate control process that, depending on the compression parameters (i.e. quality, compression ratio), will choose the amount of data to put in the final file. In addition, each image can be independently encoded in tiles, which are rectangular sub-portions of the main picture.

After this, the compressor will create a JPEG 2000 file, using a structure like the one presented in Figure 16.2.

JPEG 2000 compressed images are called codestreams. In the simplest case, a codestream is structured as a main header followed by a sequence of tile parts. The codestream file always starts with the SOC (start of codestream) marker, followed by the main codestream header (indicated by *main* in the figure above). The main header has some mandatory markers: SIZ, which specifies image information such as image and tile size, COD (coding style defaults) and QCD (quantization default), which provide default coding and quantization parameters. In the header there are also some other optional markers, such as for overriding the default parameters on a component basis, or for indicating the regions of interest. A special marker, COM (comment), can be used to add unstructured data to the codestream (comments regarding the image, classification data, author, etc.). The tile part header consist of two mandatory markers, SOT (start of tile) and SOD (start of data), and some optional markers used to override default coding parameters; the SOD marker is followed by the actual data bitstream. Finally, each JPEG 2000 codestream is closed by a two byte marker, EOC (end of codestream).

The structure of a JPEG 2000 codestream can be very helpful when sending it through a network: it can be easily parsed, to find the markers' boundary and to split the bitstream into packets that can be easily adjusted to fit network packets' size.

JPEG 2000 was chosen as the compression standard for the lossy coding of Digital Cinema sequences. Different from video compression standards such as H.264 [18] or MPEG-2 [19], the temporal redundancy existing between adjacent frames is not exploited: a DC video file is a collection of ordered codestreams, as well as some additional audio and data tracks, packaged into a format called MXF [20]. Another possible solution (not envisaged in the DCI standard) is to use a native Motion JPEG 2000 format [21].

16.2.2 Digital Cinema Initiatives System Specifications

The DCI system uses a hierarchical image structure that supports both 2K and 4K resolution files, where 2K and 4K image profiles stand for 2048 x 1080 pixels and 4096 x 2160 pixels per frame, respectively [2]. The bit depth for each one of the three color components is 12 bits, for a total of 36 bits per pixel. The DCI image structure is required to support a frame rate of 24 Frames Per Second (FPS), but it can also support frame rates of 48 FPS, for the 2K image files only. The color space used to represent image components is the XYZ color space [22], which offers a broader range of color tones over the classic RGB. The DC system uses perceptual coding techniques to achieve an image compression that is visually lossless, in order to limit transmission bandwidth or media storage usage. The JPEG 2000 final compression rate is about 6 times; each frame contains exactly one tile, 6 resolutions for each color component, and a single quality layer; the maximum codestream size is of 1,302,083 bytes per frame (corresponding to a final, encoded video bit rate of 250 Mbit/s). In particular, the data section in the codestream is structured as a sequence of three tile parts, with each part carrying exactly one color component. Since interframe coding is not performed, the video stream is simply a sequence of compressed images. The image compression reduces the required bandwidth from a maximum of 7.5 Gbit/s for a 4K video sequence to 250 Mbit/s, with a perceptually lossless method.

The DCI system uses encryption, too, in order to protect audiovisual contents from unauthorized copying and illegal distribution. In particular, the AES cipher, operating with a 128 bit key, is used to encrypt image and audio data [2].

Figure 16.3 shows how DCI movies are prepared and packaged. The video content is compressed, as specified before, and encrypted. A number of audio channels are then multiplexed into the main data stream: audio is generally left uncompressed. Subtitling and captioning in several languages are also included in the final package, as well as auxiliary data that are used during the projection and for classification purposes. The final product is called a Digital Cinema Package (DCP): the transport to the projection sites is not covered by the DCI standard specifications and it is left to what the distribution technology best offers. With this concern, security aspects are of fundamental importance in the whole system and they are carefully considered and managed by DCI.

16.3 Multicast Protocol Overview

NORM (NACK Oriented Reliable Multicast) [15] is a multicast protocol designed to provide end-to-end reliable multicast transfer. This protocol uses generic IP multicast capabilities, and

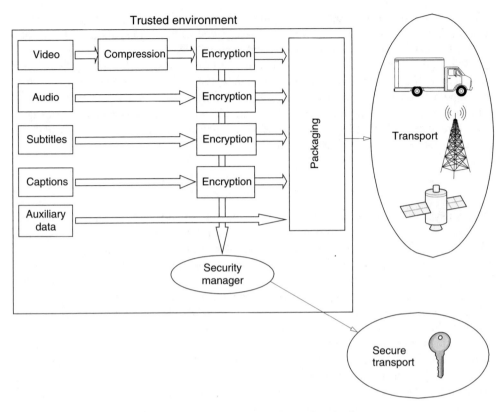

Figure 16.3   DCI encoding, encryption, and packaging process

on top of that it tries to achieve reliable transfer using NACKs (Negative ACKnowledgment). It can work both on reciprocal multicast networks (i.e. wireless or wired LANs) or unidirectional links (such as unidirectional satellite); in this way, it can satisfy all the transmission needs for D-Cinema distribution.

NORM repair capabilities are based on the use of negative acknowledgments sent from receivers to the sender upon detection of an erroneous or missing packet. The sender transmits packets of data, segmented according to a precise strategy, each of them identified by a number. When a receiver detects a missing packet, it initiates a repair request with a NACK message. Upon reception of NACKs, the sender places in a queue the appropriate repair messages, using FEC blocks. Each receiver can re-initiate a repair procedure if it does not receive repair blocks.

Feedback suppression is applied, using a random back-off algorithm; each receiver, before sending a NACK for a certain packet, waits a random interval, during which it senses the medium and checks if other receivers have requested a repair for the same packet: if so, it discards the NACK, otherwise it sends it and waits for the repair bits. When the sender receives the NACK, it places in queues the repair bits (or the entire lost packet). Feedback suppression works efficiently in this protocol, and can achieve good results [23].

NACK packets can be sent both in multicast or unicast mode, using the sender's address. In the second case, feedback suppression can be achieved using multicast advertising messages, sent from each sender, which let the receivers know which packets have a pending repair request.

A NORM sender can even autonomously add FEC parity bits to each packet, thus enabling the receiver to correct errors and recover from losses without starting a NACK procedure. FEC parity bits are created using a Reed-Solomon code [24]: the parameters n and k can be chosen in order to accommodate for variations in the channel conditions. The number of FEC bits to send can also be statically decided in advance.

Figure 16.4 shows the typical sequence of operations performed by a NORM sender and receiver during a transmission session. The sender node puts the data packets in queues, segmented according to some parameters that can be changed by the user, to satisfy particular needs: in our case, JPEG 2000 codestreams are encapsulated and an additional header is included at the end of the NORM packet header. It also periodically puts control messages in queues, such as round trip collection and rate congestion control feedback. Each receiver controls if the packet is in order and error-free: in this case, it accepts the packet and passes it to the destination application. Otherwise, it enters the NACK procedure: this consists in picking a random back-off interval, based on some parameters such as the largest round trip delay, usually supplied by the sender, and delaying NACK transmission until this interval is passed. In the meanwhile, if it receives a repair request for the same packet or the repair bits, the NACK is dropped; otherwise, it sends the NACK in multicast mode. Sending the NACK in multicast is useful for feedback suppression, as described above. When the sender receives the NACK it stops usual data transmission and sends repair packets, in the form of a bit sequence derived from the Reed-Solomon code. With this repair bits, receivers are able to recover from transmission errors. If a receiver loses a repair packet too, it can resend another NACK for the same packet after waiting for a new back-off interval.

For the programmers, an API is available (written in C++), offering access to NORM functionality in an easy and customizable way. In particular, a sender and a receiver can be created, and their behavior can be customized by setting parameters such as transmission speed, TCP port numbers to use, and Reed-Solomon code properties. The embedded congestion control mechanism can be enabled or disabled as well.

We can see that this protocol is applied not only to the off-line file transfer between the production and destination sites, but it can also represent a feasible method to broadcast live events to the theaters, in a manner which is similar to the classic streaming techniques, i.e. RTSP [25]. In fact, the widespread adoption of WLANs is fostering the diffusion of streaming protocols based on UDP, such as RTP/RTCP, which can provide an acceptable performance in case of non-guaranteed packet delivery. Recently, RTP for JPEG 2000 has been introduced and it is in the process of standardization [26].

16.3.1 Packetization Strategy and Header Format

It is important that the sender and the receiver jointly minimize the probability of retransmissions; this can be done by making them aware of the underlying codestream-based file structure. A number of fields have been added to the NORM protocol header, based on the fields utilized by the JPEG 2000 RTP streaming protocol.

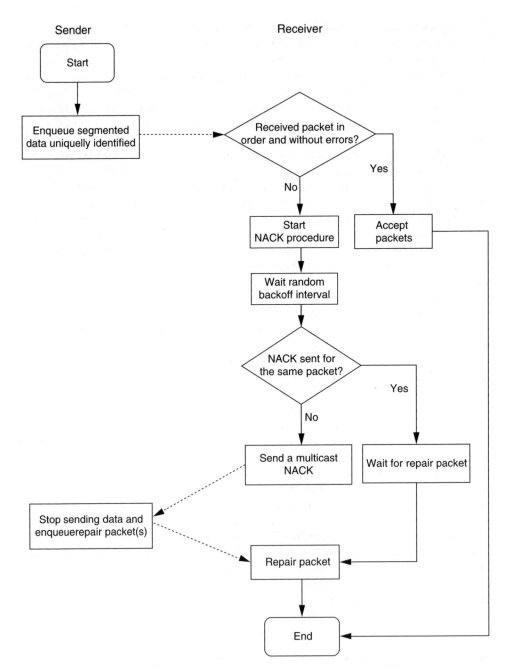

Figure 16.4   Sender and receiver sequence of operations [29]

Bits	7	15	23	31
main version	version	type	reserved
reserved	packet ID	secondary ID	
image number			
offset			
offset			
image length			

Figure 16.5   JPEG 2000 packet header format [29]

A tabular view of the additional header fields is presented in Figure 16.5; their meaning and use is described in the following:

- *main version* (8 bits): this is always set to 0xCB. If the received packet has a different value for this field, it is discarded.
- *version* (8 bits): minor version of the protocol, currently 1. Any packet with a different minor version is discarded by the receiver.
- *type* (8 bits): this indicates the kind of transfer, in particular it is related to the format of the codestream-containing file. Currently, it can be set to 0xF1 when a set of *.j2c/.j2k* files (raw codestreams) is sent, or to 0xF6 if a single *.mj2* video file (wrapped file format) is being transferred. Other types would result in packet discarding.
- *packet ID* (8 bits): if a single codestream is fragmented at the sender, it indicates the progressive number of the transmission. This is especially useful when transmitting large codestreams, which can be divided into smaller portions at the data packet boundary. This is used for packet re-ordering upon receiving;
- *secondary ID* (16 bits): this can be used to order different tracks inside a video file, or to differentiate among video, audio and synchronization data.
- *image number* (32 bits): progressive number of the image (i.e. codestream) in a movie. This is used to reconstruct the video file in the correct order, and to check for the correct receipt of packets;
- *offset* (64 bits): this represents the offset of the first data in this packet, starting from the beginning of the original file; it is directly expressed in bytes. We use 64 bits to handle files bigger than 4 GB, since DCI video content files can be as large as 250 GB. The receiver uses the value of this field to correctly put the received packet in the destination file;
- *image length* (32 bits): this denotes the total length of the codestream containing the image, expressed in bytes. It is used to check if the image has been completely received.

This header is added to each sent multicast packet for identification. This packetization strategy can send either separate codestreams or a single video file. In the first case, the content of a directory containing thousands of files (24 or 48 for each elapsed second of movie) is sent, each file having a size of 1.3 MB maximum. On the other hand, we can also send a single video file, which may be larger than 250 GB. In this case, we parse the video file and send each codestream in a separate packet.

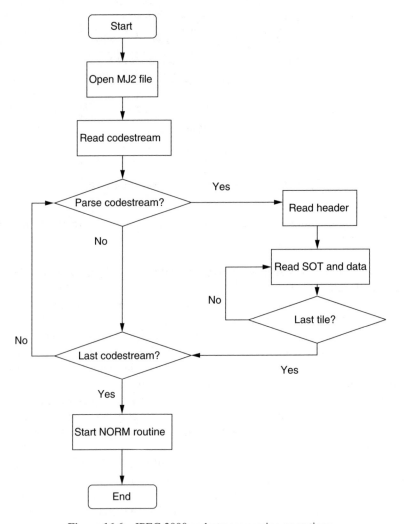

Figure 16.6   JPEG 2000 codestream parsing operations

Figure 16.6 shows the sequence of operations performed by the system before sending the codestream using the NORM protocol. An MJ2 video file, which contains codestreams and synchronization data in a box structure, is parsed to find codestreams offsets and lengths. The system allows parsing of codestream markers too. This secondary parsing could be useful in particular network situations, or if a better control over FEC data is needed (we could choose to protect with different codes parts of the codestreams).

16.4 System Architecture

Figure 16.7 shows a typical architecture of a multicast distribution system used for live events. A similar multicast system is also available in case of transmission of a stored video sequence from a production site to theatres and end-users, which is depicted in Figure 16.8. Digital

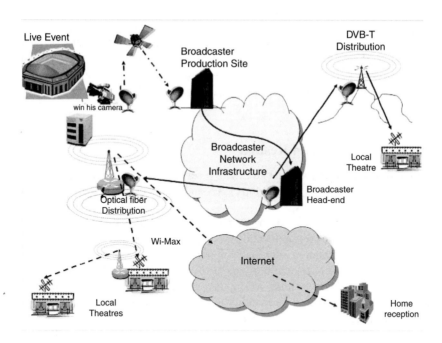

Figure 16.7   Live video distribution system architecture

cameras capture the event directly to a digital support; such cameras are already available at high resolution, that is, 2K and 4K. The captured video is transmitted to a production site using a high speed network, e.g. satellite. In case of a live event, the video content will immediately be sent to the head-end through a high speed network infrastructure and then to the final receivers: these ones could be large regional theatres, which could also act as distribution centers, sending content to local theatres or high-end users. In the last step of the distribution

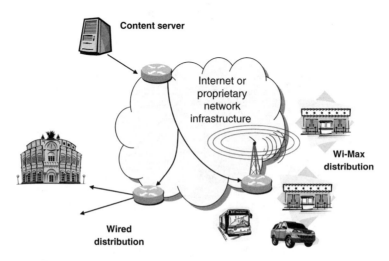

Figure 16.8   Pre-recorded movie distribution system architecture

chain, video content can be easily delivered via a wireless channel too (i.e. Mobile WiMAX), also because in rural areas a high-speed wired connection could not be available. In this case, also mobile equipment could be used to receive the live video content. If a sufficiently high transmission bandwidth is available, the content could be transferred through the Internet, using a multicast-enabled network or some tunneling system.

16.5 Test Application and Results

In our tests, we used a wired architecture, with a server acting as sender and few receivers. They were all connected using a 100 Mbps LAN and a network hub. Figure 16.9 shows the architecture of our test system.

Two separate applications were prepared for sending and receiving the data over a LAN, either wired or wireless. Each test application was written in C++, and the GUI uses wxWidgets [27], thus allowing for maximum inter-platform portability (e.g. Windows and Linux).

We tested our applications on a 100 Mbit/s LAN and an IEEE 802.11g access point, connected to the wired network. In order to simulate the presence of transmission errors in the network, we are provided with the capability to generate random errors (at a specified rate) directly at the sender side.

We used PCs with this configuration: CPU Intel Core Duo 2 E6600, 2 GB of RAM, 300 GB SATA hard disk, Windows XP Professional 64 bits.

We performed some tests by sending raw data through a wired LAN, in order to check the correct sequence of operations. The NORM multicast protocol was able to send data, up to a 15% error rate, without too much latency or overhead.

The next step was to simulate the transmission of a DC sequence; we approximated this by sending DCI-like formatted codestreams, compressed starting from a raw High Definition

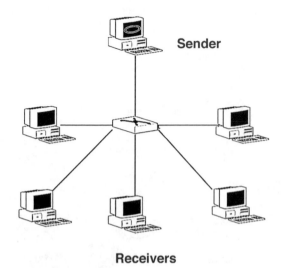

Figure 16.9   Test system architecture

Table 16.1   Measured transfer speed (in Mbps) versus error rate and FEC transmission method

Redundancy bits	Error rate			
	0%	5%	10%	15%
Sent	81.9	67.1	62.9	52.0
Stored	74.1	64.3	63.0	58.9

sequence (YCbCr color space, 4:2:0 color space subsampling, 2048x1080 pixels, 8 bits per pixel, 25 FPS) [28].

During the tests, transmission errors have been purposely inserted in the sending process: in this case, we observed a small added latency, and the average speed was reduced due to NACKing process and packet retransmissions. Indeed, a very little signaling overhead was introduced by the protocol. In order to compensate for these errors, we took advantage of the FEC capabilities of the NORM protocol. Thus, a number of repair bits were introduced in every sent packet; in this way, we managed to reduce the need for NACKs.

Table 16.1 reports some experimental results found on the 100 Mbps wired LAN. The redundancy bits column indicates the parameters of the Reed-Solomon code, used by the system. In this experiment we adopted a RS(80,64) code, where 16 parity bytes are stored in the sender for every 64 data bytes message, and they can be used in case of a NACK; however, the protocol allows sending them *a priori*, thus lowering NACK transmissions and making the data transfer more robust. The error rate columns indicate the percentage of errors introduced to simulate lossy data reception.

The results show that the *a priori* transmission of parity bytes cuts down the NACKing process with only a small decreased bandwidth.

Figure 16.10 shows a comparison of the transmission bit rate, in two different transmission scenarios. We used the same network architecture with the same packet loss rate, but in the *stored* case the sender calculates but does not send FEC bits (they are stored for an eventual future use); in the *sent* case, the application adds a small amount of parity bits to the actual data. As stated above, sending FEC parity in advance can lower transmission bit rate, but it will also lower the retransmission in case of errors. As Figure 16.10 shows, in case of high packet loss network (i.e. Wireless) the performance achieved by sending parity bits can be higher than in the normal cases. A possible development of the system is to enable it to adaptively modify the FEC rate while running; in this way we could change the parity bit percentage to accommodate for temporal variations of the channel loss rate.

16.6 Conclusion

In this chapter we presented a technique for encapsulating Digital Cinema compressed sequences into a reliable multicasting protocol, for the purpose of distribution among a main production site and the projection theaters through several transmission channels, including Mobile WiMAX. The selected compression standard, JPEG 2000, matched to the more-than-High

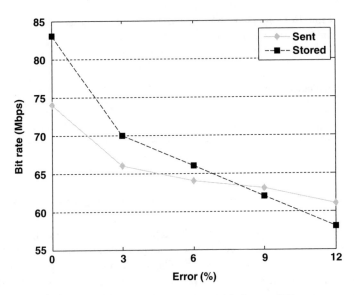

Figure 16.10   Data rate for *stored* and *sent* transmissions at different error rates

Definition quality of video sequences, deserves particular care when delivery must also minimize the probability of errors and, consequently, of retransmissions. Thus, a well crafted packetization strategy has been borrowed from the existing RTP strategy for JPEG 2000, and adapted to this particular case. The results from our simplified testbed, basically consisting of a wired network, showed very good efficiency of the protocol, even in the presence of a moderate/medium rate of transmission errors. Use of high speed wireless connections, i.e. WiMAX, is envisaged and strongly suggested in the live event distribution context. In this case, an interframe compression method should be used to lower the high bit rate demand and make the transmission feasible over a Mobile WiMAX channel. On the other hand, a pre-recorded movie will always be transmitted off-line; thus, even though the wireless channel is highly disturbed, the reliable multicast protocol (using a high FEC rate) will enable the transmission of the video content to the receivers. For future work, more sophisticated techniques for managing channel errors, relying on the tools offered by the JPWL standard, will be introduced and tested.

References

[1] T. Yamaguchi, M. Nomura, K. Shirakawa, and T. Fujii, 'SHD Movie Distribution System Using Image Container with 4096x2160 Pixel Resolution and 36 Bit Color', in *Proc. of ISCAS 2005*, Kobe, Japan, May 2005, 5918–5921.

[2] Digital Cinema Initiatives, LLC, 'Digital Cinema System Specification V1.0', July 2005. Online: available http://www.dcimovies.com.

[3] A. Ghosh, D.R. Wolter, J.G. Andrews, and R. Chen, 'Broadband Wireless Access with WiMax/802.16: Current Performance Benchmarks and Future Potential', *IEEE Commun. Magazine*, v **43**, February 2005, 129–136.

[4] Specific Targeted Research Project, IST Call 2, 'Layered Compression Technologies for Digital Cinematography and Cross-Media Conversion', *WorldScreen*, 2004. Online: available http://www.worldscreen.org.

[5] EDcine Project, IST 6th Framework Program of the European Commission. Online: available http://www.edcine.org.

[6] WiMAX Forum website, 'Mobile WiMAX – Part I: A Technical Overview and Performance Evaluation', 2006. Online: available http://www.wimaxforum.org.

[7] O. I. Hillestad, A. Perkis V. Genc, S. Murphy, and J. Murphy 'Delivery of On-Demand Video Services in Rural Areas via IEEE 802.16 Broadband Wireless Access Networks', in *Proc. of 2nd ACM Workshop on Wireless Multimedia Networking and Performance Modeling, WMuNeP* (jointly with MSWiM), October 2006.

[8] Intel Corp., 'WiMAX Potential Premieres at Sundance Film Festival', 2006.

[9] ETSI, 'Digital Video Broadcasting (DVB); Framing Structure, Channel Coding and Modulation for Digital Terrestrial Television (DVB-T) – EN 300 744 V1.5.1', June 2004.

[10] D.T. Lee, 'JPEG 2000: Retrospective and New Developments', *Proc. of the IEEE*, **93**(1), January 2005, 32–41.

[11] M.J. Lee, J. Zheng, Y.B. Ko, and D.M. Shrestha, 'Emerging Standards for Wireless Mesh Technology', *IEEE Wireless Communications*, **13**(2), April 2006, 56–63.

[12] O.I. Hillestad, A. Perkis, V. Genc, S. Murphy, and J. Murphy, 'Delivery of On-Demand Video Services in Rural Areas via IEEE 802.16 Broadband Wireless Access Networks', in *Proc. International Workshop on Modeling Analysis and Simulation of Wireless and Mobile Systems'*, Torremolinos, Spain, pp. 43–52, 2006.

[13] F. Frescura, C. Feci, M. Giorni, and S. Cacopardi, 'JPEG2000 and MJPEG2000 Transmission in 802.11 Wireless Local Area Networks', *IEEE Trans. on Consumer Electronics*, November 2003.

[14] T. Shimizua, D. Shiraia, H. Takahashia, T. Murookaa, K. Obanaa, et al., 'International Real-time Streaming of 4K Digital Cinema', *Future Generation Computer Systems*, **22**(8), October 2006, 929–939.

[15] B. Adamson, C. Bormann, M. Handley, and J. Macker, 'Negative-acknowledgment (NACK)-Oriented Reliable Multicast (NORM) Protocol', November 2004. Online: available http://cs.itd.nrl.navy.mil/work/norm.

[16] 'JPEG 2000 Image Coding System—Part 11: Wireless', Int. Standards Org./Int. Electrotech. Comm. (ISO/IEC), ISO/IEC WD2.0 15 444-11, Dec. 2003.

[17] 'JPEG 2000 Image Coding System—Part 8: Secure JPEG 2000', Int. Standards Org./Int. Electrotech. Comm. (ISO/IEC), ISO/IEC WD2.0 15 444-8, Dec. 2003.

[18] 'Advanced Video Coding for Generic Audiovisual Services', ITU-T Rec. H.264, ISO/IEC 14496-10, March. 2005.

[19] 'Information Technology – Generic Coding of Moving Pictures and Associated Audio Information: Systems', ISO/IEC 13818-1, 2005.

[20] SMPTE 377M-2004 Television – Material Exchange Format (MXF) – File Format Specification, SMPTE, 2004.

[21] 'JPEG 2000 Image Coding System—Part 3: Motion JPEG 2000', Int. Standards Org./Int. Electrotech. Comm. (ISO/IEC), ISO/IEC WD2.0 15 444-3, Dec. 2003.

[22] CIE (Commission Internationale de l'Eclairage) *Proceedings*. Cambridge: Cambridge University Press, 1931.

[23] B. Adamson and J. Macker, 'Quantitative Prediction of NACK-oriented Reliable Multicast (NORM) Feedback,' Online: available http://pf.itd.nrl.navy.mil/norm

[24] I.S. Reed and G. Solomon, 'Polynomial Codes Over Certain Finite Fields', *Journal of the Society for Industrial and Applied Mathematics*, **8**(2) June 1960, 300–304.

[25] A. Rao, R. Lanphier, and H. Schulzrinne, 'Real Time Streaming Protocol (RTSP)', RFC 2326, April 1998.

[26] S. Futemma, A. Leung, and E. Itakura, 'RTP Payload Format for JPEG 2000 Video Streams', draft-ietf-avt-rtp-jpeg2000-10, February 2006.

[27] wxWidgets, cross platform GUI library. Online: available http://www.wxwidgets.org.

[28] 'High Definition Test Sequences'. Online: available ftp://ftp.ldv.e-technik.tu-muenchen.de/pub/test_sequences

[29] P. Micanti, G. Baruffa, and F. Frescura, 'A Packetization Technique for D-Cinema Contents Multicasting over Metropolitan Wireless Networks'. *Proceeding IEEE Mobile WiMAX Symposium*, Orlando, 2007.

17

WiMAX Extension to Isolated Research Data Networks: The WEIRD System

Emiliano Guainella, Eugen Borcoci, Marcos Katz, Pedro Neves,
Marilia Curado, Fausto Andreotti, and Enrico Angori

17.1 Introduction

WiMAX, the technological driver for the broadband wireless access market, is being considered by network operators and service providers as a novel opportunity for a large number of application scenarios. In fact, the positive results in terms of performance combined with the increasing interest showed by most governments worldwide, as demonstrated by the hundreds of trials and the regulatory activities currently going on in the world, are boosting WiMAX's acceptance in the wireless market.

The standards and the technology behind the WiMAX word are many, reflecting its flexibility of use. For MAC and PHY layers, it encompasses the IEEE 802.16-2004/ETSI Hiper-MAN [1] standards, suited for fixed WLAN/WMAN segments, and the more recent IEEE 802.16e-2005/ETSI HiperMAN [2] that, incorporating the latest technological advances such as OFDMA, MIMO and beamforming, is better suited for combined fixed and mobile high speed access. This last standard is seen as a strong candidate towards the 4G network. Born to support natively the IP technology, WiMAX includes a reference network architecture ([3], [4]) proposed by the WiMAX Forum, in order to support the access technology. This architecture follows the recent trends of Next Generation Networks (NGN), following an Internet-like horizontal approach (applications independent of the transport technology) rather than a vertical approach typical of the legacy of telecom infrastructures.

This chapter describes the main outputs of the WEIRD (WiMAX Extension to Isolated Research Data networks) Integrated Project [5], funded by the European Commission and carried

Mobile WiMAX   Edited by Kwang-Cheng Chen and J. Roberto B. de Marca.
© 2008 John Wiley & Sons, Ltd

out by a consortium of 16 partners, composed of operators, manufacturers, academia and user communities. The scope of the project is to use and extend the WiMAX technology and to build four European testbeds to prototype the broadband wireless access of research communities to the next generation National Research and Education Networks (NRENs) [6]. The application scenarios considered, selected from the real needs of the user communities of the consortium and used to drive the system specifications, are in the fields of environmental and volcano monitoring, fire prevention and telemedicine. These scenarios introduced some challenging requirements, such as QoS, mobility, compatibility with legacy or non-conventional applications and reliability. They have been translated into the WEIRD system specifications, built with the aim of extending the WiMAX standards and technology, as described in the following sections [7].

The chapter is organized as follows: Section 17.2 describes the novel application scenarios driven by the research communities partners of the WEIRD consortium, Section 17.3 describes the key technologies adopted by the project beyond the state of the art, Section 17.4 describes the WEIRD's high level system architecture, Section 17.5 describes how results will be validated within the four European testbeds, and finally Section 17.6 draws the conclusion.

17.2 Novel Application Scenarios for WiMAX

As a Broadband Wireless Access (BWA) technology, WiMAX is able to provide ubiquitous Internet access, allowing end users to be connected to the Internet independent of their location. It contributes to decreasing the digital divide gap, by providing Internet access to rural areas where the deployment of a wired access solution will not be economically feasible for the telecommunication operators. A set of emergency services scenarios in which the WiMAX technology can play an important role have been identified by the WEIRD consortium: environmental monitoring, telemedicine and fire prevention.

17.2.1 Environmental Monitoring

Environmental monitoring is very important in impervious areas such as seismic and volcanic zones. An effective and reliable monitoring system must be implemented to warn of the occurrence of a natural catastrophe in these areas. Volcano monitoring is one of the applications foreseen by the WEIRD project as an example of an environmental monitoring scenario. Permanent and mobile stations, such as video cameras and sensors, are installed in the impervious area. The information collected from these stations is sent in real time to an aggregation point using Mobile WiMAX and then forwarded to the Monitoring Centre through a Fixed WiMAX backhaul. Furthermore, Mobile WiMAX is also used to allow real-time communication between mobile users, which visit the impervious area, and the Monitoring Centre. A global and simple overview of the environmental monitoring volcano scenario is shown in Figure 17.1.

This scenario requires that the Mobile WiMAX technology is able to provide QoS-aware real-time services, such as voice and video over IP communication. Moreover, fast mobility is also mandatory allowing users, as shown in Figure 17.1, to keep their connection to the Monitoring Centre while moving in the impervious area.

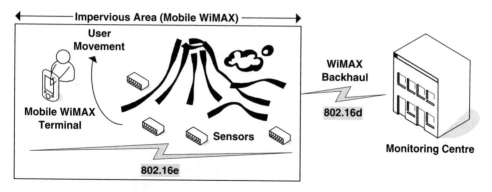

Figure 17.1 Environmental monitoring volcano scenario [26]

17.2.2 Telemedicine

E-health is one of the areas where WiMAX technologies can substantially contribute to improve the daily activities and thus enhance the quality of life. Today a large number of activities are carried out with limited success, unnecessary costs and human difficulties because of the impossibility of exchanging real-time information between different elements of the chain that are not at fixed locations. Remote diagnosis is one of the possible applications where Mobile WiMAX plays an important role. Relevant medical information obtained from a portable ultrasound device can be sent directly from an ambulance to the hospital using the Mobile WiMAX technology. As this information can be exchanged while the ambulance is on its way to the hospital, all the necessary procedures can start to be prepared beforehand. Also, a videoconference session with the medical staff from the hospital can be set up, allowing the exchange of important information about the patient's condition. This scenario is illustrated in.

The telemedicine scenario requires real-time services and applications such as voice and video over IP in a mobility environment to support real-time communication in case of emergency.

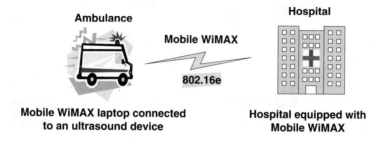

Figure 17.2 Medical information exchange while traveling scenario [26]

17.2.3 Fire Prevention

Several pilot projects have shown how the use of technologies such as video and infrared cameras can help fire detection. The main obstacles to the implementation of such systems are the costs and image quality related to GSM/GPRS communications and the difficulty of implementing radio links to transmit video in mountainous regions. The abandonment of these remote or isolated areas, most times mountainous regions, is one of the main factors that leads to a poor early fire detection system.

The *FireStation* application is one of the applications foreseen for the Fire Prevention scenario. It is related to the transmission of images and text taken from the Forest Fire Simulation System (*FireStation*) located and operated in the District Civil Protection Coordination Centre (CC) to a mobile unit in the field (PDA/Laptop). To transmit the images and the text from the District Civil Protection Coordination Centre (CC) to a mobile unit in the field (PDA/Laptop) the Mobile WiMAX technology is used. In this case, real-time data collection and transmission, as well as voice and video over IP services will be utilized.

Another relevant application for Fire Prevention is related to Fixed and Mobile Video Surveillance. In the Fixed Video Surveillance case, fixed video cameras transmit video, as well as meteorological parameters, to the CC. The collected data are transmitted to a web server and are accessed on a web page where the user is able to control the movement of the cameras. In this scenario, since fire detection and monitoring are still a human process, as they imply the presence of a system operator, it is necessary to have real-time video and camera control with the best possible resolution. For the Mobile Video Surveillance application a terrestrial mobile unit works as a *mobile watch tower* and monitors the fire. It transmits text data, such as GPS position and meteorological parameters, as well as video data (similar to the fixed cameras) to the CC. A similar application for Mobile Video Surveillance is the video and voice communication between the helicopters of the fire brigade and the CC. Helicopters can provide images from a new top-down perspective impossible to achieve with a mobile unit. This can be especially interesting as helicopters are highly mobile and can therefore be quickly ordered to move to different locations providing fast information updates under changing conditions, like a turning wind direction depicts both applications of the Mobile Video Surveillance scenario.

Figure 17.3   Mobile Video Surveillance – Terrestrial and Air Scenarios [26]

17.3 Key Technologies

The WEIRD system design is based on state-of-the-art technologies and the ongoing standard-izations processes, as well as business models and associated scenarios. In order to efficiently support the envisioned application scenarios for WiMAX, the WEIRD consortium has investigated the adoption, among others, of several key technologies and designed its system architecture to integrate them.

This section deals with physical and transport layer technologies which are being deployed in the WEIRD framework. In addition, application and control layer signaling protocols for QoS reservation are described.

17.3.1 Physical Layer Issues

In communication systems, radio-frequency (RF) signal processing functions such as frequency up-conversion, carrier modulation, and multiplexing, are performed at the BS, and immediately fed into the antenna.

The growth in high-bandwidth radio services like WiMAX has led to a renewed focus on the optimum network infrastructure able to transmit signals between base stations and antennas. As a result, there has been recent interest in Radio over Fiber (RoF) systems, which employ fiber optics to transport RF-modulated signals for wireless applications. These have potential advantages in allowing the transmission of WiMAX signals in their raw form to antennas at which no RF signal processing beyond amplification is required, which leads to a simplification of the transmission equipment. RoF technology entails the use of optical fiber links to distribute RF signals from a central location to Remote Antenna Units (RAUs). RoF makes it possible to centralize the RF signal processing functions in one shared location (Central Base Station), and then to use optical fiber, which offers low signal loss to distribute the RF signals to the RAUs. By so doing, RAUs are simplified significantly, as they only need to perform optoelectronic conversion and amplification functions.

The centralization of RF signal processing functions enables equipment sharing, dynamic allocation of resources, and simplified system operation and maintenance. These benefits can translate into significant savings in system installation and operational costs, especially in wide-coverage broadband wireless communication systems, where a high density of BS is necessary.

Among the advantages of RoF technologies there are low attenuation losses, large available bandwidth, immunity to electromagnetic interferences and eavesdropping, easy installation and maintenance and reduced power consumption. In addition, RoF systems allow multi-operator multi-service operation, and dynamic resource allocation [8].

shows two typical deployments of RoF technology. On the left, a Radio Antenna Unit (RAU) can easily provide the coverage of shadow zones without adding any new Base Station (BS). On the right, different RAUs are fed by the same carrier from a single BS, thus creating overlapping areas that are needed in order to provide coverage extension and strengthening of the received signal power with a reduced handover management. Furthermore, important reliability features can be exploited in such architecture, as a mobile station is simultaneously covered by two antennas, thus protecting the service from the eventual failure of any one of them.

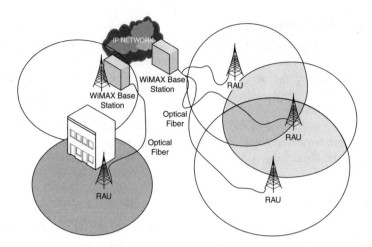

Figure 17.4 RoF deployment examples

These deployments will be very useful for those application scenarios envisaged in WEIRD, where critical customers need a reliable connection that cannot be disrupted due to, for instance, extreme weather conditions, temporary obstacles or natural disasters.

17.3.2 MAC and Service Flow Management

The IEEE 802.16d/e standard includes the Medium Access Control (MAC) layer and the Physical layer (PHY) specifications. The MAC layer is further divided in three different sublayers, the Service Specific Convergence Sublayer (CS), the Common Part Sublayer (CPS) and the Privacy Sublayer (PS), as depicted in Figure 17.5.

The MAC CS resides on top of the MAC CPS and provides the interface between the MAC layer and the network layer. It accepts and classifies the incoming Protocol Data Units (PDUs) into the appropriate connections, based on a set of packet matching criteria. As defined in the IEEE 802.16d/e standard, a connection is a unidirectional mapping between the BS and the SS MAC layers for the purpose of transporting a Service Flow's (SF) traffic. To uniquely identify a connection, a 16-bit Connection Identifier (CID) is used. The connection-oriented

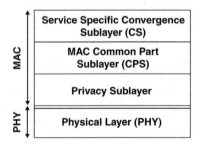

Figure 17.5 IEEE 802.16d/e protocol layering

feature is the key characteristic of the MAC layer. Therefore, all communications between the IEEE 802.16d/e nodes are based on a connection and no packets are allowed to traverse the wireless link without a specific connection previously established. After their classification, the PDUs are delivered to the MAC CPS via the MAC Service Access Point (SAP) and delivered to the peer node. The MAC CPS is responsible for the main functionalities of the system, such as connections and SFs management, bandwidth allocation, polling and access-grant mechanisms, as well as scheduling mechanisms. Encryption and authentication mechanisms are assured by the PS.

WEIRD architecture is compliant with the QoS model defined in the IEEE 802.16d/e standard. QoS is assured through the use of SFs and their associated connections. Each packet sent over the 802.16d/e air link is associated with a particular SF. A SF is a MAC-layer transport service that provides unidirectional transport of packets either to uplink packets transmitted by the SS or to downlink packets transmitted by the BS. It is characterized by a set of QoS parameters such as latency, jitter and throughput, which have to be assured over the air-link. These parameters also include the required information for the bandwidth allocation mechanism, as well as the schedulers operation in both the BS and the SS. SFs exist in both uplink and downlink direction and they are identified by a 32-bit Service Flow Identifier (SFID). Different sets of QoS parameters might be associated with a particular SF, such as:

- *ProvisionedQoSParamSet*: indicates a QoS parameter set provisioned via means outside of the scope of 802.16d/e standard, such as the operator network management system.
- *AdmittedQoSParamSet*: defines a set of QoS parameters for which the BS and the SS are reserving resources. This QoS parameter set serves in the preparation to subsequently activate the flow.
- *ActiveQoSParamSet:* specifies the set of QoS parameters that are actually being provided to the SF.

Depending on the QoS parameters set, each SF will have a status associated. The following SF statuses are defined:

- *Provisioned Service Flow*: is known via provisioning by, for example, the network management system. Its *AdmittedQoSParamSet* and *ActiveQoSParamSet* are both null.
- *Admitted Service Flow:* indicates that the SF has resources reserved by the BS for its *AdmittedQoSParamSet*, but these parameters are not yet active.
- *Active Service Flow:* resources are committed by the BS for its *ActiveQoSParamSet*, and as a result the BS is able to forward data packets through this SF for the SS.

To manage the SFs and their correspondent status, two models (*Static* and *Dynamic*) are defined in the IEEE 802.16d/e standard and supported on the WEIRD architecture. For the Static Model, the SF should pass through all the SF status until it reaches the Active status. With respect to the Dynamic Model, the SF is directly activated without traversing the remaining status. The Dynamic Model is suitable for establishing SFs for real-time applications, whereas the Static Model is appropriated for long-term QoS reservations, managed by the Network Management System.

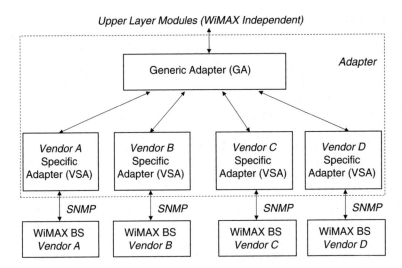

Figure 17.6   WiMAX adapter architecture

17.3.3 Low Level Hardware Transparency (Adapters)

The WEIRD architecture has been specified in order to be independent, as much as possible, of the WiMAX manufacturer equipments. To achieve this level of independency, an abstraction module, named Adapter, is provided by the WEIRD specification. Its main objective is to separate the WiMAX equipment vendor-dependent modules from the independent ones. As a result, different WiMAX vendor equipments can be seamlessly integrated and supported without requiring modifications over the remaining architecture modules and interfaces. Figure 17.6 presents a brief and simple description of the Adapter architecture and decomposition.

The Adapter module has a northbound interface with the upper layer modules from the WEIRD architecture, which are totally independent of the WiMAX equipment that is beneath, and a southbound interface with the WiMAX BS through the SNMP protocol [9]. The Adapter is further divided into a generic part, called Generic Adapter (GA), and a set of Vendor Specific Adapters (VSA) to provide a differentiated SNMP interface for each vendor WiMAX BS. The GA is responsible for processing the incoming requests and sending them to the appropriated VSA, according to the specified vendor. The VSA converts the received requests from the GA to the correspondent SNMP messages and sends them to the SNMP agent running on the WiMAX BS. Additionally, to convey the network topology information, as well as alarms notifications, SNMP Trap messages sent by the WiMAX equipments are also supported.

17.3.4 IP-Based Transport

To integrate IP over WiMAX, WEIRD develops a set of functionalities in order to provide a technology independent interface between upper-layers and WiMAX technology-dependent MAC and PHY layers.

The WEIRD transport plane will be implemented as a Convergence Layer solution including IPv4/IPv6 support, and the basic TCP and UDP transport protocols, as well as the newer ones (Stream Control Transmission Protocol) SCTP and (Datagram Congestion Control Protocol) DCCP protocols for supporting new applications and scenarios [10], [11]. Moreover, the WEIRD transport plane provides APIs for QoS management.

The functionalities of the Transport Stratum are achieved by the following layers:

- The *Transport Layer* including, TCP, UDP, SCTP and DCCP protocols to provide QoS-capabilities at transport level.
- The *Network Layer* including IPv4 and IPv6, and QoS support using DiffServ, as well as mobility protocols such as MIPv4/MIPv6, and the security functions offered by IP Security (IPSec).
- The *Convergence Layer* is the interface with the 802.16 upper layers and is composed by two entities, the QoS Convergence Layer and the Mobility Convergence Layer. The QoS Convergence Layer is responsible for the selection of 802.16 scheduling service. The Mobility Convergence Layer will monitor mobility activities in order to enhance layer 3 mobility decisions.

17.3.5 Application and Session Signaling

The application signaling denotes the end-to-end signaling used for application control. In WEIRD, this signaling should cooperate with resource reservation and allocation mechanisms at the network level, in order to provide application-driven QoS.

It is worth mentioning that the WEIRD system has the capability to support a variety of applications, Session Initiation Protocol (SIP)- and no-SIP-based, legacy and customizable, which can be split into two main categories:

- *WEIRD-aware* – SIP-based applications and other applications that can be customized for using WEIRD system services.
- *Legacy* – applications that cannot be updated but can use the WEIRD-system services through a suitable WEIRD Agent.

Regarding SIP, this is the protocol of choice for setting up multimedia communication sessions between two or more endpoints. SIP defines various signaling methods that are used services such as instant messaging, presence, and real-time communication using audio and video. SIP is a lightweight, transport-independent, text-based protocol [12].

Table 17.1 WEIRD applications and QoS support

	Customizable	Legacy
SIP-based	SIP/SDP extension(s) could be needed to support QoS-assured model.	COTS SIP clients, with (at least) QoS-enabled model.
not-SIP-based	Use of WEIRD API on client side.	Use of WEIRD agent on client side.

For session-based SIP applications, the WEIRD architecture aims to support two different QoS models: QoS-assured and QoS-enabled.

According to the QoS-assured model, a call can be established only if the requested/required level of QoS can be set; in other words the QoS setup becomes a precondition for calls. In SIP-based applications, QoS preconditions are media stream specific, therefore they are specified in SDP messages as attributes of the media. In a QoS-enabled model, the availability of QoS resources does not affect the success of a call; it only affects the effective level of QoS associated with the call.

Customizable SIP-based applications could be enhanced with particular SIP/SDP extensions in order to support a QoS-assured model, while Commercial-Off-The-Shelf (COTS) SIP applications can also be used in WEIRD with a QoS-enabled model.

In both cases, the reservation process is driven by a suitable SIP Proxy in the signaling path, leaving the SIP endpoints to participate in the precondition verification mechanism (if used). This WEIRD SIP Proxy, located on the core service network side, will be used for service description information (SDI) extraction from the SIP/SDP messages, resource reservation triggering toward the WEIRD control plane and interaction with AAA services as well. This approach is compliant with that used in the IMS/TISPAN framework [13], [14], [15], where QoS is triggered by an Application Function (AF) using a standardized interface (Gq/Gq′) based on the Diameter protocol to communicate with the resource control.

In order to support signaling for other kind of applications (not-SIP based), WEIRD provides specifically designed software running on the client side.

If it is possible to adapt the application, WEIRD-awareness is obtained by integrating the source code with a suitable Application Programming Interface, the WEIRD API. The purpose of this interface is to let applications trigger a reservation from the client side for the resource control plane.

On the other hand, legacy applications will use WEIRD system services configured via the same API by a suitable Agent with two levels of functionality. Assuming that the application flow identifiers are known in advance, the basic version of this Agent is an off-line pre-provisioning module used for configuration and activation of the channel before the application starts. The enhanced WEIRD Agent is an on-demand resource provisioning module which configures the channel once it gathers information about the application flow identifier(s).

Reference [23] summarizes, for each kind of WEIRD application, how QoS provisioning will be supported.

17.3.6 Resource-Oriented Signaling

WEIRD System adopts Next Steps in Signaling (NSIS), a signaling framework being developed by the Internet Engineering Task Force (IETF) for the purpose of installing and maintaining flow states in the network 0. NSIS is mainly based on the most known RSVP signaling protocol, and it reuses its mechanisms, whenever possible, since they have already been widely tested. However, NSIS avoids unnecessary complexity by removing capabilities such as multicast support, becoming, thus, simpler and more scalable than RSVP.

The NSIS architecture is composed of two layers: one for signaling transport and the other for signaling applications. This separation allows the use of the same signaling transport protocol for the support of all signaling applications.

The NSIS transport layer protocol, known as General Internet Signaling Transport (GIST), is responsible for the transport of signaling messages between network entities. When a signaling message is ready to be sent, it is passed to the GIST layer along with the flow information associated with it. The role of the GIST layer is to send the message to the next NSIS Entity on the path.

The signaling layer contains specific functionality of signaling applications and may comprise several NSIS Signaling Layer Protocols, generically known as NSLPs. The main signaling protocols being developed at the NSIS IETF WG are the QoS-NSLP 0 and the Network Address Translation (NAT) and Firewall (FW) NSLP 0.

The main NSIS features include support for on-path and off-path signaling, transparent signaling through the network, signaling aggregation for a set of flows, while supporting scalability, flexibility, and security. NSIS allows per-flow signaling and aggregated flows signaling based on the use of the DSCP field or tunnels. While NSIS works on a hop-by-hop basis between NSIS-aware nodes, it is transparent when NSIS is not supported by some network entities.

In WEIRD, NSIS is used for QoS signaling, comprising thus the GIST and QoS-NSLP layers. To be able to establish and maintain resource reservations, QoS NSLP defines four messages types: RESERVE, QUERY, RESPONSE and NOTIFY. Each message contains three parts: Control information, QoS specification (QSPEC), and Policy objects. In WEIRD, the QSPEC is modified in order to contain authentication information which will be used by WEIRD AAA functions, as described in the next section.

17.3.7 AAA Framework

With respect to the Authentication, Authorization, and Accounting (AAA) services, the WEIRD approach will rely on Diameter [19]. The Diameter protocol was derived from the RADIUS [20] protocol with many improvements in different aspects, and is generally believed to be the next generation AAA protocol. Together with SIP, it is widely used in the IMS architecture for IMS entities to exchange AAA-related information. The various WEIRD applications that require AAA functions can define their own extensions on top of the Diameter base protocol, and can benefit from the general capabilities provided by the Diameter base protocol. Moreover, for applications which do not rely on SIP signaling, the NSIS protocol transports authentication information to be handled by Diameter.

Three logical AAA layers are identified in the WEIRD system:

1. *SIP Session Signaling layer* – performs authentication and authorization during the SIP registration and session setup. It also records accounting information in each SIP session.
2. *QoS Reservation layer* – evaluates authentication and authorization QoS requests on behalf of applications (both SIP- and no-SIP based).
3. *WiMAX Access Network layer* – can be used to cope with the AAA framework supported by the specific equipment vendor.

Configuration and administration of user profiles, identities and credentials are provided in WEIRD by means of a Web Server.

17.4 System Architecture

This section presents the WEIRD system architecture and its relationships with standard and future generation architectures for networks and services.

17.4.1 Recent Architecture Standards and Trends

The recent architectural efforts made by several standardization organizations and groups (IETF, IEEE 802.x, ETSI, WiMAX Forum, ITU-T TISPAN, 3GPP, etc.) are focused on enhancing the existent architectures or defining new ones, targeting (among others), two major objectives: support of seamless connection of different heterogeneous lower layer technologies (wireline/wireless, mobile/fixed) and of unification and integration of the logical framework for high level services and applications. 'All IP' is a unifying paradigm to allow uniform but flexible processing at higher layers. Horizontal architectural decomposition in multiple planes offers better possibility to control and manage the network and services entities.

Given the objectives of this chapter, a very short review is first presented, on IEEE 802.16 L1/L2 and WiMAX Forum network architecture, [1–4]. The IEEE 802.16 architectural stack comprises the PHY and MAC layer only, mainly for *Data Plane* and *Control Plane*. The standards also define the Service Access Points for interaction between the Data and Control Plane and the *Management Plane*. The WiMAX Forum extends the 802.16 architecture in its *Network Reference Model* (NRM), [3], [4] seen as a logical representation of the WiMAX network architecture. The NRM objective is to allow multiple implementation options for a given functional entity, and yet achieve interoperability (among different realizations of entities) based on the definition of communication protocols and data plane treatment between functional entities, with overall aim of achieving E2E functions (e.g. QoS assurance, security or mobility management).

The NRM introduces the *Network Control and Management System* which may include the following services: AAA; Service Flows and Connection Identifiers Management; RF Transmission and Synchronization; BS Scheduling and coordination; Paging; Security; Mobility Management; Gateway and Router Services; Network Management; Inter-working; Data Cache; Multimedia Session Management; Media Independent Handover Function.

The NRM identifies functional entities and normative reference points (RP) over which interoperability is achieved. It defines the functionalities of: *Subscriber Station/Mobile Station (SS/MS), Access Service Network (ASN),* and *Connectivity Service Network (CSN)*. The NRM defines the interfaces between entities as Reference Points (RP), R1, R2, . . . R8. Each entity, SS/MS, ASN and CSN represents a functional grouping. The functions of one entity may be realized in a single physical device or distributed over multiple devices (functions grouping and distribution is an implementation choice). The functional entities on either side of an RP represent a collection of control and bearer (data) plane end-points. Interoperability will be verified based only on protocols exposed across an RP, which would depend on the E2E function or capability realized (based on the usage scenarios supported by the overall network).

As seen in Section 17.3.5, in IETF architectures a notable trend is the enhancement of Control Plane functions, e.g. by developing the signaling NSIS framework and also a strong effort is seen in the domain of high level services signaling SIP/SDP, security and AAA framework and protocols.

Recognizing IP as a basis of integration in NGN, the telecom-originated organizations defined the NGN as an enhanced IP-based network and services architecture. The starting point was that new applications have new needs, that are not fulfilled by the traditional IP architecture. Consequently, the ITU-T Standardization Sector (ITU-T) Rec.Y.2001 [21] identified a number of key features necessary in an NGN. ITU-T Rec.Y.2011 [22] provides a general framework for the architecture.

The main NGN step beyond traditional telecommunications architectures is the shift from separate vertically integrated application-specific networks to a single network capable of carrying all services. Circuit switching migrates to IP packet switching, but still NGN is capable to offer services with quality comparable to the telecom one. Additionally, sophisticated applications and services are possible, by introducing the IP Multimedia Subsystem (IMS) based on SIP framework. The transport envisaged by NGN is assured by multiple broadband QoS-enabled transport technologies while the high level service-related functions are independent on this transport. Such an approach enables flexible access for users to networks and competing service providers and/or services of their choice. It also supports generalized mobility that will allow consistent and ubiquitous provision of services to users.

17.4.2 WEIRD Overall Multi-plane Architecture

17.4.2.1 High Level Description

The WEIRD system aims to be a part of full multi-domain network architecture, allowing fixed and mobile access in new scenarios. Among other targets, WEIRD aims to contribute to developing end-to-end QoS enabled services. The WEIRD system can support different business models; each business actor can offer the others high-level services or connectivity services, in the access and/or core transport. The architecture allows organizational and technical independence (in terms of management and control) of the entities managing the network domains: Network Access Provider (NAP), Core Network Service Provider (NSP), etc.

The WEIRD system is built upon a networking infrastructure, with three components: *Customer Premises Equipment (CPE)*, *Access Service Network (ASN)* and *Connectivity Service Network (CSN)*. Each one can be managed by a different business entity. Note that a WEIRD customer may be a user/provider or might have both roles. They may be located within CPE or linked to the CSN, in case of *Application Service Providers (ASP)*. The infrastructure includes the mobility support. has been simplified and abstracted, to emphasize the WiMAX Forum, interfaces R1, R2, . . . , R8 0. The CPE can be composed of single-user IEEE 802.16 Subscriber Stations (SS) or Multiple users SSs (MSS), in case that an SS offers access to LANs/WLANs having several users/hosts. An ASN may control, and aggregate several BSs, based on a wireline or wireless IP infrastructure. The ASN is linked through an ASN Gateway (ASN-GW) to the CSN. The ASN-GW plays here both the data gateway role, and also the control role for *ASN*. In a mobile environment the CSN may be the Home CSN or Visited CSN respectively. Connectivity with other networks may be realized via IP backbone. Application entities' clients and/or servers can exist in the CPE side or in CSN networks. In the WiMAX Forum model, also direct interfaces between different ASNs can exist (denoted by R4).

One important goal is to control and offer end-to-end QoS-enabled services. WEIRD should achieve and control QoS in its scope: WiMAX segment and ASN. To this aim, it defines appropriate interfaces with CPE and CSN and runs QoS-oriented signaling to these interfaces. The

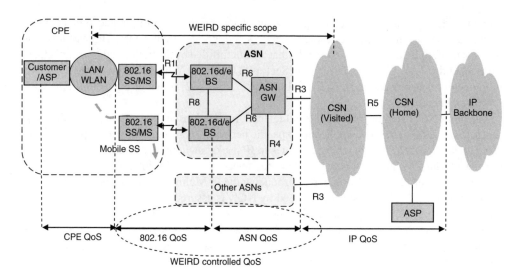

Figure 17.7   Simplified WEIRD overall network infrastructure

WEIRD system offers different levels of QoS to the high-level services/applications mapping them on appropriate IEEE 802.16 classes of services (Unsolicited Grant Service (UGS), real-time Polling Service (rtPS), extended real time Polling Service (ertPS), non-real-time Polling Service (nrtPS) and Best Effort (BE)), in WiMAX segment. This architecture supports different applications, capable or not to signal their QoS requirements (SIP/non-SIP-based applications, legacy, etc.) by offering appropriate APIs, as explained in Section 17.3.5.

The overall WEIRD architecture is structured as a multi-plane (Figure 17.8) and it is fully described in [23], [24]. Vertically there are two 'macro-layers', or 'strata', i.e., *Application and Service Macro-Layer/Stratum* and *Transport Macro-Layer/Stratum*. Horizontally, there are three ('parallel') planes: *Management (MPl), Control (CPl)* and *Transport/Data Plane (DPl)*. This structuring aims to decouple the applications and high level services from transport technologies, in order to support heterogeneity of the core and access network technologies [22], [25].

The *Applications and Service Stratum* includes the layers and functions for management, control and also operations on data. These operations are independent *of network transport*. An application generally contains a graphical user interface (GUI), a media module and signaling modules. Some applications are QoS signaling-capable (based on SIP or other protocols). Legacy applications are supported by a specially defined WEIRD agent, capable to signal their requirements. The WEIRD API Interface adapts the applications data and control flows to the Transport Stratum.

Transport Macro-Layer/Stratum performs management and control of resources/traffic, as well as data operations in order to transport the information flow through various networking infrastructures. The MPl performs medium- and long-term management functions: for high-level service management at the Application and Service Layer macro-layer, and respectively resource and traffic management at the Transport macro-layer. It also provides coordination between all the planes. The CPl layers perform short-term control actions. In the Services and

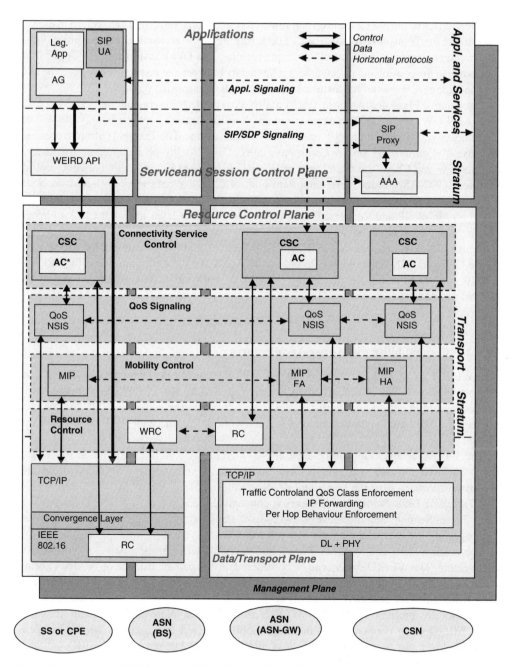

Figure 17.8 WEIRD high level overall architecture (control plane details)

Notes: UA – User Agent; SIP – Session Initiation Protocol; API – Application Programmer Inteface; CSC – Connectivity Service Controller; AC – Admission Control; AC* – this will exist only in an SS if it manages multiple users, to control the CPE segment resource allocation; NSIS – Next Step in Signaling Modules; MIP – Mobile IP; RC Resource Controller; WRC – WiMAX Resource Controller.

Applications Stratum the CPl sets up and releases connections, restores a connection in case of a failure; in the Transport Stratum, the CPl performs the short-term actions for resource control and traffic engineering and control, including routing. The DPl transfers the user/application data but also the control and management related data between the respective entities. The DPl may include functions and mechanisms to act upon the transported packets.

The Control Plane architecture, 'horizontally' covers the following entities: SS/MS, ASN (BS, ASN-GW) and CSN. The Application and Service Stratum contains mainly the session signaling (e.g., SIP), including SIP agents and AAA functions. The Transport Stratum contains the 'layers': *Connectivity Service Control* as a layer of blocks with specific internal structure for SS, SN-GW and CSN. (the main focus of WEIRD is on WiMAX and ASN network control, therefore CSC-ASN is the most important control block); *QoS signaling*, based on NSIS signaling as QoS messages vehicle; *Mobility Control*, including micro and macro mobility based on Mobile IP and Resource Control which is the lower layer having the task to install resources in the network segments.

Figure 17.8 does not include the RC for CPE network and CSN because these are specific to the CPE and CSN technologies. In case of WIMAX the RC communicate with WRC via SNMP in order to install Service Flows in WiMAX segment. A detailed description of the control architecture is given in [24].

17.4.2.2 Compatibility with Other Architectures

The WEIRD architecture is open and agrees with the current developments in IP and telecom architectures. The WEIRD system is *entirely based on the TCP/IP stack*, being easily interconnected in end-to-end multi-domain chains. Its signaling protocols NSIS for resource control, the SIP-based signaling and AAA framework are in line with IETF recent developments.

The WEIRD architecture can be seen also as a NGN-style architecture due to its vertical split structured into two 'macro-layers', or 'strata', i.e., *Application and Service Stratum* and *Transport Stratum*. Also the horizontal decomposition in *Management (MPl), Control (CPl)* and *Transport Data Planes (DPl)* makes it conceptually compliant to NGN. This approach frees the applications and high-level services from transport technologies, in order to support heterogeneity in the core and access network technologies.

The IEEE 802.16 and WiMAX Forum architectural stacks are the basis of the WEIRD architecture. The WEIRD system can be considered on one part an instance of WiMAX Forum stack, but actually is more than WiMAX forum specification, because it is extended vertically with NSIS, SIP/SDP and AAA framework.

The mobility problem is solved by WEIRD architecture at micro or macro scale (the latter using MIPv4/v6 in cooperation with micro-mobility solved at L2 layer). Media Independent Handover can be supported by our architecture. Last but not least, WEIRD architecture allows the decoupling of logical resource management from each domain own policy of network routes dimensioning, due to the overlay concept of resource management. This is based on traffic trunks concepts, treated as logical links intra and inter-domain. The mapping on these trunks on physical paths is accomplished by traffic engineering blocks where the WEIRD system gives complete freedom to be defined and implemented at the will of each operator.

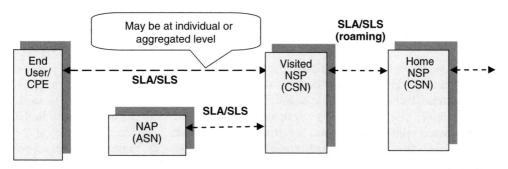

Figure 17.9 Business SLA/SLS relationships – example

17.4.2.3 End-to-end Aspects

The WEIRD architecture is responsible for QoS control in WiMAX and ASN segments only. It reserves resources in the access network, including the WiMAX channel. The resource allocation is done on the following segments: SS – BS and BS – (ASN-GW). The core network resources can be in a simpler approach supposed as over-provisioned, or in a dynamic approach signalled between domain managers. Therefore the resource control and allocation in the (ASN-GW) – CSN and CSN – far-end segments are beyond the WEIRD scope.

The current phase of the WEIRD architecture supposes that static Service Level Agreements (SLAs) are established between business actors, as follows:

- *WiMAX SS/MS/CPE and Home NSP*: – allows the WiMAX subscriber to have access to a suite of WiMAX services and enable accurately billing for these services by the Home NSP.
- *(Visited) NSP and NAP*: – authorizes NSP to use a given NAP's coverage area (or a part of it).
- *NSPs* – this is a Roaming Agreement: – establishes roaming agreements between NSPs.

In this approach, logical pipes (with a given bandwidth) can be statically reserved (e.g. in terms of bandwidth, etc., for different classes of services). The reservation is done based on some forecasting information on future calls amount. After SLA/SLS conclusion, the actual allocation of resources can be done (depending on SLA contract clauses, immediately or later) i.e. they are installed through management actions by the Resource Manager of each segment. The allocation can be done per-individual flow basis or at aggregated level in the zone SS – BS and usually at aggregated level in the zones: BS – (ASN-GW); (ASN-GW) – CSN.

17.4.2.4 Scalability and Reliability Issues

All scenarios considered by WEIRD pose stringent requirements in terms of reliability. Indeed, people's life and natural resources could be threatened in those environments and therefore, they ultimately depend on the ability of the system to convey reliably information to the appropriate destination. By adding redundancy (or diversity) to the system architecture, reliability could be radically enhanced. Such system redundancy is fundamental as functional parts of the network are potentially under direct risk of being damaged, e.g., the network will deployed in a geographic area prone to volcanic eruptions, fires, harsh meteorological conditions, seismic activity, etc. Several approaches are identified and the WEIRD architecture can support them, in order to increase end-to-end reliability. The most promising include:

- *Path redundancy:* The design is such that there is always more than one available path (or route) from any terminal to the BS. Typically implementing two paths is a good engineering compromise. Paths (routes) could be always active or dynamic path selection can be used. In the former case more hardware is required but signal diversity can be exploited. In the latter case, a primary path (route) is active, and the secondary path comes into play if the primary path fails. In general, paths can be implemented using the same or different access technologies
- *Macro diversity:* The network is designed in such a way that: (a) the signal from each terminal can always be received by two BSs; (b) the signal from each terminal can always be received by two distributed antennas (connected to the same BS by RoF) with overlapping patterns; and (c) the signal from each terminal can always be received using multiple beams from the same or multiple base stations
- *Composite cellular/ad hoc networks:* Terminals are also allowed to communicate with each others (using WiMAX or other access technology). If direct access from a given node to the BS is not possible (e.g., due to a blocking obstacle, channel fading conditions or system failure), an additional node can be used as a relaying/repeating station to deliver the signal to the destination node.

17.4.3 Functional Description

17.4.3.1 QoS and Resource Management and Control

In WEIRD, the support of application QoS requirements is done dynamically, through the use of a set of signaling and management approaches, including off-line and online procedures. The aim is to offer support for high level services, (i.e. reserving and then allocating the necessary network level resources).

The architecture allows efficient resource management and control. The medium–long term resource provisioning can be done by the management plane, thus preparing in advance the resources to be used in the future by the high level services. The provisioning can be defined for several classes of services offering different QoS levels. From the granularity point of view, the provisioning can be done:

- at the individual or aggregated level (not per individual flow) in the SS-BS zone of the chain usually at aggregated level in the zones BS- (ASN-GW), (ASN-GW)-CSN and inside or between CSNs.

Then, individual resources for different flows can be dynamically established at request.

The off-line provisioning of resources is made a priori, (when there are not yet explicit requirement for resources from the end users), by the Resource Management module. The resource provisioned beforehand can be allocated to the service flows which enter the network. At this stage, the on-line resource control mechanisms take charge in order to allow WEIRD to support the QoS requirements of applications.

The triggering of WEIRD signaling for QoS support can be made in different ways, as explained in Section 17.3.5, depending on the type of application used (WEIRD-Aware, Legacy). The individual requests for WiMAX and ASN resources are received at the ASN-GW either from NSIS or SIP signaling. In the first case, resource reservation can be triggered by legacy applications which can be modified in order to request for resource reservation through the WEIRD API, or by the WEIRD agent acting in place of the legacy application which cannot be changed. In the second case, resource reservation is triggered by SIP, either by having a WEIRD SIP proxy which can interpret the application requirements or by an application signaling, which is able to depict its QoS requirements on SDP to be interpreted by the WIRD SIP proxy.

After the QoS requests arrive at the ASN-GW, the CSC module activates the admission control module to check whether it is possible to admit them. The CSC-ASN will admit the requests which are still available, taking into consideration the pre-provisioned resources by the Resource Manager. The criterion for admission is based on the amount of resources that were pre-provisioned. It is also possible that the request is completely 'new' in terms of its scope; in such a case the AC applied by the CSC-ASN, if successful, will determine installation of new service flows and new pipes in ASN.

Once the AC provides a positive response, the resources are allocated on the WiMAX channel, and NSIS is used to transmit the reply about the QoS request to the SS/MS.

If the application triggering is done through SIP, NSIS is used for the end-to-end signaling, from the ASN-GW, through the CSN and up to the destination network.

17.4.3.2 Mobility Management

Mobility management is highly relevant to the WEIRD Project, in particular in its second phase where mobile WiMAX will be implemented in the dynamic scenarios that rely on mobile terminals roaming in wide areas. Mobility management is required to track the location of the terminals and ultimately deliver information to them, regardless of their geographical position. Mobility management can be approached from host mobility, session mobility and network mobility standpoints. The main aim in WEIRD is to provide support for terminal mobility, which includes the handover between base stations and networks. Thus both location and handover management are considered here. The WEIRD system supports some common mobility management protocols such as MIPv6, HIP or HMIPv6. Triggering mechanism with dynamic channel modifications can be used for enhancing the mobility management. Note that channel state information is required for mobility and handover management.

17.4.3.3 Network Management and Monitoring

shows a simplified picture of the Management Plane (MPl). It performs the classical network management functions (CNMS) and the medium-long term resource management (RM). NMS is composed of two subsystems: *Conventional Network Management Systems* (CNMS) having 'classical' functions such as network static provisioning, network monitoring, alarm collection and management; *Resource Manager* which is responsible for managing reservation and allocation of connectivity resources in the ASN and WiMAX segments.

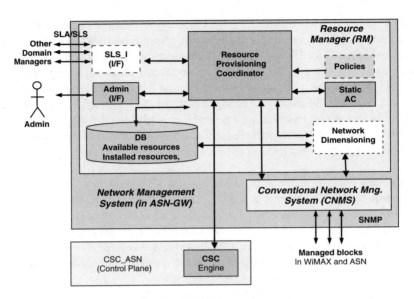

Figure 17.10   WEIRD management plane

As presented in Section 17.4.3.1, resource pre-provisioning is done by management actions, thus preparing in advance (based on some forecasting information on future calls amount) the resources to be used in the future by the high level services. The provisioning can be defined for several classes of services offering different QoS levels. This is the main role of the RM part of the NMS and it comes completely under the WEIRD scope. As described above, individual (per call) resources allocations for different flows can be dynamically established at request by the Control Plane, while taking into account the pre-provisioning done previously by RM and also the limits fixed by the RM.

The proposed architecture is flexible in the sense that it allows extensions (currently not in the scope of WEIRD): agreements between domain managers can be established by (SLA/SLS), on the amount of resources to be provisioned within each domain or between domains. Also a Network Dimensioning module can map the physical topology and link capacities information on a logical map of traffic trunks, described as a matrix of virtual pipes, independent of network infrastructure. This allows general algorithms to be defined for AC and in general for resource control in an independent manner with respect to networking technology. Details of these are given in [24].

17.5 Validating Results: Four European Testbeds

One of the key tasks of WEIRD project is to validate and demonstrate the developed results by implementing project-specific testbeds. In fact, four testbeds have been implemented and deployed in Finland, Italy, Portugal and Romania, each one built around a specific scenario and employing specific technologies suitable for that scenario and its associated applications.

The testbeds are used to assess the enhancements done at the different layers, including improvements on the physical layer (range and coverage extension, reliability, etc.), data link layer, QoS support, handover performance and access control mechanisms at the convergence layer. Moreover, a number of new technologies needed to enlarge the capabilities and operating scenarios of WiMAX are validated in the testbeds, for instance, Radio over Fibre (RoF). The testbeds not only will provide wireless connectivity in their corresponding scenario but they will also be interconnected with each other through their national high-speed data networks and the pan-European GÉANT 2 research network. As the testbeds will have different profiles and employ specific technologies, the scale of the testing and technology validation will be unusually large. Thus, the project is not only developing novel solutions aimed at enlarging the application domain of WIMAX but also implementing, testing and validating these solutions. This is considered one of the most valuable and unique assets of the WEIRD project. In addition to the testing to be carried out locally at each testbed, some applications like video and voice over IP will be evaluated through the interconnecting network.

Table 17.2 Key mobile demonstrations and related scenarios of WEIRD test-beds

WEIRD test-bed	Demonstration	WEIRD test-bed	Demonstration
Italy (3.5 GHz)	Demonstration: – Mobility capabilities – Full management functionality – LOS/NLOS – RoF Scenario: Telemedicine	Portugal (3.5 GHz)	Demonstration: – Mobile video surveillance – A/V streaming – LOS – Real time data collection and transmission Scenario: Fire prevention
Romania (3.5 GHz)	Demonstration: – Mobility capabilities – Full management functionality – LOS/NLOS – VoIP Scenario: video conf., surveillance, e-learning	Finland (3.5 GHz)	Demonstration: – Mobile capabilities – Multimedia transmission with error resilient coding – Range extension - Reliability Scenario: Environmental monitoring

WEIRD test-bed	Demonstration	Scenario
Italy (3.5 GHz)	Mobility capabilities Full management functionality LOS/NLOS RoF	Telemedicine
Portugal (3.5 GHz)	Mobile video surveillance A/V streaming LOS Real time data collection and transmission	Fire prevention
Romania (3.5 GHz)	Mobility capabilities Full management functionality LOS/NLOS VoIP	Video conferencing, surveillance, e-learning
Finland (3.5 GHz)	Mobile capabilities Multimedia transmission with error resilient coding Range extension Reliability	Environmental monitoring

Among the performance figures and capabilities to be tested in each particular testbed, we highlight achievable data throughput, coverage, handover performance, QoS, security, reliability, compatibility between WiMAX equipment as well as between WiMAX and other access technologies (WiFi, 2G, 3G). Table 17.2 summarizes the main WiMAX related demonstrations of the four testbeds. The considered scenarios impose strict requirements on the technical solutions and testbeds validating them. Among them we underline broadband transmissions in NLOS radio channels (all scenarios), wide coverage and very long ranges (volcano monitoring), very high mobility (fire prevention and monitoring) and high reliability (all scenarios). Fulfilling such requirements, identified as one of the key technical challenges of the project, would certainly contribute to broaden the usability of WiMAX technology.

depicts a block diagram of a typical WEIRD testbed, in particular where wireless transfer of multimedia and environmental data is tested. Data is generated from a variety of sources, weather stations, seismic monitors, video cameras, etc. The media server does the rate adaptation and streaming, connecting this IP-based information into the core network. In the considered example the WiMAX access network provides the multimedia data stream to the terminals in the downlink direction. In the uplink direction, the example shows how environmental information could be monitored and transferred to other testbeds.

17.6 Conclusion

WiMAX as a Next Generation access network, both for fixed and mobile connectivity is a promising technology for a large number of real application scenarios in the academic,

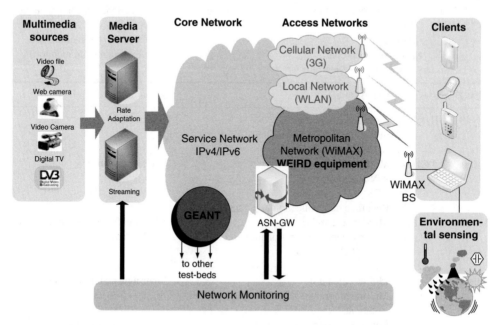

Figure 17.11 Example of a WEIRD test-bed conveying multimedia and environmental data over a WiMAX network

scientific and research community. The analysis of the new applications requirements highlighted the necessity to smoothly integrate, from the beginning, the WiMAX technology into the Next Generation Network architecture in order to incorporate the existing features and to be compliant with the infrastructure that will be present in the market in the near future.

This chapter described how the WEIRD project is facing this challenging problem by defining and specifying complete and open system architecture to fulfil the new applications' requirements in terms of mobility and Quality of Service. The architecture definition was followed by system design and prototype implementation. The WEIRD system accommodates several types of applications. While the applications might not be IMS compliant, as in the case of some of the scenarios described, some enhancements on the client and network side will permit the communication of the user requirements in terms of QoS and mobility introducing new functional modules such as the WEIRD Agent and API, the NSIS signaling modules and the Adapter.

In June 2007, the first prototype of the WEIRD system was released for evaluation in the testbeds, while the final results of the tests and trials will be ready for the end of the project, expected in June 2008.

Acknowledgments

The authors would like to thank all the partners of the WEIRD consortium for their valuable suggestions.

References

[1] 'Air Interface for Fixed Broadband Wireless Access System', IEEE STD 802.16 – 2004, October 2004.

[2] 'Air Interface for Fixed Mobile Broadband Wireless Access System', IEEE P802.16e/D12, February, 2005.

[3] 'WiMAX End-to-End Network System Architecture (Stage 3: Detailed Protocols and Procedures)', WiMAX Forum, August, 2006.

[4] 'WiMAX End-to-End Network Systems Architecture (Stage 2: Architecture Tenets, Reference Model and Reference Points)', WiMAX Forum, March 2006.

[5] WEIRD Website and Public Deliverables: http://www.ist-weird.eu

[6] European Research Network Gèant and National Research Networks Official Website: http://www.geant2.net/

[7] E. Guainella, E. Borcoci, M. Katz, P. Neves, M.Curado, F. Andreotti, and E. Angori, 'WiMAX Extension to Isolated Research Data Networks', paper presented at IEEE Mobile WiMAX World Summit, March 2007, Orlando, FL. *Proceeding IEEE Mobile WiMAX Symposium*, Orlando, 2007.

[8] F. Martínez, J. Campos, A. Ramírez, V. Polo, A. Martínez, D. Zorrilla, and artí, 'Transmission of IEEE802.16d WiMAX Sgnals over Rdio-over-fibre IMDD Lnks', First WEIRD Workshop on WiMAX, Wireless and Mobility, May 22nd, 2007.

[9] D. Levi, P. Meyer, and B. Stewart, 'Simple Network Management Protocol (SNMP) Applications', STD 62, IETF RFC 3413, December 2002.

[10] R. Stewart., Q. Xie, K. Morneault, C. Sharp, H. Schwarzbauer, T. Taylor, I. Rytina, M. Kalla, L. Zhang, and V. Paxson, 'Stream Control Transmission Protocol', RFC 2960, October 2000.

[11] E. Kohler, M. Handley, and S. Floyd, 'Datagram Congestion Control Protocol (DCCP)', IETF, RFC4340, March 2006.

[12] J. Rosenberg, H. Schulzrinne, G. Camarillo, A. Johnston, J. Peterson, R. Sparks, M. Handley, and E. Schooler, 'SIP: Session Initiation Protocol', RFC 3261, June 2002.

[13] TISPAN project official website: http://www.etsi.org/tispan/

[14] 3G Partnership Project official website: http://www.3gpp.org

[15] M. Poikselka et al., *The IMS IP Multimedia Concepts and Services in the Mobile Domain*.Chichester: John Wiley & Sons, Ltd., 2005.

[16] R. Hancock, G. Karagiannis, J. Loughney, and S. Van den Bosch, 'Next Steps in Signaling (NSIS): Framework', RFC 4080, June 2005

[17] J. Manner, G. Karagiannis, and A. McDonald, NSLP for Quality-of-Service Signaling, IETF Internet-draft <draft-ietf-nsis-qos-nslp-12.txt>, work in progress, 2006.

[18] M. Stiemerling, H. Tschofenig, C. Aoun, and E. Davies, NAT/Firewall NSIS Signaling Layer Protocol (NSLP), IETF Internet-draft <draft-ietf-nsis-nslp-natfw-13.txt>, work in progress, 2006.

[19] V. Fajardo, J. Arkko, and J. Loughney, Diameter Base Protocol, IETF Internet-draft draft-dime-rfc3588bis-03.txt, 2007.

[20] C. Rigney, A. Rubens, W. Simpson, and S. Willens, 'Remote Authentication Dial in User Service (RADIUS)', RFC 2865, June 2000

[21] ITU-T Rec. Y.2001, 'General Overview of NGN'.

[22] ITU-T Rec. Y.2011, 'General Principles and General Reference Model for Next Generation Network'.

[23] WEIRD Consortium, 'D2.3 System Specification', deliverable, May 2007. www.ist-weird.org

[24] WEIRD Consortium, 'D3.1 Preliminary Implementation Description', deliverable, May 2007. www.ist-weird.org

[25] K. Knightson, N. Morita and T. Towle, 'NGN Architecture: Generic Principles, Functional Architecture, and Implementation', *IEEE Communications Magazine*, October 2005, 49–55.

[26] E. Guainella, E. Borcoci, M. Katz, P. Neves, M. Curado, F. Andreotti, and E. Angori, 'WiMAX Technology Support for Applications in Environmental Monitoring, Fire Prevention and Telemedicine'.

18

Business Model for a Mobile WiMAX Deployment in Belgium

Bart Lannoo, Sofie Verbrugge, Jan Van Ooteghem, Bruno Quinart, Marc Casteleyn, Didier Colle, Mario Pickavet, and Piet Demeester

18.1 Introduction

Only at the end of the nineties, broadband access networks (mainly including DSL and cable networks, but also Fiber To The Home) were commonly deployed by telecom operators. The number of broadband connections is still expanding rapidly, from about 100 million subscribers at the end of 2003 to 247 million subscribers in June 2006 [1]. Besides broadband Internet connectivity, mobile phones are responsible for another booming market, with a worldwide increase of approximately 1 billion users between end 2003 and June 2006 (currently 2.4 billion users) [2]. When considering the Belgian situation, in June 2006, Belgium counted 2.1 million broadband Internet subscribers (with a market share of 65% for DSL and 35% for cable networks) [1] and 9.3 million mobile phone users [3], corresponding to a penetration of about 20.3% and 88.5% respectively. Moreover, an indisputable new trend is the ultimate desire to have broadband Internet access anywhere and anytime.

It may thus be clear that there exists a great potential for broadband services on mobile terminals. In this respect, a promising new technology that gained a lot of attention the last few years in the telecom industry is WiMAX (based on the IEEE 802.16 standards). Today, two important WiMAX profiles are defined: a Fixed and Mobile version. Fixed WiMAX makes use of the IEEE 802.16-2004 standard (ratified in June 2004) and has the potential to bring broadband Internet access to the millions of people worldwide who are not connected to a wired network infrastructure. However, due to the high penetration of copper (100%) and coax (84%) and the potential higher bit rates of both DSL and cable networks, Fixed WiMAX does not look attractive for the Belgian market. With the introduction of Mobile WiMAX, based on the IEEE 802.16e-2005 revision (approved in December 2005), a new important feature is added: mobility. Since Mobile WiMAX combines the possibilities of Fixed WiMAX with

Mobile WiMAX   Edited by Kwang-Cheng Chen and J. Roberto B. de Marca.
© 2008 John Wiley & Sons, Ltd

mobility, it is expected that mainly Mobile WiMAX will be used in the future. On the Belgian market, it could be a competitive technology next to UMTS and WiFi.

Currently, in Belgium, only some pre-WiMAX networks are being deployed in a few large cities and the coastal area. The operators are Clearwire Belgium and Mac Telecom, both have a license in the 3.4-3.6 GHz band (25 MHz duplex each), which leaves no room for other operators in that frequency band; unless the frequency bands that are now used for video transmission from helicopters of the VRT (the public broadcasting company) are released. Further, in the 10.15-10.65 GHz band, Mac Telecom, Clearwire Belgium and EVONET Belgium already have a license. In the 27.5-29.5 GHz band, only Mac Telecom has a license. Spectrum is still available for two licenses for the entire Belgian territory in this frequency band [4]. Recently, the Belgian telecom regulator BIPT decided to offer new licenses in the 2.6 GHz band by the end of 2007, which can then also be used for the rollout of a WiMAX network.

To further investigate the potential of Mobile WiMAX, a complete business model has been worked out in this chapter, including different rollout and business scenarios. A basic model was already presented in [5], and now it is further extended with enhanced adoption models, new rollout schemes, more accurate technical parameters and updated cost figures, leading to more realistic results. To perform a well-founded study, we have developed a planning tool that takes into account a lot of technical features of Mobile WiMAX. An overview of the most important characteristics of Mobile WiMAX is given in Section 18.2 of this chapter, followed by a section that provides the reader with a detailed explanation of the planning tool itself. Section 18.4 elaborates our business model and considers different business and rollout scenarios that can be applied onto the Belgian market. Section 18.5 presents the most important results, together with a sensitivity analysis to indicate the most influencing parameters within our model.

18.2 Technical and Physical Aspects of Mobile WiMAX

Before creating a planning tool to dimension a Mobile WiMAX network, a lot of technical and physical aspects have to be taken into account. This section gives an overview of a WiMAX network and the related equipment, and next, the most important physical aspects of Mobile WiMAX are described.

18.2.1 Network and Equipment

Mobile WiMAX typically uses a cellular approach, comparable to the exploitation of a GSM network. At the operator side, throughout the area that needs to be served, several sites or base stations (BS) have to be installed. An important feature of a WiMAX system is the use of advanced antenna techniques such as the built-in support of Multiple Input Multiple Output (MIMO) techniques and beam forming using smart antennas (also indicated with the term Adaptive Antenna System or AAS). Next to the above techniques, the capacity per base station can also be increased by installing several sectors on one site, each containing one (or more, in case of MIMO and/or AAS) sector antenna(s). One sector can then provide services to multiple simultaneous users. The antennas themselves are preferably placed at a certain altitude so that their signal is not being blocked by adjacent buildings. The installation of WiMAX base stations and especially the pylons is a determining cost factor in a WiMAX deployment. Site sharing with currently available mobile networks can also be included, depending on the business case. However, it is very important to properly dimension the network by calculating the required number of base stations and their optimal placing.

The base stations are then connected to a WiMAX Access Controller through a backhaul network. The WiMAX Access Controller, connected to the backbone network of the operator, is responsible for, among other things, the access control and accounting. It also guarantees the assignment of IP addresses to the users and mobility, by coordinating the handovers. This mobility is performed by using Mobile IP.

At the client side, the wireless signals are captured and interpreted by a subscriber station (SS), also commonly known as Customer Premises Equipment (CPE). This CPE can be compared with a modem in DSL or cable broadband connections. However, it has to capture a wireless signal and it has thus an attached or integrated antenna. Since the wireless signals, which are transmitted to and from the base station, get severely degraded due to attenuation loss, best performance is reached by using a roof-mounted outdoor CPE. The signal is then brought to the end user computer using in-house Ethernet cabling or a WiFi access point. The end-user could also try to use an indoor CPE. This simplifies the installation process and no roof works are required, but comes at the cost of degraded network performance. If too many objects or walls are located between CPE and base station, communication can fail.

18.2.2 Physical Aspects

A lot of physical aspects of Mobile WiMAX are of importance to deploy a WiMAX network [6–8]. The physical layer modulation of Mobile WiMAX is based on Scalable Orthogonal Frequency Division Multiple Access (SOFDMA). The channel bandwidth is divided into smaller subcarriers which are orthogonal with each other, generated by the Fast Fourier Transform (FFT) algorithm. There are three types of OFDM subcarriers: data subcarriers (for data transmission), pilot subcarriers (for various estimation and synchronization purposes) and null subcarriers (used as guard bands and DC subcarrier). Data and pilot subcarriers are divided into subsets of subcarriers, called subchannels. Subchannels are the smallest granularity for resource allocation and can be assigned to individual users. The physical layer is well adapted to the non-line-of-sight (NLOS) propagation environment in the 2–11 GHz frequency range and it is fundamentally different from the Code Division Multiple Access (CDMA) modulation used in the UMTS technologies. Another feature which improves performance is adaptive modulation, which is applied to each subscriber individually and can be dynamically adapted according to the radio channel capability. If the signal-to-noise ratio (SNR) is high enough, 64-QAM can be used, but with a decreasing SNR 16-QAM or QPSK is applied. WiMAX also provides flexibility in terms of carrier frequency and channel bandwidth. In Europe, the 3.5 GHz licensed band and the 5.8 license free band are the most important ones for Fixed WiMAX at the moment. Also the 2.5 GHz band is investigated, and this one is preferred for Mobile WiMAX [7,8]. Concerning the channel width, channels from 1.25 MHz to 20 MHz are possible. For Mobile WiMAX, channel bandwidths of 1.25 MHz, 5 MHz, 10 MHz and 20 MHz are specified. Note that SOFDMA allows scaling the number of OFDM subcarriers with the channel bandwidth to keep the carrier spacing unchanged. Mobile WiMAX uses Time Division Duplexing (TDD) as duplex mode, which means that downlink and uplink use the same frequency, but at a different time.

18.3 Technical Model and Planning Tool

In order to investigate the feasibility of a WiMAX rollout, one has to be able to asses the number of base stations that will be needed in a specific area, dependent on the offered services and the number of active users. This is possible with the developed planning tool based on an accurate technical model. The tool takes into account the major technical characteristics of Mobile WiMAX together with the desired service specifications. It also has a certain degree of flexibility to introduce adaptations like e.g. new hardware.

To start, we discuss the calculation of the link budget, which indicates to what extent the signal may weaken. Then, a propagation model is proposed to determine the range, by taking into account the link budget. Based on this range, we illustrate the calculation of the cell coverage area. In a next step, we calculate the bit rate per cell sector and finally, the cell areas and bit rates are combined to estimate the required number of base stations.

18.3.1 Link Budget

The link budget depends on several parameters, which are discussed in this paragraph. Different parameter values can be chosen in the planning tool, and for both the downlink and the uplink a separate link budget is calculated. We also indicate which values are selected for the business modeling study.

18.3.1.1 Base Station

We consider a base station (BS) with three sectors, and there is a choice from three BS profiles: "Standard BS", "BS with 2×2 MIMO" and "BS with 2×2 MIMO and 2 element AAS". Most base stations which are now entering the market, belong to the category "BS with 2×2 MIMO" (which is considered in our study). Every profile contains the values for six different parameters required for the link budget calculation. Table 18.1 gives an overview of them, where DL and UL stands for downlink and uplink respectively, and T_x and R_x for transmitter and receiver. The used values are mainly deduced from [8] and [9], but they can easily be substituted when e.g. new hardware comes on the market. Note that additionally a BS feeder loss of 0.5 dB is taken into account.

Table 18.1 Base station parameters

	Standard BS	BS with 2×2 MIMO	BS with 2×2 MIMO and AAS
DL T_x power	35 dBm	35 dBm	35 dBm
DL T_x antenna gain	16 dBi	16 dBi	16 dBi
Other DL T_x gain	0 dB	9 dB	15 dB
UL R_x antenna gain	16 dBi	16 dBi	16 dBi
Other UL R_x gain	0 dB	3 dB	6 dB
UL R_x noise figure	5 dB	5 dB	5 dB

Table 18.2 CPE parameters

	Portable CPE	Mobile CPE
UL T_x power	27 dBm	27 dBm
UL T_x antenna gain	6 dBi	0 dBi
Other UL T_x gain	0 dB	0 dB
DL R_x antenna gain	6 dBi	0 dBi
Other DL R_x gain	0 dB	0 dB
DL R_x noise figure	7 dB	7 dB

18.3.1.2 Customer Premises Equipment (CPE)

With regard to the CPE, we can choose from two profiles: "Portable CPE" and "Mobile CPE". The first type is comparable with e.g. a usual cable modem: they are installed indoors, have their own power supply and are usually connected via an Ethernet cable to the computer. They do not guarantee any form of mobility. Solutions with PCMCIA cards and receivers integrated in e.g. a laptop belong then to the second type (which is used in our study). Every profile contains again six parameters (Table 18.2) [8].

18.3.1.3 Receiver Sensitivity

The receiver sensitivity is defined by the thermal noise, the receiver SNR, the noise figure (Table 18.1 and Table 18.2) and the implementation loss.

Thermal Noise
The thermal noise is dependent on the channel bandwidth and can be estimated as (in dBm): $-174 + 10\log_{10}(\Delta f)$, where Δf is the bandwidth in hertz over which the noise is measured. As physical bandwidth (BW), there is a choice from 1.25 MHz, 5 MHz, 10 MHz and 20 MHz, where today 10 MHz is the most standard value (and also used in our study). For the calculation of the thermal noise, the bandwidth Δf has to be scaled to the effectively used bandwidth. So the value of BW has to be multiplied by the ratio between the number of used subcarriers (N_{Used}) and the total number of OFDM subcarriers or FFT size (N_{FFT}), and the sampling factor (n). For each bandwidth, the model contains different values for N_{FFT} and N_{Used} (Table 18.3). Note that N_{Used} is equal to the sum of the number of data subcarriers (N_{Data}) and pilot subcarriers (N_{Pilot}), together with the DC carrier. Table 18.3 also shows N_{Data} (used to determine the bit rates,

Table 18.3 Parameters per channel bandwidth

BW	N_{FFT}	N_{Used}	N_{DataDL}	N_{DataUL}	$N_{SubChDL}$	$N_{SubChUL}$
1.25 MHz	128	85	72	56	3	4
5 MHz	512	421	360	280	15	17
10 MHz	1024	841	720	560	30	35
20 MHz	2048	1681	1440	1120	60	70

Table 18.4   Parameters per modulation scheme

Modulation scheme	SNR CC (AWGN, BER 10^{-6})	SNR CTC (AWGN, BER 10^{-6})	Data bits per symbol
QPSK 1/2	5 dB	2.5 dB	1
QPSK 3/4	8 dB	6.3 dB	1.5
16-QAM 1/2	10.5 dB	8.6 dB	2
16-QAM 3/4	14 dB	12.7 dB	3
64-QAM 1/2	16 dB	13.8 dB	3
64-QAM 2/3	18 dB	16.9 dB	4
64-QAM 3/4	20 dB	18 dB	4.5

Section 18.3.4) and the number of subchannels (N_{SubCh}, used to calculate the subchanneling gain, Section 18.3.1.4).

The sampling factor n determines the subcarrier spacing (in conjunction with the bandwidth and used data subcarriers), and the useful symbol time. This value is set to 28/25 for channel bandwidths that are a multiple of any of 1.25, 1.5, 2 or 2.75 MHz (which is applicable in our case). The thermal noise is then determined by (18.1).

$$\left(\frac{Thermal}{noise}\right) = -174 + 10\log_{10}\left(BW \times n \times \frac{N_{used}}{N_{FFT}}\right) \tag{18.1}$$

Receiver SNR

The receiver SNR depends on the modulation scheme and the corresponding values are shown in Table 18.4, for two different forward error correction (FEC) methods (convolution code (CC) [6], used in our study, and convolution turbo code (CTC) [10]) in an additive white Gaussian noise (AWGN) channel at a bit error rate (BER) of 10^{-6}. As WiMAX adaptively selects the modulation scheme per user, the appropriate SNR value used in the link budget calculation is dynamically adapted. The modulation scheme also defines the number of data bits per symbol, but this parameter only influences the bit rate per sector which will be discussed further in this chapter (Section 18.3.4).

Implementation Loss

The implementation loss includes non-ideal receiver effects such as channel estimation errors, tracking errors, quantization errors, and phase noise. The assumed value is 2 dB [8].

18.3.1.4   Uplink Subchanneling Gain

In the uplink direction, it will hardly occur that data is sent over all subcarriers simultaneously. To set off this effect, an uplink subchanneling gain is taken into account, based on the number of used subchannels per user and defined by (18.2) [11]. $N_{SubChUL}$ is already given in Table 18.3 and $N_{UsedSubChUL}$ is based on the number of subchannels required for the offered uplink data rate per user, and will also depend on the modulation scheme (Section 18.3.4).

$$\left(\frac{Uplink}{subchanneling\ gain}\right) = -10\log_{10}\left(\frac{N_{UsedSubChUL}}{N_{SubChUL}}\right) \tag{18.2}$$

Table 18.5   Urban corrections

Urban type	Correction
Rural	+5 dB
Suburban	0 dB
Urban	−3 dB
Dense urban	−4 dB

18.3.1.5 Margins

To calculate the link budget, we have to consider several margins, such as the fade margin, the interference margin and an urban correction factor.

Fade Margin
Fading covers the effect of the variation of the signal strength during the time on a fixed location. In contrast to shadowing which takes into account the variation of the signal strength between different locations on the same distance from the transmitter, fading is not incorporated in the propagation model. The fade margin in our model is fixed at 10 dB [12].

Interference Margin
Due to co-channel interference (CCI) in frequency reuse deployments, users at the cell edge or the sector boundaries may suffer degradation in connection quality. The assumed interference margin is 2 dB for DL and 3 dB for UL respectively [7].

Urban Correction
Buildings obstruct the transmitted electromagnetic signals. Since the used propagation model does not sufficiently take into account this effect, an extra correction on the link budget is added. The different possibilities are summarized in Table 18.5.

18.3.1.6 Link Budget Calculation

With the data discussed in the previous sections, it is possible to calculate the link budget, which is specified for the downlink by (18.3). The uplink situation is similar, only the uplink subchanneling gain still has to be added.

$$
\begin{pmatrix} DL\ link \\ budget \end{pmatrix} = \begin{bmatrix} \begin{pmatrix} DL\ T_x \\ power \end{pmatrix} + \begin{pmatrix} DL\ T_x \\ antenna \\ gain \end{pmatrix} + \begin{pmatrix} Other \\ DL\ T_x \\ gain \end{pmatrix} + \begin{pmatrix} DL\ R_x \\ antenna \\ gain \end{pmatrix} + \begin{pmatrix} Other \\ DL\ R_x \\ gain \end{pmatrix} \\ - \begin{pmatrix} Thermal \\ noise \end{pmatrix} - \begin{pmatrix} R_x \\ SNR \end{pmatrix} - \begin{pmatrix} DL\ R_x \\ noise \\ figure \end{pmatrix} - \begin{pmatrix} Implementation \\ loss \end{pmatrix} \\ - \begin{pmatrix} Interference \\ m\ \arg in \end{pmatrix} - \begin{pmatrix} Fade \\ m\ \arg in \end{pmatrix} + \begin{pmatrix} Urban \\ correction \end{pmatrix} \end{bmatrix}
$$

$$(18.3)$$

18.3.2 Propagation Model

Departing from the link budget, we can calculate the range of a base station by using an appropriate propagation or path loss model. Path loss (*PL*) is the reduction in power density of an electromagnetic wave as it propagates through space. We have used the Erceg-Greenstein model [13], which is also applied by the IEEE 802.16 working group [14]. This propagation model is based on extended experimental measurements in the US. Beyond some close-in distance d_0 the path loss can be written as (18.4).

$$PL = \left[A + 10\gamma \log_{10} \left(\frac{d}{d_0} \right) + s + \Delta PL_f + \Delta PL_h \right] \tag{18.4}$$

where:
- d = range, and $d \geq d_0$, with $d_0 = 100$ m
- $A = 20\log_{10} \left(\frac{4\pi d_0}{\lambda} \right)$ which is the link budget at $d = d_0$ (with λ = wavelength)
- $\gamma = a - bh_b + c/h_b$ (with h_b = base station height)
- s = shadow fading variation
- ΔPL_f = frequency correction term
- ΔPL_h = receiver antenna height correction term

To calculate the range distance d, we have to determine the other parameters and to equate the *PL* with the link budget from (18.3). The wavelength λ depends on the carrier frequency which is already discussed in Section 18.2.2. The parameters a, b and c are constants, depending on the terrain type and specified in Table 18.6. The shadowing effect s follows a lognormal distribution with mean 0, and a standard deviation depending on the terrain type, which is also specified in Table 18.6. In the planning tool, the user has to choose a percentage p which results in a value s_p that we will use for s ($s \leq s_p$ with probability p). We will use a percentage of 90% for p (s_{90} also depicted in Table 18.6). The terrain types are divided in next three categories:

- Type A: hilly, moderate tree
- Type B: intermediate
- Type C: flat, light tree

The model is valid for a base station height h_b between 10 m and 80 m. In Belgium, the current GSM pylons e.g. have a height between 20 m and 40 m. Finally, without the terms

Table 18.6 Parameters for different terrain types

Parameter	Type A	Type B	Type C
A	4.6	4	3.6
B	0.0075	0.0065	0.0050
C	12.6	17.1	20
std dev s	10.6 dB	9.6 dB	8.2 dB
s_{90}	13.6 dB	12.3 dB	10.5 dB

ΔPL_f and ΔPL_h, the above formula is only valid for frequencies close to 2GHz and for receiver antenna heights close to 2 m. That is why these correction terms are introduced. They are equal to:

- $\Delta PL_f = 6 \log_{10}(f/2000)$
- $\Delta PL_h = -10.8 \log_{10}(h/2)$, for Type A and B
  $\quad\quad -20 \log_{10}(h/2)$, for Type C

where f is the carrier frequency (in MHz) and h is the receiver antenna height between 2 m and 10 m.

To conclude this section, the user of the planning tool can choose the terrain type (Type C), base station height h_b (30m), receiver height h (2m) percentage p (90%) for shadowing s_p, and the carrier frequency f (2.5 GHz). The values used in our business modeling study are indicated between brackets. With the mentioned parameter values, the obtained Mobile WiMAX ranges vary from 1800 m (rural) and 1150 m (urban) for QPSK 1/2 to 780 m (rural) and 500 m (urban) for 64-QAM 3/4 in the downlink, and from 1210 m (rural) and 770 m (urban) for QPSK 1/2 to 750 m (rural) and 480 m (urban) for 64-QAM 3/4 in the uplink.

18.3.3 Cell Area

Mobile WiMAX uses a cellular network structure and we consider a hexagonal cell area, defined as $3 \times d^2 \times \sin(\pi/3)$, with d the coverage range as indicated in Figure 18.1.

18.3.4 Bit Rate per Sector

Both the channel bandwidth and the modulation scheme have an important influence on the bit rate, and these two parameters were already discussed above (Table 18.3 and Table 18.4). Besides, the bit rate is also determined by the guard time, the overhead and the TDD down/up ratio. The guard time is intended to overcome multi path effects and in the planning tool the user can select a fraction from a particular set, specified in the standard (1/8 is used in our study). The overhead is defined as the percentage of time that no data is sent and is the time used for e.g. initialization and synchronization, and it also covers the headers (we assume an overhead of 20%). Finally, the ratio between the downlink and uplink time is defined by a TDD down/up ratio parameter (fixed at 3:1 in our model). The downlink bit rate is then given

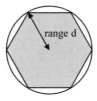

Figure 18.1   Illustration of cell area calculation [5]

by (18.5), and for a 10 MHz channel, this results in bit rates ranging from 4.2 Mbps (QPSK 1/2) to 18.9 Mbps (64-QAM 3/4). The uplink bit rate is similarly calculated.

$$\left(\begin{array}{c} DL \\ bit\ rate \end{array}\right) = \left[\begin{array}{c} BW \times n \times \dfrac{N_{DataDL}}{N_{FTT}} \times \left(\begin{array}{c} data\ bits \\ per\ symbol \end{array}\right) \\ \times \dfrac{1 - overhead}{1 + guard\ time} \times (TDD\ down/up\ ratio) \end{array}\right] \tag{18.5}$$

18.3.5 Required Number of Sites and Sectors

The final goal of the planning tool is to deliver the number of sites and sectors required to cover a particular region, and this information will then be used to formulate different business scenarios. The area of the region, the user density and the desired downlink and uplink bit rate per user are additional input parameters. Operators also take into account that not every user simultaneously uses his connection, and for this purpose a parameter for simultaneous usage (overbooking) is introduced, which defines the percentage of the users that effectively use the service (we assume 5%). As already mentioned, WiMAX dynamically selects the best possible modulation scheme per user, which is illustrated in Figure 18.2.

18.3.6 Planning Tool: Graphical User Interface

Figure 18.3 shows the graphical user interface of the planning tool, and the main blocks are discussed below. Channel specifications, including the carrier frequency and channel bandwidth, can be adapted in block A. Further this block contains some protocol parameters which influence the effective bit rate, such as the guard time, protocol overhead and the ratio between down- and uplink traffic. A last parameter in block A is the used FEC method. In block B and C the hardware specifications of the base station (cf. Table 18.1) and CPE (cf. Table 18.2) are defined. Block D covers the terrain specifications and margins, incorporating three terrain types (cf. Table 18.6) and four urban types (cf. Table 18.5). Further, this block specifies the BS and CPE heights, the shadowing variation, the fade margin, the interference margin and the implementation loss. All the above parameters have their influence on the link budget and propagation range.

Block E contains the service specifications such as the desired download and upload speed per user and a factor that covers the simultaneous usage since not all users will utilize their

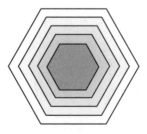

Figure 18.2 Range of the different modulation schemes, indicated by different colors. The lighter the color, the less data bits per symbol (cf. Table 18.4) [5]

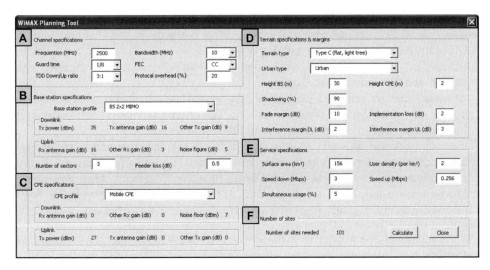

Figure 18.3 Graphical user interface of the planning tool

connection at the same time. Finally, the surface area and user density are also very important to calculate the required number of sites in a specific area. The final result is then shown in block F.

Note that the selected values in Figure 18.3 were used during the calculations in this chapter, with exception of the urban type and surface area (dependent on the region), the user density (increasing during the years) and the user bit rates (dependent on the offered service).

18.4 Business Model

In collaboration with Belgacom, the former Belgian incumbent telecom operator, a generic business model has been worked out. The model is then applied for a specific case that investigates the rollout of Mobile WiMAX in Belgium. The considered time period is 2007 till 2016, but the results can easily be projected to any other time period.

18.4.1 Model Input Parameters

This section describes the defined business and rollout scenarios, and the market forecast.

18.4.1.1 Business Scenarios

Three different business scenarios are defined for offering Mobile WiMAX in [5]:

- "Nomadicity pack": the operator offers its current customers the possibility to make use of wireless Internet via Mobile WiMAX for an extra monthly fee (comparable with a hotspot subscription). The bandwidth connection will be limited as this is just a complement to the fixed line broadband service (512 kbps downstream, 128 kbps upstream) and the emphasis lies on the mobility of the user.

- "Stand alone wireless broadband": WiMAX completely replaces the fixed line broadband connection. It offers a comparable bandwidth (3 Mbps downstream, 256 kbps upstream), but additionally combines it with the mobility of WiMAX.
- "All services": this scenario is a combination of the two previous services together with a "prepaid" and "second residence" offer:
  - "Prepaid": the user can buy a prepaid card that grants him a limited number of hours for making use of the Mobile WiMAX network. The offered service is comparable to the "nomadicity pack" (512 kbps downstream, 128 kbps upstream).
  - "Second residence": this service is an extra option to existing customers of the operator, and it can be considered as a kind of "nomadicity pack" for heavy Internet users. It offers a bandwidth similar to the "stand alone wireless broadband" service (3 Mbps downstream, 256 kbps upstream), but at a reduced tariff. It is mainly intended for users that need a second connection e.g. customers with studying children, or with a holiday cottage.

As the additional cost to offer an extra service over the WiMAX network is limited, and thanks to the high differentiation of the offered services, the "all services" scenario is by far the best option [5]. This study will focus on the last business scenario that includes the four described services.

18.4.1.2 Rollout Scenarios

Next to the three business scenarios, there are also considered three different rollout scenarios in [5], ranging from a limited rollout in the ten most important Belgian cities during two years (*Urban*) and an extension to the Belgian coast in the same time period (*Extended Urban*) to a nationwide rollout performed in only three years (*Nationwide*). After a profound update of some technical parameters however, we have to conclude that the mentioned nationwide scenario from [5] is not anymore economic feasible. With a cell diameter of maximum 2 to 3 km (based on the calculations in Section 18.3.2), a complete nationwide rollout involves too high investments to cover the rural areas. In this way, the nationwide scenario is reduced to a nationwide urban rollout, which corresponds to all areas with minimum 1000 inhabitants/km^2. In the evaluation, we will consider three rollout speeds for the nationwide urban scenario (fast in three years, moderate in five years and slow in eight years). Together with the urban and extended urban scenario from [5], in total five scenarios are evaluated (Table 18.7). The urban and extended urban scenarios coincide with the first two years of respectively the 5-year and 3-year nationwide urban scenario.

Table 18.7   Rollout scenarios

Scenario	Area (% of Belgium)	Population (% of Belgium)	Rollout period	Description
Urban	4%	25%	2 years	10 most important cities
Extended Urban	5%	27%	2 years	Urban + Belgian coast
Nationwide Urban 3Y			3 years	
Nationwide Urban 5Y	8%	36%	5 years	Areas with more than 1000 inh./km^2
Nationwide Urban 8Y			8 years	

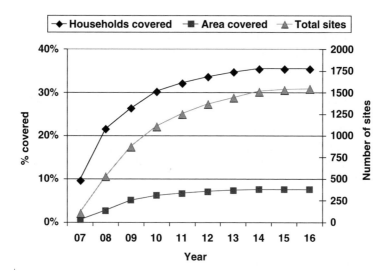

Figure 18.4   Households and area covered in a nationwide urban rollout during 8 years, related to the number of base stations

It is supposed that the nationwide urban scenarios are following a gradual scheme, starting in the largest cities and moving on to the less populated ones. To illustrate this, Figure 18.4 shows a complete rollout scheme for the 8-year nationwide urban scenario. It is clearly noticeable that the percentage of covered households (final value of 36%) increases much faster than the covered areas (final value of 8%). Note that Belgium counts 10,511,382 inhabitants on a 32,545 km^2 territory [3].

Figure 18.4 also depicts the required number of base station, which is the outcome of the planning tool and depends on the used scenario parameters (both business and rollout scenarios).

18.4.1.3 Market Forecast

The most crucial part in the model is associated with the market forecast - residential as well as business users are considered. For predicting the number of residential customers, the analysis starts from the total number of Belgian households. Taking into account the number of broadband connections, a forecast can be made for the targeted number of customers. Business customers are also interested in these services, especially the "nomadicity pack" and "second residence" service for offering mobile subscriptions to their employees.

We assume that "stand alone wireless broadband" will not be successful due to the fact that the foreseen bandwidth will not be sufficient enough as primary broadband connection when triple play services with e.g. large video streams will be offered. Prepaid cards will replace the current WiFi hotspot service. "Second residence" and "nomadicity pack" will initially grow at the same rate, but as we assume that the former is intended for a smaller population of people with a second domicile, the latter will become slightly more popular after a while. Maybe the fact that more devices will be equipped with wireless cards (e.g. PDA, cell phone, etc.) might lead to a larger usage.

A market forecast based on the Gompertz model [15] is made for the take-up of the four offered services. The Gompertz model is given by (18.6), and it is determined by three adoption parameters that have to be predicted. Separate parameters are defined for the different services and the different user groups (residential and business users). Note that the used curve for "stand alone wireless broadband" will somewhat differ from the Gompertz curve, since we assume a decrease in the take rate after some years. Besides, for the areas that are not covered with WiMAX from year one, the Gompertz curve will be shifted in time. However, this time shift will be a little smaller (maximum one and a half years smaller) than the difference in rollout time so that a faster adoption will be modeled in the areas that are later covered. This can be motivated by the fact that the WiMAX service will already be better known in the rest of the country after some years.

$$y(t) = C \cdot \exp[-\exp(-b(t-a))] \tag{18.6}$$

where:
- C = the saturation point (i.e. maximum adoption percentage)
- a = the infliction point (i.e. year between a progressive and degressive increase)
- b = the take rate (i.e. indication of the slope of the maximum increase)

In comparison with [5], we have adapted a few adoption percentages that were somewhat overestimated. Especially the figures for the business users are reduced, which has its impact on the total adoption of the "nomadicity pack" and the "second residence" service. On the other hand, for prepaid cards, we have slightly increased the usage in the first years (by changing the a and b parameters from the Gompertz model), on the assumption that the barrier to buy a prepaid card is much lower than the barrier to take a monthly subscription.

Finally, by combining the previous considerations, we obtain Figure 18.5 and Figure 18.6 for the adoption in the 3-year and 8-year nationwide urban rollout respectively (the residential and business users for "nomadicity pack" and "second residence" are merged). The difference

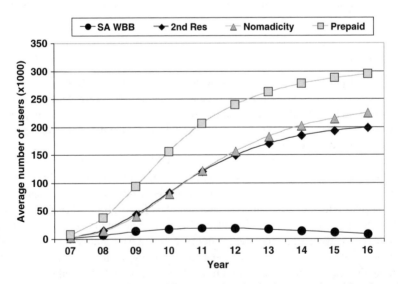

Figure 18.5   Adoption curves for the different services in the 3-year nationwide urban rollout

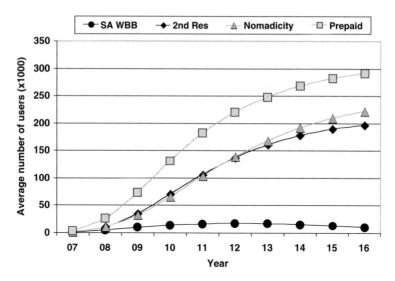

Figure 18.6 Adoption curves for the different services in the 8-year nationwide urban rollout

between the two rollout scenarios is clearly noticeable. For a WiMAX rollout that is only finished after eight years (instead of three), the curves will increase more slowly, but the final values are still situated in the same range. Note that the prepaid cards are depicted per sold prepaid card of three hours (and not per individual user), which explains their high number.

18.4.2 Costs

The costs are split into capital (CapEx) and operational expenditures (OpEx).

18.4.2.1 Capital Expenditures (CapEx)

Capital expenditures (CapEx) are the long term costs which can be depreciated. CapEx contain the rollout costs of the new WiMAX network, listed in Table 18.8. After ten years, the number of needed base stations varies from respectively ca. 1000 and 1200 for the urban and extended urban rollout to ca. 1600 for the nationwide urban rollouts. A site sharing of 90% is assumed in urban areas (in less populated areas this would be lower), as regulation declared that pylons for e.g. GSM or UMTS must be shared between operators. For the remaining 10%, new sites

Table 18.8 Detailed CapEx costs

Capex	Costs	Depreciation
Cost site	40,000 €	20 years
Cost WiMAX equipment main unit	15,000 €	5 years
Cost WiMAX equipment sector unit	6,000 €	5 years
Cost core equipment	10% of WiMAX equip.	5 years
Cost normal backhaul (CapEx part)	5,000 €	5 years

will be built, equipped with a pylon if required (also possible on the rooftop of a building) and a WiMAX base station. Owned pylons can also be let to other operators, which will result in revenues for the operator. The equipment cost per base station contains the WiMAX main unit & sector units, as well as backhaul costs for connecting to the backhaul network. In addition, an investment must take place in central infrastructure (core equipment) such as WiMAX Access Controllers, routers or network operation centre infrastructure. Equipment is renewed every five years (economic and technical lifetime).

18.4.2.2 Operational Expenditures (OpEx)

The operational expenditures (OpEx) contain the yearly returning costs. OpEx are mostly underestimated and determine in a large extend the total costs of networks. Therefore a thorough analysis is essential as all important factors must be taken into account. A model that can be used for this analysis is described in [16]. The most important network OpEx are (Table 18.9) the WiMAX spectrum license, operations & planning (depends on the growth of the network), maintenance (WiMAX standard and core equipment), costs made for owning and leasing the sites of the pylons and backhaul traffic costs. OpEx specifically related to the service contains marketing costs (making the users familiar with the service), sales & billing and helpdesk.

18.4.3 Revenues

Starting from the forecasted number of users, we can calculate the total revenues per service. Assumptions have been made about the tariffs of the different services (Table 18.10). For the "nomadicity pack", a premium tariff of €13 incl. VAT is set, which is competitive compared to hotspot services. The "second residence" service is priced at €20 per month incl. VAT, which is higher than the previous one which reflects the higher bandwidth connection. These two services are vouching for 80% of the overall revenues. The other two services are relatively less important. Prepaid cards are offered at €9 for 3 hours. The price remains the same in the upcoming years but the duration of the card will enlarge. The "stand alone wireless broadband"

Table 18.9 Detailed OpEx costs

OpEx	Costs	Comments
WiMAX spectrum license	1,250,000 €	Yearly payments
Network operations		Related to the roll out of the network
Network planning		Related to the roll out of the network
Maintenance WiMAX st. equip.	7%	% of CapEx
Maintenance core equip.	10%	% of CapEx
Light backhaul – OpEx part	500 €	Per base station per year
Normal backhaul – OpEx part	3,000 €	Per base station per year
Lease and maintenance own sites	5,000 €	Per base station per year (own sites)
Cost shared sites	6,000 €	Per base station per year (leased sites)
Marketing	1,500,000 €	Maximum (First years, % of this number based on the covered households)
Sales & billing	10%	% of revenues
Helpdesk		Based on the number of calls per user

Table 18.10   Overview of the tariffs per service

Service	Tariff (incl. VAT)	Offered bandwidth	
		Downstream	Upstream
Nomadicity pack	13 €/month	512 kbps	128 kbps
Second residence	20 €/month	3 Mbps	256 kbps
Prepaid	9 €/3-hour card	512 kbps	128 kbps
Stand alone wireless broadband	60 €/month	3 Mbps	256 kbps

service is priced at €60 incl. VAT per month, which is hard to compete with current fixed cable or DSL broadband connections. The usage of this last service will decline after a few years (cf. Figure 18.5 and Figure 18.6) as more bandwidth is requested by the users, which cannot be guaranteed by WiMAX at this stage.

18.5   Economic Results for a Mobile WiMAX Rollout in Belgium

Based on the model input parameters, together with the costs and revenues from the previous sections, the five rollout scenarios from Table 18.7 are extensively compared to each other. A static cash flow and net present value (NPV) analysis, together with an extensive sensitivity analysis are presented in this section.

18.5.1   Static Analysis

The results of the cash flow analysis are shown in Figure 18.7, for the three rollout scenarios that are most different from each other. In the first three to four years, costs for rolling out WiMAX base stations will generally dominate the result as revenues cannot compensate the

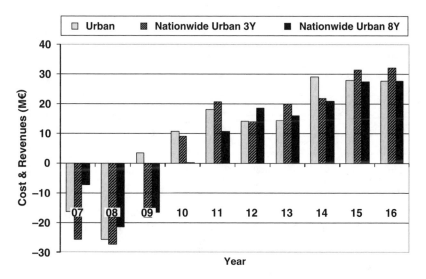

Figure 18.7   Cash flow analysis for the three most different rollout scenarios

investments. After this period extra investments are still required to satisfy the user needs or to cover the rest of the nationwide cities in the 8-year rollout. However, from now on, the number of users has increased to create enough revenues to cover this.

Figure 18.7 indicates some differences between a fast 3-year rollout and a more gradual 8-year rollout. The former requires very high investments in the first years, which involve a high financial risk for the operator. From year 4 the costs are more and more related to the increasing customer base: on the one hand OpEx which will more and more determine the total costs and at the other hand new investments to meet the needs of the customers. Renewing of equipment is important from year 6, which reflects in a small decrease of the cash flows in year 6 and 7. As can be seen in Figure 18.7, the 3-year nationwide urban rollout generates a positive cash flow from year 4. (The same is valid for the urban rollout, where the costs and revenues are already balanced in year 3). Concerning the 8-year nationwide urban rollout, during the first two years, its cash flows are less negative than in the other scenarios. So, the yearly investments and related risks are much smaller than in case of a fast rollout. However, it now takes a year longer to generate a clearly positive cash flow.

Next to the above cash flow analysis, a net present value (NPV) analysis is more suited to assess the financial feasibility of long-term projects. Figure 18.8 shows the results of the NPV analysis for the five proposed rollout scenarios (discount rate is set at 15%). As could be expected from the cash flow analysis, the NPV reaches its minimum in year 3 or 4, with the lowest NPV at that time (-63 M€) for the 3-year nationwide urban scenario. This again confirms the large financial risk for such a rollout. After approximately eight years, the NPV of both urban scenarios becomes positive, while the three nationwide urban scenarios do not show a positive NPV before year 10 (i.e. a discounted payback period of respectively eight and ten years). Note that the high investments in the 3-year nationwide urban rollout are still noticeable in the NPV after ten years. So, this NPV analysis indicates that a slow or moderate

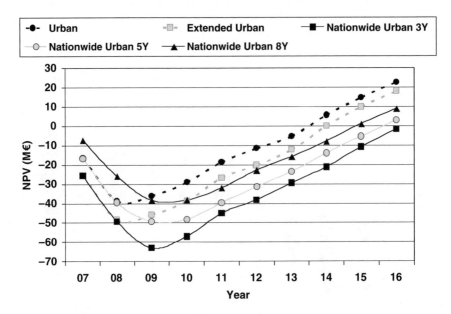

Figure 18.8   NPV analysis of the five different rollout scenarios (discount rate = 15%)

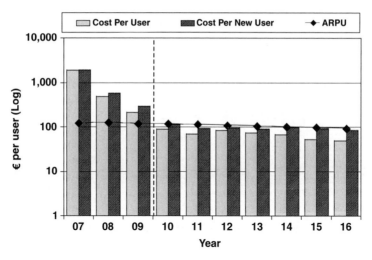

Figure 18.9   ARPU vs. Cost per User in the 3-year nationwide urban rollout

rollout speed is more suitable than a fast rollout and that the high investment costs to cover the less populated cities are not yet compensated after ten years.

Although the NPV analysis clearly shows that the best strategy for an operator consists of a (slow) rollout limited to the big cities, in some cases, an operator might decide to extend its target area. The main reason to move to a nationwide urban rollout is to create a higher customer base. As can be derived from the cash flow analysis (Figure 18.7), the cash flow in year 10 has the highest value for the 3-year nationwide urban rollout, which will involve that the NPV will rise faster in the years afterwards (on the assumption that the network is still sufficient for the user needs in this year or can be upgraded with limited extra investments). Furthermore, it could be possible that a faster rollout will lead to a higher adoption, while a slower rollout has the opposite effect. So, in some cases it is difficult to deduce commonly valid conclusions from the NPV analysis. To estimate the influence of the different parameters however, we will perform an extensive sensitivity analysis in Section 18.5.2.

To end this section, Figure 18.9 and Figure 18.10 depict the average revenue and cost per user for the 3-year and 8-year nationwide urban rollout respectively. The average revenue per user (ARPU) steadies around 100€ per user annually from the beginning, with a small decline over the years since the prices will slightly decrease for offering the same service. Note that the small increase between the first and second year originates from the higher take-up of prepaid cards at the beginning, which is the service with the lowest ARPU. The cost per user graphs can be split up. The first graph, average cost per user, is calculated based on the fact that all users must pay for the extra investments (form of cross subsidizing). From this graph, it is clear that the 3-year rollout generates a positive cash flow after four years and the 8-year rollout after five years, which is completely in line with Figure 18.7. The second graph shows the cost per new user where CapEx and OpEx are separately allocated for new and existing users. The year that the whole considered area is covered can be derived from this graph (indicated by the dashed line). While the network is expanded, the cost per new user is still higher than the ARPU. This can be explained by the fact that the first year a new area is covered, the number of users is still too low to compensate the new investments.

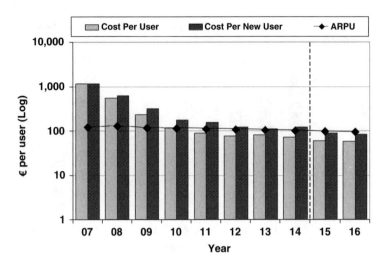

Figure 18.10   ARPU vs. Cost per User in the 8-year nationwide urban rollout

18.5.2 Sensitivity Analysis

We have set several parameters in our model for which we are uncertain whether the values are realistic or not. Adoption parameters, CapEx and OpEx costs and the service tariffs are the most important ones. Therefore, we have performed a sensitivity analysis in which we let fluctuate the respective parameter values around an average value, according to a well-defined distribution (Gaussian, Uniform or Triangular). The sensitivity analysis is done by using the Crystal Ball tool [17], and for every rollout scenario 100,000 runs with varying parameters were performed to get a realistic view of the uncertain outcome (especially NPV results after five and ten years in this case). The relative influences of the diverse parameters are shown in Figure 18.11, Figure 18.12 and Figure 18.13. The depicted percentages are a measure of the impact of a varying input parameter on the end result. Note that the sum of all percentages per scenario (from the three figures together) is 100%.

The parameters related with the market forecast are definitely the most uncertain ones in the business model (total influence of approximately 40% after ten years). Figure 18.11 gives each time two kind of parameters corresponding to the Gompertz model (18.6): a maximum adoption percentage (max %, related to the parameter C) and an indication of the adoption speed (adopt., related to the parameters a and b which are supposed to be correlated with each other). A first important trend is that the adoption speed is very important in the beginning years (cf. NPV after five years), while in the following years the maximum adoption logically becomes more and more important. Further, the "nomadicity pack" and the "second residence" service are the most important services, together with the "stand alone wireless broadband" service in the first years. The importance of the latter is decreasing due to a declining user interest, as mentioned in Section 18.4.1.3.

The tariff setting also greatly influences the end results and becomes more and more important during the years (Figure 18.12), i.e. when more customers make use of the WiMAX network (total influence of more than 25% after ten years). The importance of the different services is again clearly noticeable, and especially the second residence service is a determining service to generate a positive business case.

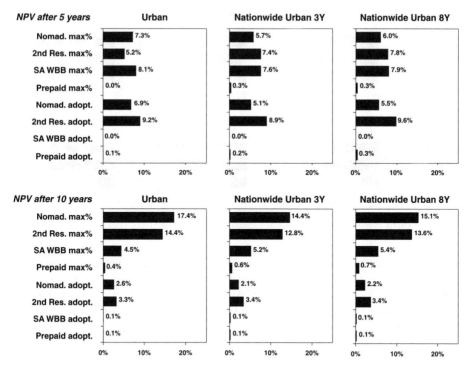

Figure 18.11   Sensitivity results for the market forecast parameters (for three rollout scenarios)

When considering the costs (Figure 18.13), then the CapEx to install the complete WiMAX network (especially the high number of base stations) is a very important factor in the first years of the business case. Since this is a non-recurring cost, its influence becomes less important after ten years. In a later stadium, OpEx, dominated by the operational costs of the WiMAX sites, becomes more influential than CapEx. The other OpEx represent the general OpEx costs (such as help desk, marketing and sales & billing) and the WiMAX license.

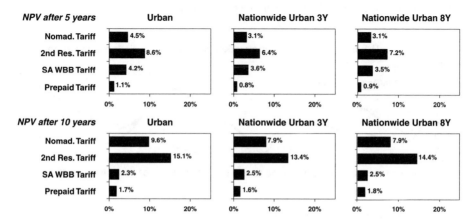

Figure 18.12   Sensitivity results for the service tariffs (for three rollout scenarios)

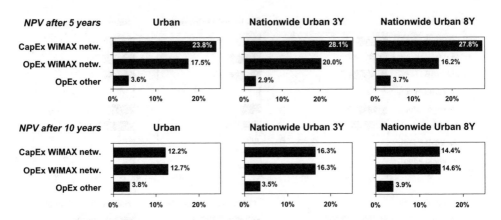

Figure 18.13   Sensitivity results for the different cost factors (for three rollout scenarios)

In Figure 18.14, a trend analysis of the forecasted NPV can be observed for the three most different rollout scenarios. The nationwide urban scenarios have a more varying outcome after ten years than the urban one (with a range of 89 M€ and 102 M€ vs. 76 M€). Further, after ten years, 99% of the Crystal Ball runs give a positive NPV outcome for the urban scenario, meaning that a positive business case for Mobile WiMAX should be possible in the big cities. For the 8-year nationwide urban scenario, a positive NPV is still reached in 78% of the runs, while this is dropped to only 42% for the 3-year rollout. So, the 3-year nationwide urban rollout not only shows a higher financial risk due to the high negative cash flows during the first years, but it also has the highest uncertainty and risk profile.

18.6 Conclusion

Mobility becomes a very important topic when discussing the rollout of new access networks, and Mobile WiMAX may possibly offer an appropriate solution. First, this chapter presents a planning tool to make an accurate calculation of the number of required base stations for a WiMAX network. The calculation is based on the physical characteristics of Mobile WiMAX together with some specific rollout and service parameters.

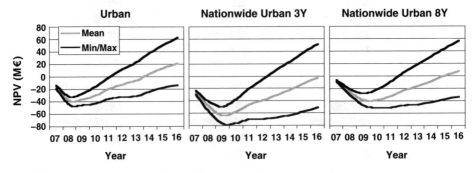

Figure 18.14   NPV trend analysis over ten years, for three rollout scenarios

Further, a business model has been created for the rollout of WiMAX, and several rollout scenarios for a WiMAX deployment in Belgium are considered. The results of our model indicate that a full nationwide rollout in Belgium is not feasible with the current technology (using MIMO 2x2). Although a rollout limited to the urban areas is well able to generate a positive business case for the operator, it is clear that a WiMAX rollout outside the big cities remains a risky project. So, a moderate rollout speed, which will probably be tuned to the user adoption, is recommended. As this analysis was based on a number of uncertain parameter values, we have conducted a sensitivity analysis, which indicates that the most determining factors are related to user forecast & service pricing and the high number of required base stations. Especially for the latter factor, it is important to remark that an evolving Mobile WiMAX technology, with increasing ranges (e.g. by using AAS) can greatly improve the business case. While this study clearly shows that a business case is no longer feasible outside the urban areas, this can be totally different if the covered area per base station would increase.

Acknowledgements

The authors would like to thank Jeffrey De Bruyne and Wout Joseph (both from Ghent University) for their useful support regarding the technical aspects of WiMAX.

References

[1] Point Topic: Global broadband statistics, http://www.point-topic.com.

[2] GSM Association, "GSM subscriber statistics", http://www.gsmworld.com/news/statistics/.

[3] Statistics Belgium, http://statbel.fgov.be.

[4] BIPT, Belgian Institute for Postal services and Telecommunications, "Radio Communications: Frequencies, Wireless local loop", http://www.bipt.be.

[5] B. Lannoo et al., "Business scenarios for a WiMAX deployment in Belgium", in Proceedings of IEEE Mobile WiMAX 2007 conference, Orlando, USA., Mar. 2007. *Proceeding IEEE Mobile WiMAX Symposium*, Orlando, 2007.

[6] IEEE Std. 802.16e - 2005, Amendment to IEEE Standard for Local and Metropolitan Area Networks, "Part 16: Air interface for fixed broadband wireless access systems - Physical and Medium Access Control Layers for Combined Fixed and Mobile Operations in Licensed Bands", Feb. 2006.

[7] WiMAX Forum, "Mobile WiMAX – Part I: A Technical Overview and Performance Evaluation", Aug. 2006.

[8] WiMAX Forum, "WiMAX System Evaluation Methodology", Jan. 2007.

[9] L. Nuaymi, "WiMAX: Technology for Broadband Wireless Access", Wiley, Jan. 2007.

[10] WiMAX forum, "WiMAX Forum[TM] Mobile System Profile Release 1.0 Approved Specification", May 2007.

[11] WiMAX Forum, "Simulation Results for Subchannelization", Nov. 2002.

[12] ITU-R Recommendation P.530-10, "Propagation data and prediction methods required for the design of terrestrial line-of-sight systems", 2001.

[13] V. Erceg et al., "An empirically based path loss model for wireless channels in suburban environments," IEEE JSAC, vol. 17, no. 7, Jul. 1999, pp. 1205–1211.

[14] IEEE 802.16 Working Group, "Channel models for fixed wireless applications", IEEE, New York, Jun. 2003.

[15] K. Vanston and R. Hodges, "Technology forecasting for telecommunications", Telektronikk 4.04, 2004.

[16] S. Verbrugge et al., "Modeling operational expenditures for telecom operators", in Proceedings of ONDM2005, Milan, Italy, Feb. 2005, pp. 455–466.

[17] Crystal Ball, http://www.crystalball.com.

Index